AF595456

Analysis, Design and Fabrication of Micromixers

Analysis, Design and Fabrication of Micromixers

Editor

Kwang-Yong Kim

MDPI • Basel • Beijing • Wuhan • Barcelona • Belgrade • Manchester • Tokyo • Cluj • Tianjin

Editor
Kwang-Yong Kim
Department of Mechanical
Engineering
Inha University
Incheon
South Korea

Editorial Office
MDPI
St. Alban-Anlage 66
4052 Basel, Switzerland

This is a reprint of articles from the Special Issue published online in the open access journal *Micromachines* (ISSN 2072-666X) (available at: www.mdpi.com/journal/micromachines/special_issues/micromixer).

For citation purposes, cite each article independently as indicated on the article page online and as indicated below:

LastName, A.A.; LastName, B.B.; LastName, C.C. Article Title. *Journal Name* **Year**, *Volume Number*, Page Range.

ISBN 978-3-0365-1368-3 (Hbk)
ISBN 978-3-0365-1367-6 (PDF)

Contents

About the Editor

Kwang-Yong Kim

Kwang-Yong Kim is currently an Inha Harlim professor at the Department of Mechanical Engineering and was the dean of the Engineering College of Inha University, Incheon, Korea. He served as the associate editor of ASME *Journal of Fluids Engineering*, as the co-editor-in-chief of the *International Journal of Fluid Machinery and Systems*, as the editor-in-chief of the *Transactions of Korean Society of Mechanical Engineers*, as the president of Korean Society for Fluid Machinery, and as the chairman of the Asian Fluid Machinery Committee. He is a fellow of the Korean Academy of Science and Technology, a member of National Academy of Engineering of Korea, a fellow of the American Society of Mechanical Engineers (ASME), and an associate fellow of the American Institute of Aeronautics and Astronautics (AIAA).

 *micromachines*

Editorial

Editorial for the Special Issue on Analysis, Design and Fabrication of Micromixers

Kwang-Yong Kim 

Department of Mechanical Engineering, Inha University, Incheon 22212, Korea; kykim@inha.ac.kr;
Tel.: +032-860-7317

During the last couple of decades, there have been rapid developments in analysis, design, and fabrication of micromixers. Micromixers are essential components of micro-total analysis system (μ-TAS) and lab-on-a-chip, and have various chemical and biological applications [1]. Due to the small scale, flows of the liquid samples in micromixers are laminar, and thus have a difficulty in mixing. Therefore, efficient mixing is a critical goal in the design of micromixers. A variety of passive or active designs are used to enhance the mixing. Design optimization is an efficient tool to maximize the mixing by optimizing the geometric and/or operating parameters using systematic algorithms. Both numerical and experimental methods have been used to analyze the flow and mixing in micromixers. However, due to the micro scale, quantitative experimental measurement of mixing is quite limited in micromixers, and special techniques are also required for the fabrication.

The current special issue includes 12 research papers [2–13]. The topics of the papers are focused on design of micromixers [2–7,9,12,13], fabrication [8,9,12], and analysis [11]. Some of them propose novel micromixer designs [2,3,5,6], which include a slight modification of an existing micromixer [6]. Most of them deal with passive micromixers, but two papers [5,11] report the studies on electrokinetic micromixers. Fully three-dimensional (3D) micromixers were investigated in some cases [2–4,13]. Hossain et al. [4] applied optimization techniques to the design of a 3D micromixer. Raza et al. [13] performed a review of recently developed passive micromixers and a comparative analysis of 10 typical micromixers. Martinez-Lopez et al. [8], Wang et al. [9], and Sabotin et al. [12] focused on the fabrication techniques for micromixers, and two of them [8,9] used 3D print techniques.

The review paper of Raza et al. [13] has a unique feature that is different from other reviews on micromixers. They classified passive micromixer designs developed so far into five types: 2D designs using serpentine, spiral, curved helical channels, 2D designs with split-and-recombination (SAR) structures, 3D design with serpentine and/or SAR structures, 3D design with patterned grooves, and 3D designs with SAR two-layer crossing channels. Along with a general review of micromixers in each type, they performed a quantitative comparison among 10 representative micromixers selected from the five types with their own calculations. For this comparison, they calculated the mixing indexes and pressure drops of theses micromixers under the same axial length and Reynolds numbers. Eight Reynolds numbers in a range of Re = 0.01–120 were tested, which have been selected from three ranges: low-Re (Re < 1), intermediate-Re (1 < Re < 40), and high-Re (Re > 40) ranges. Even through a number of micromixer designs have been developed, there has not been any quantitative comparison of these micromixers under the same operating conditions. Therefore, this review report is expected to contribute to the selection of effective micromixers for specific applications.

Martinez-Lopez et al. [8] characterized photopolymers and stereolithography processes for manufacturing 3D print molds and polydimethylsiloxane castings of micromixers. Validation of the soft tooling approach was performed for an asymmetric SAR micromixer with different cross sections under various flow conditions (10 < Re < 70). Wang et al. [9] designed a jet mixer with arrays of micro-nozzles and fabricated it using 3D printing

Citation: Kim, K.-Y. Editorial for the Special Issue on Analysis, Design and Fabrication of Micromixers. *Micromachines* **2021**, *12*, 533. https://doi.org/10.3390/mi12050533

Received: 1 May 2021
Accepted: 6 May 2021
Published: 7 May 2021

technology. The design has two opposite arrays of micro-nozzles to enhance the mixing performance with chaotic advection. The mixing efficiency of the mixer was measured experimentally and analyzed comparatively with a conventional Y-shaped micromixer. Sabotin et al. [12] constructed a simple technical model of micro electrical discharge machining (EDM) milling. They performed an experiment by machining different microgrooves into corrosive-resistant steel and tested the following parameters: material removal rate, electrode dressing time, machining strategy, electrode wear, and electrode wear control time. The technical model was demonstrated through a bottom grooved micromixers (BGM) design. The mixing performances of different BGM designs were evaluated numerically to find an optimum design.

Different methods have been applied to the analysis of mixing. Most of the works used only computational fluid dynamics (CFD) for the analysis [2–6,10,13]. However, some of the works employed experimental analysis [8,9] or both experimental and CFD analyses [7,12]. On the other hand, Kim et al. [11] performed a theoretical analysis of an electromagnetic micromixer and compared the results with experimental measurements. They suggested the physical foundation for the electromagnetic coupling in the micromixer, and derived a correlation between the mechanical velocity and electrical driving point impedance. Exceptionally, Granados-Ortiz and Ortega-Casanova [10] analyzed the thermal mixing between two parallel laminar flows with different temperatures in a simple 2D channel with an obstacle located at the center. They performed unsteady 2D Navier-Stokes analysis to analyze the mixing caused by the vortex shedding behind the obstacle.

I would like to express my appreciation to all the experts who contributed to this special issue. I also thank all the reviewers of the submitted papers. This special issue will be continued as Analysis, Design and Fabrication of Micromixers II. I hope for further up-to-date technologies of micromixers to be reported in the next special issue.

Conflicts of Interest: The author declares no conflict of interest.

References

1. Afzal, A.; Kim, K.-Y. *Analysis and Design Optimization of Micromixers, SpringerBriefs in Computational Mechanics*; Springer Nature Singapore: Singapore, 2021.
2. Okuducu, M.B.; Aral, M.M. Toward the Next Generation of Passive Micromixers: A Novel 3-D Design Approach. *Micromachines* **2021**, *12*, 372. [CrossRef] [PubMed]
3. Naas, T.T.; Hossain, S.; Aslam, M.; Rahman, A.; Hoque, A.S.M.; Kim, K.-Y.; Islam, S.M.R. Kinematic Measurements of Novel Chaotic Micromixers to Enhance Mixing Performances at Low Reynolds Numbers: Comparative Study. *Micromachines* **2021**, *12*, 364. [CrossRef] [PubMed]
4. Hossain, S.; Naas, T.T.; Islam, F.; Kaseem, M.; Bui, P.D.H.; Bhuiya, M.M.K.; Aslam, M.; Kim, K.-Y. Enhancement of Mixing Performance of Two-Layer Crossing Micromixer through Surrogate-Based Optimization. *Micromachines* **2021**, *12*, 211. [CrossRef] [PubMed]
5. Wang, C. Liquid Mixing Based on Electrokinetic Vortices Generated in a T-Type Microchannel. *Micromachines* **2021**, *12*, 130. [CrossRef] [PubMed]
6. Wu, C.-Y.; Lra, B.-H. Numerical Study of T-Shaped Micromixers with Vortex-Inducing Obstacles in the Inlet Channels. *Micromachines* **2020**, *11*, 1122. [CrossRef] [PubMed]
7. Hsu, C.-W.; Shih, P.-T.; Chen, J.M. Enhancement of Fluid Mixing with U-Shaped Channels on a Rotating Disc. *Micromachines* **2020**, *11*, 1110. [CrossRef] [PubMed]
8. Martínez-López, J.I.; Cervantes, H.A.B.; Iturbe, L.D.C.; Vázquez, E.; Naula, E.A.; López, A.M.; Siller, H.R.; Mendoza-Buenrostro, C.; Rodríguez, C.A. Characterization of Soft Tooling Photopolymers and Processes for Micromixing Devices with Variable Cross-Section. *Micromachines* **2020**, *11*, 970. [CrossRef] [PubMed]
9. Wang, Y.; Zhang, Y.; Qiao, Z.; Wang, W. A 3D Printed Jet Mixer for Centrifugal Microfluidic Platforms. *Micromachines* **2020**, *11*, 695. [CrossRef] [PubMed]
10. Granados-Ortiz, F.-J.; Ortega-Casanova, J. Mechanical Characterisation and Analysis of a Passive Micro Heat Exchanger. *Micromachines* **2020**, *11*, 668. [CrossRef] [PubMed]
11. Kim, N.; Chan, W.X.; Ng, S.H.; Yoon, Y.-J.; Allen, J.B. Understanding Interdependencies between Mechanical Velocity and Electrical Voltage in Electromagnetic Micromixers. *Micromachines* **2020**, *11*, 636. [CrossRef] [PubMed]
12. Sabotin, I.; Tristo, G.; Valentincic, J. Technical Model of Micro Electrical Discharge Machining (EDM) Milling Suitable for Bottom Grooved Micromixer Design Optimization. *Micromachines* **2020**, *11*, 594. [CrossRef] [PubMed]
13. Raza, W.; Hossain, S.; Kim, K.-Y. A Review of Passive Micromixers with a Comparative Analysis. *Micromachines* **2020**, *11*, 455. [CrossRef] [PubMed]

Review

A Review of Passive Micromixers with a Comparative Analysis

Wasim Raza [†], Shakhawat Hossain [†,‡] and Kwang-Yong Kim *

Department of Mechanical Engineering, Inha University, Incheon 22212, Korea; wasimkr@live.in (W.R.); msfk3742@yahoo.com (S.H.)

* Correspondence: kykim@inha.ac.kr; Tel.: +82-32-872-3096; Fax: +82-32-868-1716

† These authors contributed equally to this work.

‡ Current address: Department of Industrial and Production Engineering, Jashore University of Science and Technology, Jessore 7408, Bangladesh

Received: 2 March 2020; Accepted: 24 April 2020; Published: 27 April 2020

Abstract: A wide range of existing passive micromixers are reviewed, and quantitative analyses of ten typical passive micromixers were performed to compare their mixing indices, pressure drops, and mixing costs under the same axial length and flow conditions across a wide Reynolds number range of 0.01–120. The tested micromixers were selected from five types of micromixer designs. The analyses of flow and mixing were performed using continuity, Navier-Stokes and convection-diffusion equations. The results of the comparative analysis were presented for three different Reynolds number ranges: low-*Re* ($Re \leq 1$), intermediate-*Re* ($1 < Re \leq 40$), and high-*Re* ($Re > 40$) ranges, where the mixing mechanisms are different. The results show a two-dimensional micromixer of Tesla structure is recommended in the intermediate- and high-*Re* ranges, while two three-dimensional micromixers with two layers are recommended in the low-*Re* range due to their excellent mixing performance.

Keywords: passive micromixers; comparative analysis; Navier-Stokes equations; mixing index; pressure drop; mixing cost

---

1. Introduction

Microfluidics is becoming more important in many chemical and biological applications [1–3]. Diffusion instead of turbulence governs the mixing of fluid species at the micrometer scale, and the mixing process is prolonged. Consequently an enhanced fluid mixing capability is essential for the design of micromixers. Micromixers are necessary parts of lab-on-a-chip (LOC) devices and micro-total analysis systems (μ-TAS) [4–10]. In various microfluidic applications, the mixing capability of micromixers may affect the overall performance of the entire systems. For example, fast mixing of cells, reagents, and organic solutions are essential in many bioengineering and biochemical systems [11–13]. A notable example of the mixing of reagents or organic solutions is nanomaterial synthesis [14,15]. Efficient mixing significantly improves the detection sensitivity and reduces the analysis time [16,17].

Based on the mixing mechanisms, micromixers generally come under two categories: active and passive types. In active micromixers, flow perturbation is created using external energy sources, such as magnetic fields, electric fields, and ultrasonic vibration. Additionally, active micromixers require a control mechanism, which makes the entire system more complex and creates difficulties in fabrication and operation. However, acoustofluidic devices [18–20] showed promising prospects for lab-on-a-chip applications and in the bio-medical diagnostic field. Sharp-edge-based acoustic active micromixers [20,21] with serpentine-like channels were recently fabricated in a simple manner and used for versatile applications. Furthermore, an active acoustic-based micromixer [18] also achieved combined pumping and mixing in a single device; in other words, it required no external source of fluid pumping.

Contrarily, no external energy source except for pumping of fluids is required in passive micromixers, and the mixing is performed by molecular diffusion and chaotic advection. Chaotic advection is generally created by modifying the microchannel geometry to reduce the diffusion length and maximize the interfacial area between the fluids by manipulating or reorganizing the fluid flow. Thus, compared with active micromixers, passive micromixers are economical, convenient, and can easily be incorporated into LOC systems [10,22–24]. Passive micromixers are also used as 'concentration gradient generators' for biological and pharmaceutical applications [25,26]. Recently, various passive mixers have been proposed to accomplish effective mixing. Chaotic flows induced by repeated perturbations of the fluid flow can significantly increase the performance of micromixers [27–29], and are generated by several different microchannel geometries: two-dimensional (2D) structures [30–41], three-dimensional (3D) serpentine structures [42–49], patterned groove structures [29,50–58], 2D and 3D split-and-recombination (SAR) structures [59–81], and two-layer crossing channels [82–89].

Several review articles have introduced a variety of passive micromixers with their mixing mechanisms [6,24,90–95]; these micromixers have various dimensions and working conditions (e.g., Reynolds number). However, quantitative comparisons of their mixing performance are rarely found in the literature, even though designers would benefit from the information on mixing performance of different micromixers under the same geometric and working conditions. Such information will assist in the selection of suitable designs to meet specific requirements for different processes [92].

Quantitative comparisons have been performed only for specific types of passive micromixers. Falk and Commenge performed a comprehensive study of conventional T-type micromixers and micromixers based on the concept of SAR or multi-lamination through the Villermaux-Dushman test reaction [96]. They analyzed and compared the mixing efficiencies of the micromixers while considering Reynolds number and power dissipation per unit mass of the liquid. Viktorov et al. presented a comparative analysis of three passive micromixers (tear-drop, Y-Y, and H-C micromixers) in a wide range of Reynolds numbers [97]. They conducted a numerical simulation and experimental analysis to evaluate the mixing performance at Reynolds numbers ranging from 1 to 100. Bošković et al. analyzed and compared the characteristics of the residence time in three different passive micromixers (micromixers with 3D serpentine structure, staggered herringbone grooves, and split and recombine (SAR) structure) [98]. The microstructures with similar channel designs were analyzed across a wide range of Reynolds numbers ($0.3 \leq Re \leq 110$).

As mentioned above, a variety of micromixer designs have been developed so far, but no one has reported a quantitative evaluation of the micromixers operating at the same conditions, which is necessary for the selection of effective micromixers under different flow conditions in various microfluidic applications. Therefore, in the present work, a review of a wide range of existing passive micromixers is presented and a comparative analysis of selected micromixers was performed under the same working fluids, Reynolds number, and axial channel length for a quantitative comparison among them. For the comparative analysis, ten micromixers were selected among the high-performance micromixers found in the literature, which cover five typical micromixer designs. These micromixers achieved efficient mixing with different mixing mechanisms or their combinations, as explained in Section 2. The numerical analyses of mixing and fluid flow were performed using Navier-Stokes and advection-diffusion equations for momentum and mass transports, respectively. The comparison was performed in a Reynolds number range of 0.01–120.

2. Mixing Mechanisms of Micromixer Types and Selected Micromixers

In the last two decades, a number of passive micromixers involving different microchannel designs have been proposed, as introduced in the previous section. The microchannel designs can be categorized into five types, as shown in Table 1. For the quantitative comparison in this work, ten representative micromixers (M-1 to M-10) were selected as shown in Table 2, and Figures 1–10 show that their schematics. M-1 to M-4 are 2D planar designs, and M-5 to M-10 are 3D designs. To compare the mixing performances at an equal axial length, which refers to the length of the channel in the x-direction between the start of the mixing unit and the micromixer exit, the number of mixing units in each micromixer was changed from the original number as indicated in each figure caption. The dimensions of the mixing unit in each micromixer were the same as those in its original design. The mixing capability was evaluated at each micromixer exit (5050 μm (L_t) downstream of the start of the mixing unit) for the comparison.

Table 1. Types of micromixer designs.

Type	Micromixer Design	Mixing Mechanism	Selected Micromixer
1	2D designs using serpentine, spiral, curved helical channels [30–41]	Inertial force (secondary flow, Dean vortex)	M-1 [40], M-2 [41]
2	2D designs with SAR structures [68–79]	Inertial force, SAR	M-3 [68], M-4 [77]
3	3D design with serpentine and/or SAR structures [42–49,59–67,80,81]	Inertial force, chaotic mixing, multi-lamination	M-5 [45], M-6 [59], M-7 [81]
4	3D design with patterned grooves [29,50–58]	Inertial force, chaotic mixing	M-8 [58]
5	3D designs with SAR two-layer crossing channels [82–89]	Chaotic mixing, multi-lamination	M-9 [84], M-10 [87]

Table 2. Selected micromixers.

Micromixer Designation	Micromixer	Geometry (Type in Table 1)	Designers [Ref.]	Year
M-1	Curved micromixer	2D serpentine (1)	Hossain et al. [40]	2009
M-2	Curved micromixer with grooves	2D serpentine (1)	Alam and Kim [41]	2012
M-3	Modified P-SAR (planar SAR) micromixer with dislocation sub-channels	2D SAR (2)	Li et al. [68]	2013
M-4	Modified Tesla micromixer	2D SAR (2)	Hossain et al. [77]	2010
M-5	3D serpentine micromixer	3D serpentine (3)	Ansari and Kim [45]	2009
M-6	3D serpentine SAR micromixer	3D serpentine SAR (3)	Hossain and Kim [59]	2015
M-7	Improved serpentine laminating micromixer	3D serpentine SAR (3)	Park et al. [81]	2008
M-8	Barrier embedded chaotic micromixer	3D grooves (4)	Kim et al. [58]	2004
M-9	Chaotic micromixer with two-layer crossing microchannels	3D SAR (5)	Xia et al. [84]	2005
M-10	Chaotic micromixer with two-layer serpentine crossing microchannels	3D serpentine SAR (5)	Hossain et al. [87]	2017

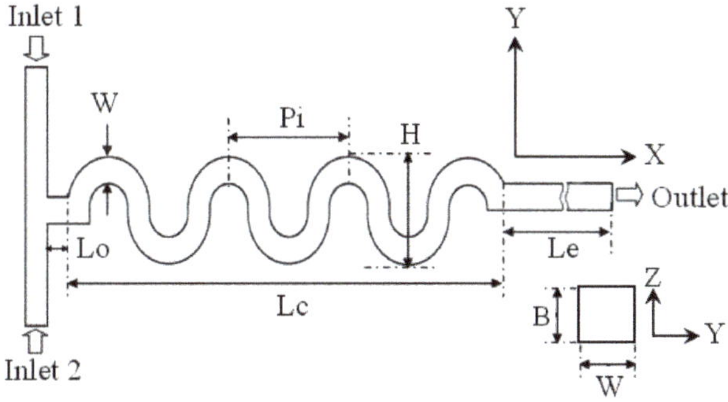

Figure 1. Curved micromixer: M-1 (no. of mixing units used: 8). Reproduced with permission from [40].

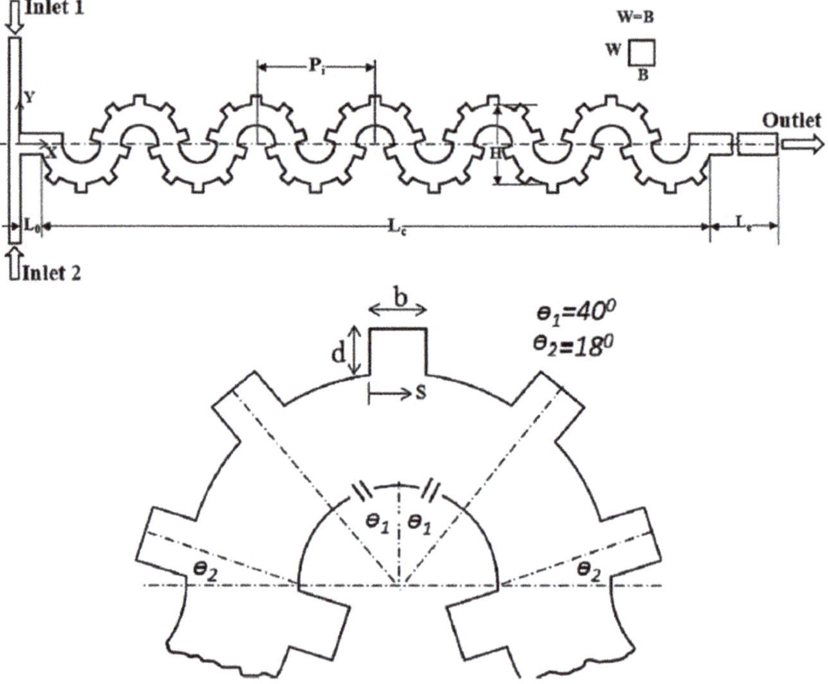

Figure 2. Curved micromixer with rectangular grooves: M-2 (no. of mixing units used: 8). Reproduced with permission from [41].

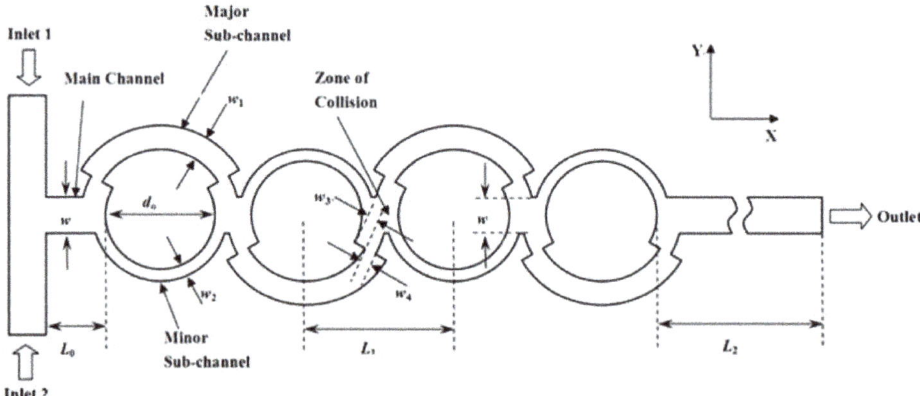

Figure 3. Modified P-SAR micromixer with dislocation sub-channels: M-3 (no. of mixing units used: 4). Reproduced with permission from [68].

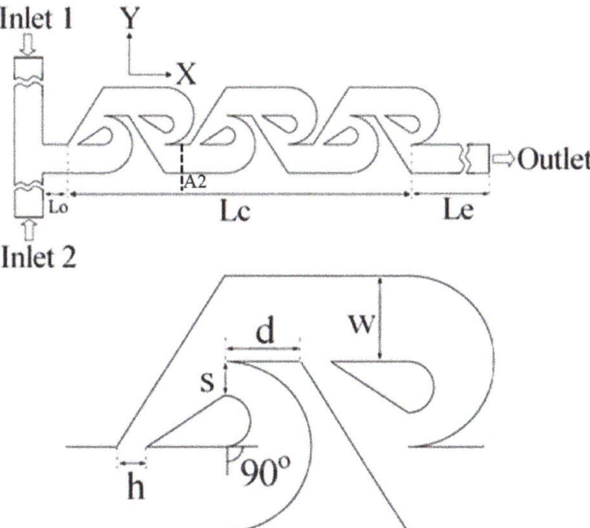

Figure 4. Modified Tesla micromixer: M-4 (no. of mixing units used: 6). Reproduced with permission from [77].

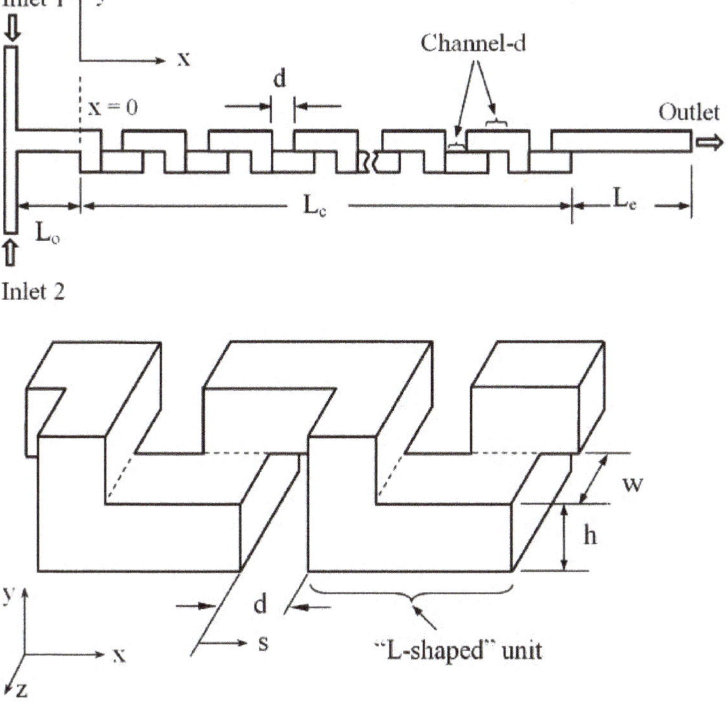

Figure 5. 3D serpentine micromixer: M-5 (no. of mixing units used: 8). Reproduced with permission from [45].

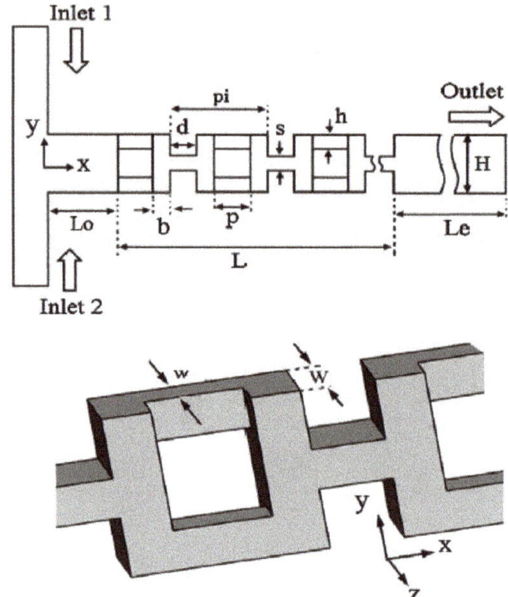

Figure 6. 3D serpentine SAR micromixer: M-6 (no. of mixing units used: 18). Reproduced with permission from [59].

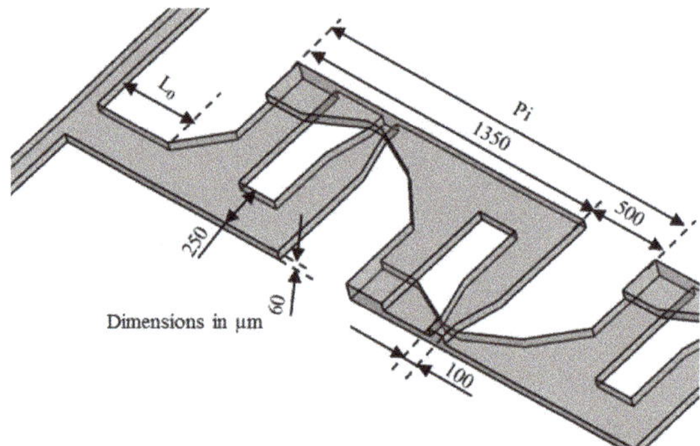

Figure 7. Improved serpentine lamination micromixer: M-7 (no. of mixing units used: 2) [81].

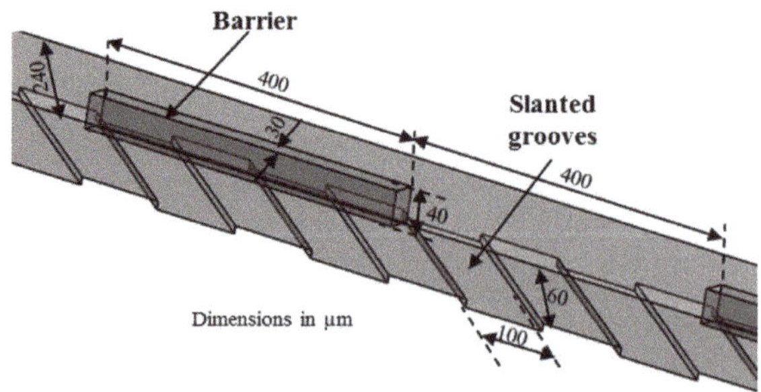

Figure 8. Barrier embedded chaotic micromixer: M-8 (no. of mixing units used: 6) [58].

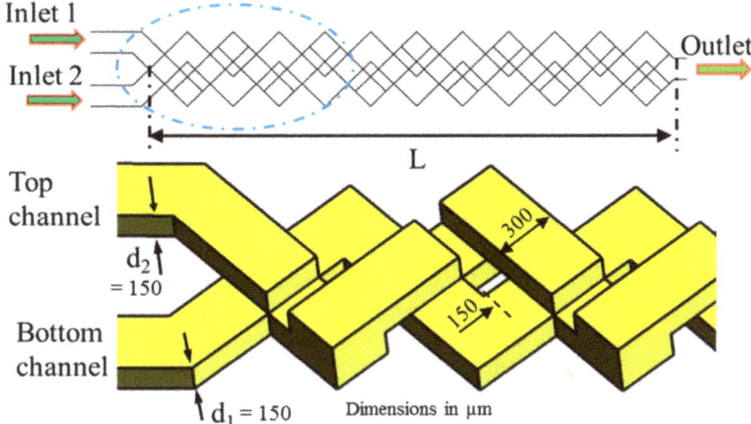

Figure 9. Chaotic micromixer with two-layer crossing microchannels: M-9 (no. of mixing units used: 8). Reproduced with permission from [89].

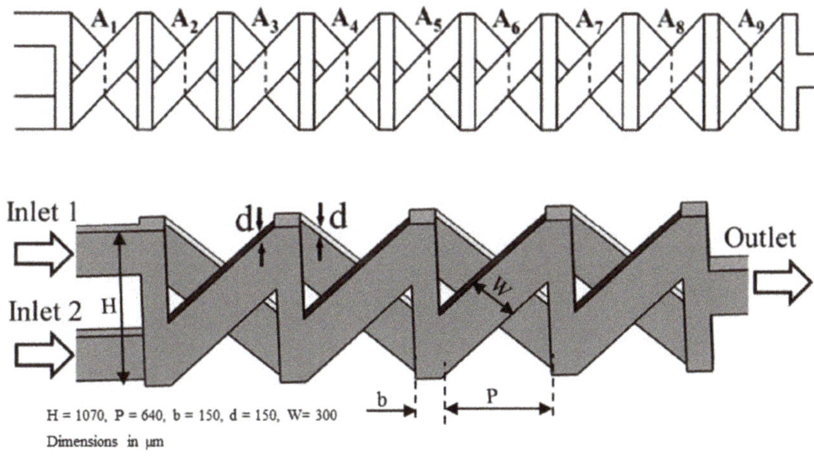

Figure 10. Chaotic micromixer with two-layer serpentine crossing microchannels: M-10 (no. of mixing units used: 6). Reproduced with permission from [87].

2.1. 2D Designs Using Serpentine, Spiral, and Curved Helical Channels (Type 1)

2D planar micromixer designs have an advantage of simplicity in the fabrication compared to the complex 3D designs. Mixing in 2D serpentine, spiral, and curved helical channels [30–41] mainly depends on the advection caused by the secondary flow or Dean vortices created by the inertia force. The performance of this type (type 1 in Table 1) of micromixers improves as the Reynolds number increases due to the dependence of secondary flow/Dean vortices on Reynolds number [78]. Hossain et al. [40] and Alam and Kim [41] conducted numerical investigations of mixing in 2D planar micromixers, M-1 (Figure 1) and M-2 (Figure 2), respectively. Hossain et al. [40] estimated the mixing performance of three serpentine passive micromixers (square-wave, zigzag, and curved shape microchannels) across a wide range of Reynolds numbers (0.267–267). Alam and Kim [41] introduced rectangular grooves on the sidewalls of a curved serpentine channel to evaluate mixing in a Reynolds number range of 0.5–90.

2.2. 2D Designs with SAR Structures (Type 2)

A variety of planar SAR micromixers (type 2 in Table 1) involving multi-lamination and inertial flow have been developed [68–79]. The SAR micromixers generate multi-laminating flow patterns successively with the three underlying flow mechanisms of splitting, recombination, and rearrangement [6]. Dean and expansion vortices are generated through curved channels and expansion-contraction at high Reynolds numbers [70,71,76].

Ansari et al. [76] proposed and analyzed a micromixer using asymmetrical splits and collisions of fluid streams. The lowest mixing performance was obtained with uniform sub-channel widths representing balanced collision over a range of Reynolds numbers. Induced Dean vortices at the interfaces in the curved sub-channels and SAR were found to enhance the mixing performance. Xia et al. [70] designed an asymmetric SAR micromixer with a fan-shaped cavity to achieve efficient mixing by the synergistic effect of expansion vortices and Dean vortices in the fan-shaped cavity along with unbalanced collision in the recombination zone. The selected micromixer, M-3 [68] shown in Figure 3, represents a planar asymmetric SAR (P-ASAR) design with outward protruded sub-channels, which is an improved form of a previous micromixer [76]. In this micromixer, the mixing performance was enhanced by the synergistic effect of unbalanced inertial collisions, expansion vortices, and Dean vortices on mixing.

Xia et al. [74] designed and fabricated a 2D planner micromixer consisting of a series of gaps and baffles in a simple microchannel. A sudden contraction provided by a gap accelerates the fluid stream and produces symmetrical expansion vortices. The accelerated fluid stream is separated by a baffle, and the same flow pattern is repeated. The synergistic effects of abrupt contraction and expansion, multiple SAR, and multiple secondary vortices increase the interfacial area of the fluids, resulting in excellent mixing performance. Hong et al. [72] proposed an innovative micromixer with a 2D modified Tesla structure (M-4) shown in Figure 4, which takes advantage of the Coanda effect. The splitting and reuniting of the fluids effectively reduce the diffusion path between the fluid streams. The structure causes chaotic flow by the collision of the fluid streams on redirection, and improves mixing. Hossain et al. [77] improved the efficiency of a modified Tesla micromixer through an optimization.

2.3. 3D Design with Serpentine and/or SAR Structures (Type 3)

Several micromixers involving 3D serpentine and/or SAR structures [42–49,59–67,80,81] (type 3 in Table 1) have been proposed. The 3D serpentine path induces a stirring flow at each bend and generates a secondary flow to enhance mixing [43,45]. The SAR structure provides lamination that decreases the diffusion path and enhances mixing at low Reynolds numbers [65]. The secondary flow combines with the axial flow, and creates chaos by stretching and folding the fluid interface [47].

Ansari and Kim [45] investigated the effects of flow and geometric parameters on the mixing performance of an L-shaped 3D serpentine micromixer (M-5) shown in Figure 5. Hossain and Kim [59] proposed a 3D serpentine SAR micromixer composed of O- and H-shaped units (M-6) shown in Figure 6. The O- and H-structures split and recombine the fluid streams repeatedly. Continuous splitting and recombining of the fluid streams generate chaotic mixing. Kim et al. [80] proposed a chaotic-mixing-based serpentine lamination micromixer (SLM) composed of F-shaped mixing units. The SLM structure combines two general chaotic mixing mechanisms: SAR induced by the mixing segments and chaotic advection induced by the overall 3D serpentine channel path. Park et al. [81] proposed a geometrical modification of SLM to improve the mixing efficiency. Figure 7 shows the improved SLM (ISLM) (M-7), where the original F-shaped mixer is altered at the recombination region. The reduced cross-sectional area enhances the vertical lamination by enhancing the local advection, consequently improving the mixing performance.

2.4. 3D Design with Patterned Grooves (Type 4)

Patterned grooves on the channel wall were also used to promote mixing in microchannels [29,50–58] (type 4 in Table 1). The grooves induce 3D helical flow in the microchannels, which promotes mixing. Kim et al. [58] designed a micromixer with rectangular barriers on the top of the slanted grooves (M-8) shown in Figure 8. The periodically located barriers on the top wall are capable of creating velocity fields characterized by two elliptic points and a hyperbolic point alternately within the helical motion produced by the grooves at the bottom, which results in enhancement of chaotic mixing.

2.5. 3D Designs with SAR Two-Layer Crossing Channels (Type 5)

3D micromixers with SAR two-layer crossing channels (type 5 in Table 1) have achieved remarkable mixing performance [82–89]. The crossing channels produce chaotic flow due to repeated splitting, stretching, rotating, and folding processes. A saddle-shaped flow structure is produced despite low Reynolds number. Hence, the chaotic flow inside the crossing channels does not depend on the fluid inertia [84].

Xia et al. [84] designed a chaotic micromixer (M-9) using 3D X-shaped crossing channels (TLCCM) shown in Figure 9. The micromixer exhibited an outstanding mixing efficiency of 96% at a low Reynolds number (*Re* = 0.2). Hossain et al. [89] further analyzed the responses of the flow structure and mixing performance to the variations in geometric parameters of TLCCM.

Recently, Hossain et al. [87] designed a chaotic micromixer with two-layer serpentine crossing microchannels (M-10) based on the mixing mechanism of TLCCM (Figure 10) to improve the mixing capability at low Reynolds numbers. This micromixer consists of two layers of serpentine channels. The bottom and top layers contain a series of N- and inverse N-shaped segments, respectively. The fluid streams are interconnected at the vertical sections and the intersections of the crossing channels. The fluid flow in successive mixing modules produces chaotic advection through continuous splitting, recombination, enlarging, and folding of the fluid streams.

3. Analysis Methods

The flow inside the microchannel was assumed to be steady, incompressible, laminar, and Newtonian. The continuity (Equation (1)), momentum (Equation (2)), and convection-diffusion (Equation (3)) equations were solved numerically for the flow and mixing analysis. The CFD software ANSYS CFX 15.0® [99] employs a coupled solver and finite volume technique to discretize Equations (1)–(3):

$$\nabla \cdot \vec{V} = 0 \tag{1}$$

$$\left(\vec{V} \cdot \nabla\right)\vec{V} = -\frac{1}{\rho}\nabla P + \nu \nabla^2 \vec{V} \tag{2}$$

$$\left(\vec{V} \cdot \nabla\right)C = D\nabla^2 C \tag{3}$$

where $\vec{V}$, C, D, ρ, P, and ν are the velocity, dye concentration, diffusion coefficient, density, pressure, and kinematic viscosity, respectively.

A finite-volume-based commercial code ANSYS CFX 15.0® [99] has been used to solve the governing differential equations. Tetrahedral meshes were generated using ICEM CFD® for the grid systems. The boundary conditions were the uniform velocity profiles at the inlets, zero static pressure at the outlets, and no-slip conditions at the channel walls. Dye-water solution and water, both having the properties of water at 25 °C (dynamic viscosity: 8.8×10^{-4} kg/m·s; density: 997 kg/m^3) were used as the working fluids. The numerical solution was assumed to be converged as root-mean-squared residual values for momentum and mass fraction reach a value less than 1.0×10^{-6}. The diffusivity constant of the solution was fixed to be 1.0×10^{-10} m^2/s. Reynolds number was calculated using the hydraulic diameter and average velocity at the inlets.

The numerical methods same as used in the present work, were validated comparing with experimental results for a variety of micromixers in previous works [40,41,45,54,55,57,59,69,75–77,79,87–89].

The mixing index was estimated by calculating the variance of concentration on a particular transverse plane normal to the fluid flow. The mass fraction variance was determined as:

$$\sigma = \sqrt{\frac{1}{N}\sum_{i=1}^{N}(c_i - \bar{c}_m)^2} \tag{4}$$

where N denotes the number of data points on the cross-sectional plane, $\bar{c}_m$ is the optimal mass fraction, and c_i is the mass fraction at a point i. The mixing index was defined as:

$$M = 1 - \frac{\sigma}{\sigma_{\max}} \tag{5}$$

where $\sigma_{\max}$ is the maximum variance over the data range. The mixing index ranges from zero (wholly separated fluid streams) to unity (complete mixing).

The other mixing performance parameter, mixing cost (*MC*) [100] that takes pressure drop into account along with the mixing index was defined as follows: where *N* denotes the number of data points on the cross-sectional plane, c_m is the optimal mass fraction, and c_i is the mass fraction at a point *i*. And, the mixing index was defined as:

$$\text{mixing cos t} \equiv \frac{M}{\Delta P} \tag{6}$$

where ΔP denotes the pressure drop. A high value of *MC* indicates an efficient micromixer.

4. Results and Discussion

4.1. Grid Refinement Test

Optimal numbers of computational grid nodes were determined after performing exhaustive grid refinement tests at *Re* = 40 for all the micromixers. Mixing index at the exit was selected as an indicator for choosing the optimal grids. The number of grid nodes from 3.86×10^5 to 2.55×10^6 were tested in the grid refinement tests. The optimum numbers of nodes that were selected for the ten micromixers varies from 1.55×10^6 to 2.21×10^6, as listed in Table 3.

Table 3. Optimum numbers of grid nodes selected through grid refinement tests at *Re* = 40.

Micromixer	Optimum Number of Meshes	Number of Finer Meshes	% Deviation of Mixing Index between the Optimum and Finest Meshes	Micromixer	Optimum Number of Meshes	Number of Finer Meshes	% Deviation of Mixing Index between the Optimum and Finest Meshes
M-1	1.81×10^6	2.04×10^6	1.39	M-6	2.07×10^6	2.34×10^6	0.10
M-2	2.07×10^6	2.41×10^6	1.16	M-7	2.19×10^6	2.55×10^6	0.54
M-3	1.55×10^6	2.08×10^6	0.94	M-8	2.21×10^6	2.53×10^6	1.05
M-4	2.16×10^6	2.33×10^6	0.20	M-9	1.90×10^6	2.03×10^6	0.10
M-5	1.79×10^6	1.96×10^6	0.10	M-10	1.61×10^6	1.80×10^6	0.11

4.2. Quantitative Comparisons in Different Reynolds Number Ranges

The examined range of Reynolds numbers (*Re* = 0.01–120) was divided into three sub-ranges: low-*Re* (*Re* ≤ 1), intermediate-*Re* (1 < *Re* ≤ 40), and high-*Re* (*Re* > 40) ranges. One of the primary purposes of the present comparative analysis was to find efficient micromixers in each *Re* range.

Evaluated values of mixing index (M), pressure drop (ΔP), and mixing cost (MC) for each micromixer are presented in Tables 4–6, respectively.

Table 4. Mixing index at the exit.

Micromixer	Mixing Index at the Exit							
	Low-*Re* Range			Intermediate-*Re* Range		High-*Re* Range		
	$Re = 0.01$	$Re = 0.1$	$Re = 1$	$Re = 20$	$Re = 40$	$Re = 60$	$Re = 80$	$Re = 120$
M-1	0.560	0.250	0.224	0.572	0.793	0.979	0.990	0.998
M-2	0.554	0.244	0.221	0.694	0.858	0.974	0.997	0.997
M-3	0.268	0.141	0.125	0.241	0.422	0.657	0.828	0.890
M-4	0.465	0.255	0.203	0.883	0.999	0.999	0.999	0.999
M-5	0.546	0.361	0.377	0.999	0.999	0.999	0.999	0.999
M-6	0.649	0.506	0.472	0.999	0.999	0.999	0.999	0.999
M-7	0.904	0.594	0.537	0.567	0.733	0.856	0.909	0.963
M-8	0.310	0.238	0.226	0.250	0.284	0.310	0.337	0.401
M-9	0.939	0.909	0.905	0.934	0.972	0.987	0.998	0.996
M-10	0.970	0.926	0.915	0.901	0.929	0.995	0.973	0.996

Table 5. Pressure drop through micromixer.

Micromixer	Pressure Drop (Pa)							
	Low-*Re* Range			Intermediate-*Re* Range		High-*Re* Range		
	$Re = 0.01$	$Re = 0.1$	$Re = 1$	$Re = 20$	$Re = 40$	$Re = 60$	$Re = 80$	$Re = 120$
M-1	2.95×10^{0}	2.95×10^{1}	2.95×10^{2}	6.77×10^{3}	1.63×10^{4}	2.81×10^{4}	4.18×10^{4}	7.70×10^{4}
M-2	2.87×10^{0}	2.87×10^{1}	2.87×10^{2}	6.64×10^{3}	1.62×10^{4}	2.80×10^{4}	4.15×10^{4}	7.68×10^{4}
M-3	3.58×10^{-1}	3.72×10^{0}	3.73×10^{1}	8.30×10^{2}	1.95×10^{3}	3.39×10^{3}	5.14×10^{3}	9.72×10^{3}
M-4	1.03×10^{0}	1.03×10^{1}	1.04×10^{2}	3.53×10^{3}	1.20×10^{4}	2.58×10^{4}	4.50×10^{4}	9.70×10^{4}
M-5	7.94×10^{-1}	7.94×10^{0}	7.96×10^{1}	2.44×10^{3}	7.12×10^{3}	1.39×10^{4}	2.19×10^{4}	4.32×10^{4}
M-6	1.67×10^{1}	1.67×10^{2}	1.68×10^{3}	5.32×10^{4}	1.57×10^{5}	3.10×10^{5}	5.15×10^{5}	1.09×10^{6}
M-7	2.46×10^{0}	2.46×10^{1}	2.47×10^{2}	5.27×10^{3}	1.16×10^{4}	1.95×10^{4}	2.90×10^{4}	5.49×10^{4}
M-8	2.09×10^{0}	2.12×10^{1}	2.12×10^{2}	4.29×10^{3}	8.73×10^{3}	1.33×10^{4}	1.79×10^{4}	2.75×10^{4}
M-9	1.78×10^{-1}	1.78×10^{0}	1.78×10^{1}	4.42×10^{2}	1.16×10^{3}	2.21×10^{3}	3.59×10^{3}	7.37×10^{3}
M-10	1.63×10^{-1}	1.63×10^{0}	1.63×10^{1}	3.90×10^{2}	9.94×10^{2}	1.84×10^{3}	2.93×10^{3}	5.90×10^{3}

Table 6. Mixing cost.

Micromixer	Mixing Cost, MC (Pa^{-1})							
	Low-*Re* Range			Intermediate-*Re* Range		High-*Re* Range		
	$Re = 0.01$	$Re = 0.1$	$Re = 1$	$Re = 20$	$Re = 40$	$Re = 60$	$Re = 80$	$Re = 120$
M-1	1.90×10^{-1}	8.47×10^{-3}	7.58×10^{-4}	8.45×10^{-5}	4.85×10^{-5}	3.48×10^{-5}	2.39×10^{-5}	1.30×10^{-5}
M-2	1.93×10^{-1}	8.48×10^{-3}	7.67×10^{-4}	1.05×10^{-4}	5.29×10^{-5}	3.47×10^{-5}	2.40×10^{-5}	1.30×10^{-5}
M-3	7.51×10^{-1}	3.79×10^{-2}	3.36×10^{-3}	2.90×10^{-4}	2.16×10^{-4}	1.94×10^{-4}	1.61×10^{-4}	9.15×10^{-5}
M-4	4.50×10^{-1}	2.47×10^{-2}	1.96×10^{-3}	2.50×10^{-4}	8.36×10^{-5}	3.87×10^{-5}	2.22×10^{-5}	1.03×10^{-5}
M-5	6.88×10^{-1}	4.55×10^{-2}	4.74×10^{-3}	4.09×10^{-4}	1.40×10^{-4}	7.18×10^{-5}	4.57×10^{-5}	2.32×10^{-5}
M-6	3.89×10^{-2}	3.03×10^{-3}	2.82×10^{-4}	1.88×10^{-5}	6.38×10^{-6}	3.22×10^{-6}	1.94×10^{-6}	9.20×10^{-7}
M-7	3.67×10^{-1}	2.41×10^{-2}	2.18×10^{-3}	1.07×10^{-4}	6.29×10^{-5}	4.40×10^{-5}	3.13×10^{-5}	1.76×10^{-5}
M-8	1.48×10^{-1}	1.12×10^{-2}	1.07×10^{-3}	5.81×10^{-5}	3.25×10^{-5}	2.34×10^{-5}	1.88×10^{-5}	1.46×10^{-5}
M-9	5.28×10^{0}	5.11×10^{-1}	5.08×10^{-2}	2.12×10^{-3}	8.38×10^{-4}	4.47×10^{-4}	2.78×10^{-4}	1.35×10^{-4}
M-10	5.95×10^{0}	5.68×10^{-1}	5.60×10^{-2}	2.31×10^{-3}	9.34×10^{-4}	5.41×10^{-4}	3.32×10^{-4}	1.69×10^{-4}

Figure 11 shows the general trend of the mixing index variation with Reynolds number for the curved micromixer, M-1 [40]. At low Re ($Re \leq 1$), low velocity of the fluid stream causes a long residential time of the fluids within the microchannel, which provides sufficient time for diffusive mixing. The mixing deteriorates rapidly as Reynolds number increases due to the reduction in the residential time, reaching a minimum at around Re = 1, where the residential time is inadequate, and the transverse flow is still ineffective for generating the secondary flow. Thus, the mixing index remains at a low level. However, beyond this Re, the residence time reduces further, but the secondary flow becomes active. Thus, mixing starts to increase with Reynolds number. The specific variation of the mixing index with Reynolds number strongly depends on micromixer configuration, but the trends are similar for different passive micromixers.

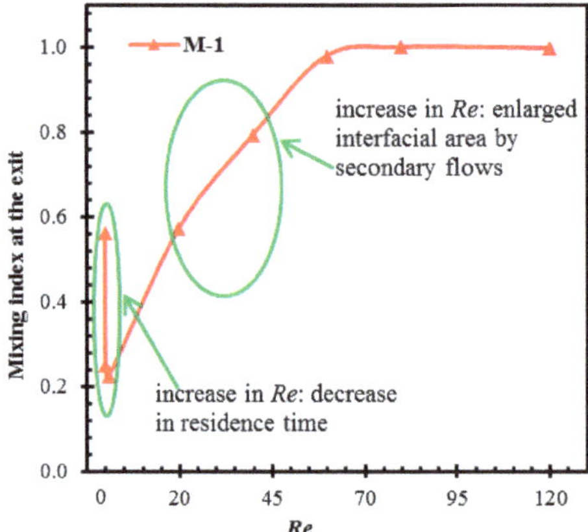

Figure 11. Variation of the mixing index at the exit with Reynolds number in a curved micromixer [40].

4.2.1. Mixing in Low-*Re* Range (Re ≤ 1)

In the low-*Re* range, mixing in passive micromixers is limited by molecular diffusion. Therefore, the mixing in this range mainly depends on the residential time of the working fluids in the micromixer. Mechanical stirring is not an effective method for enhancing mixing [36] because the secondary flow is hardly induced in this range. Thus, it is very challenging for researchers to design an efficient micromixer at this low-*Re* range.

It is found from Table 4 that, among the 2D micromixers (M-1 to M-4), the mixing indices at the exits of the curved micromixer (M-1) and the curved micromixer with rectangular grooves (M-2) are higher than the others at $Re = 0.01$ (M = 0.560 and 0.554, respectively). Interestingly, in this Reynolds number range, M-1 shows similar mixing performance as M-2. This is because, at low Reynolds numbers, mixing depends on molecular diffusion, and thus, geometric modification of the 2D planar micromixer is not effective in enhancing the mixing. The mixing index strongly depends on the time for which the working fluids remain in the micromixer. Hence, the mixing indices of these micromixers decrease as the Reynolds number increases up to 1. This trend does not only apply to the 2D micromixers but also to most of the tested micromixers.

The SAR micromixers (M-3 and M-4) generally show lower mixing than the micromixers with 2D serpentine structures (M-1 and M-2) in this range. However, the pressure drops are much lower in SAR micromixers, as shown in Table 5. It is observed that M-4 shows a 16.9% lower mixing index at the exit with a 65% lower pressure drop, as compared to M-1 at $Re = 0.01$. Among the 2D mixers, M-1, M-2, and M-4, showing high mixing indices, M-4 represents the highest *MC* values in the low-*Re* range, as shown in Table 6. M-3 shows the worst mixing performance among the tested micromixers in this range.

As shown in Table 4, the 3D micromixers using SAR with two-layer crossing channels (M-9 and M-10) achieve remarkable mixing performance at low Reynolds numbers. M-9 and M-10 show mixing indices over 0.90 in the entire low-*Re* range, while M-7 is only the micromixer which shows a mixing index over 0.90 (at $Re = 0.01$) in this *Re* range among the remaining micromixers. This is due to the generation of saddle-shaped flow structure in M-9 and M-10 at all Reynolds numbers, while in M-7, mixing decreases with the increase in Reynolds number due to diffusion dominant mixing. M-8 shows the lowest mixing indices among the 3D micromixers, which are even lower than those of some 2-D micromixers. This indicates that the mixing relying on elliptic and hyperbolic points generated through alternating barriers above the groove on the walls will require longer channel length for complete mixing.

Surprisingly, M-9 and M-10 also show the least pressure drops among the tested micromixers in the low-*Re* range, as shown in Table 5. Therefore, M-9 and M-10 show the highest MC values among the tested micromixers in the whole low-*Re* range (Table 6). In all the tested micromixers, MC decreases with increasing Reynolds number in this *Re* range. The micromixer M-7, which shows a not-much-lower mixing index than M-9 at $Re = 0.01$, shows about a 13-times-higher pressure drop with a 93% lower *MC* value at the same *Re*. Thus, the structure of 3D SAR with two-layer crossing channels is proved to be efficient in both enhancing mixing and reducing pressure drop in the low-*Re* range.

Figure 12 shows the developments of the mixing for M-4, M-9, and M-10 at $Re = 0.01$. The 2D micromixer (M-4) shows relatively slow development of mixing compared to the 3D micromixers (M-9 and M-10). Although M-9 and M-10 attain almost similar mixing at the exit, M-10 shows much faster development rate near the inlet of the micromixer. M-9 and M-10 reach a mixing index over 0.80 within 50% of the total length. M-9 attains the mixing index equal to the mixing index at the exit of M-4, within 35% of its length. This emphasizes that the 3D micromixer with an even slow development rate outperforms the 2D micromixer.

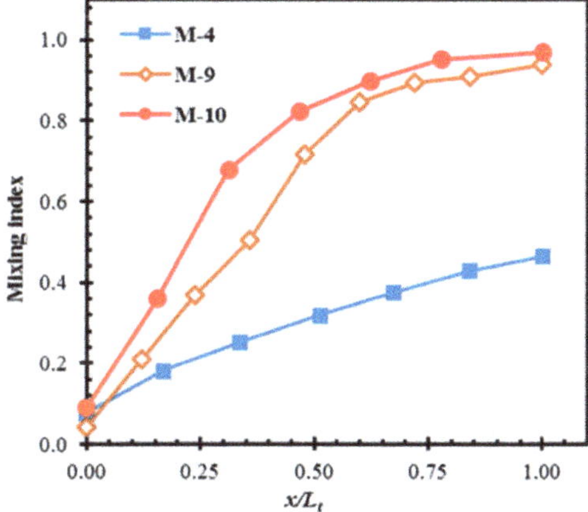

Figure 12. Developments of mixing along the length of the micromixers at $Re = 0.01$.

4.2.2. Mixing at Intermediate Reynolds Numbers ($1 < Re \leq 40$)

In the intermediate-*Re* range, the fluid inertia starts to increase with Reynolds number, and generates secondary flows that play a dominant role in mixing enhancement. Hence, an increase in Reynolds number results in an increase in mixing index in most of the cases shown in Table 4.

Among the 2D micromixers, the mixing index at the exit of the Tesla micromixer (M-4) is highest at $Re = 20$ and 40, and M-4 achieves almost perfect mixing ($M = 0.999$) at $Re = 40$. In this micromixer, with the increase in Reynolds number, the collision of fluid streams on rejoining is strengthened, which enhances chaos in the flow and hence the mixing [72,77]. It is also observed that the M-2 shows higher mixing than M-1 at these Reynolds numbers. This is due to the increase in secondary flow caused by the grooves in M-2 [41]. At $Re = 20$ and 40, M-2 shows 21.3% and 8.2% higher mixing indices than M-1, and M-4 shows 27.2% and 16.5% higher mixing indices than M-2, respectively. It is also observed that M-4 shows much lower pressure drops than M-1 and M-2. M-4 shows 46.8% and 26.3% lower pressure drops with 139.1% and 58.2% higher *MC* values than M-2 at $Re = 20$ and 40, respectively. Among the 2-D micromixers, M-3 shows the lowest mixing indices (Table 4) in the intermediate-*Re* range, but shows the highest MC values (Table 6) due to the lowest pressure drops (Table 5). Mixing in M-3 depends upon the unbalanced inertial collision. Hence, the lowest mixing indices in M-3 can be attributed to insufficient inertial force to cause an effective collision of the fluid streams that enhances chaos in the recombination zone. It also highlights that the flow instability in the inertia-based micromixers occurs at different Reynolds numbers depending upon the microchannel designs. Hence, these micromixers can be used in different *Re* range depending upon the efficient mixing range.

Among the 3D micromixers, most of the micromixers perform well in intermediate-Reynolds number range except M-7 and M-8. This indicates that locally accelerated advection due to the narrowing of flow path along with SAR mechanism in M-7 and the alternating velocity field creating elliptic and hyperbolic points due to the barrier and groove configuration in M-8 are not as effective as mixing mechanisms for other micromixer designs. Especially, M-5 and M-6 show nearly perfect mixing (M = 0.999) in the whole intermediate-*Re* range. In this *Re* range, stretching and folding of the fluid interfaces is developed due to the transverse flows induced by inertial forces. It causes enlargement of the fluid interfacial area for diffusion, thereby promotes mixing [45,59]. The 3D swirling flow along with the vortical flow due to serpentine channel is present in M-7 as discussed in [81], showing 26.7% lower mixing index than M-5 and M-6 at *Re* = 40.

The structures of 3D SAR with two-layer crossing channels (M-9 and M-10) achieve over 90% mixing in the intermediate-*Re* range even though it does not show the best mixing indices among the tested cases as in the low-*Re* range. Although M-5 and M-6 show the same mixing indices, the pressure drops in M-6 are approximately 22 times higher as compared to M-5 at *Re* = 20 and 40, and thus M-5 has much higher MC values than M-6. M-10 shows 9.8% and 7.0% lower mixing indices and 84.0% and 86.0% lower pressure drops compared to M-5 at *Re* = 20 and 40, respectively. M-10 shows the highest MC values among the tested micromixers, which are 5.6 times and 6.6 times higher than those of M-5 at *Re* = 20 and 40, respectively.

The developments of mixing in M-4, M-5, M-6, M-9, and M-10 at *Re* = 20 and 40 are shown in Figure 13a,b, respectively. At *Re* = 20, M-4, M-9, and M-10 show similar development rates, while M-5 and M-6 show higher development rates near the micromixer inlet. At *Re* = 40, in addition to M-5 and M-6, M-4 also shows a higher rate of development as compared to M-9 and M-10. M-5 and M-6 achieve almost complete mixing at 50% and 65% of the total length, respectively, at *Re* = 20. At *Re* = 40, M-6 also shows the highest development rate and attains almost complete mixing around 30% of the length. M-4 achieves a mixing index over 0.80 in less than 25% of the length and almost complete mixing at 65% of length, whereas M-9 and M-10 achieve a mixing index over 0.8 within 60% of their lengths at *Re* = 40. M-4, M-5, and M-6 show higher increases in the development rate with the increase in Reynolds number than M-9 and M-10. The non-dependency of mixing on the inertia in M-9 and M-10 is consistent with the observation in [84].

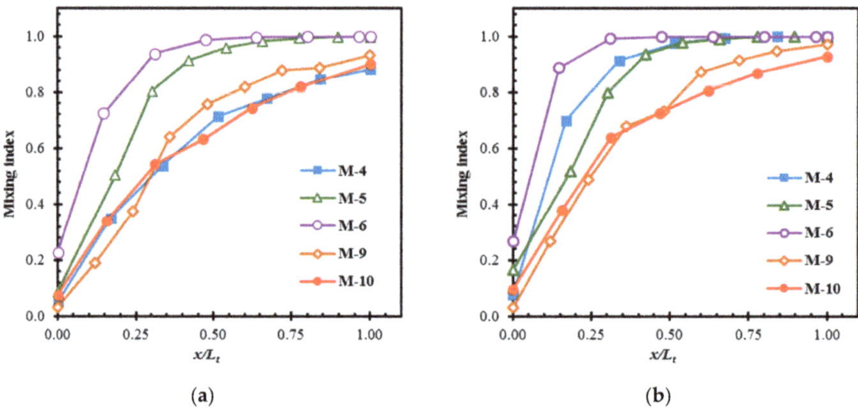

Figure 13. Developments of mixing along the length of the micromixers: (**a**) *Re* = 20 and (**b**) *Re* = 40.

4.2.3. Mixing at High Reynolds Numbers (Re > 40)

At high Reynolds numbers, transverse flows (Dean vortices or secondary flow) become stronger, and mixing is established at a faster rate by stretching and folding of the interfacial area. Increasing the inertial force of the fluids is one of the easiest ways to enhance mixing at high-*Re* range. At the highest Reynolds number (*Re* = 120), almost all the micromixers except M-8 show excellent mixing performance. And, M-4, M-5, and M-6 attain almost complete mixing in the whole high-*Re* range.

The quantitative comparison in Table 4 demonstrates that even simple 2D serpentine micromixers (M-1 and M-2) show more than 97% mixing at *Re* = 60, and more than 99% mixing at *Re* = 80 and 120. M-3 having asymmetrical SAR structure reaches a mixing index of 0.89 at *Re* = 120. From Table 5, it is observed that M-1 and M-2 show 7.1% and 20.6% reduction in pressure than those of M-4 at *Re* = 80 and 120, respectively. Hence, M-4 shows 7.4% and 20.5% lower *MC* values than those of M-2 at *Re* = 80 and 120, respectively, in Table 6. However, it shows 11.6% higher *MC* than that of M-2 at *Re* = 60. M-3 shows the highest MC values among the 2-D micromixers due to the lowest pressure drops.

The 3D micromixer M-7 also reaches a mixing index over 0.9 at *Re* = 80. There are minimal variations among the mixing indices of the 3D micromixers in this *Re* range if M-7 and M-8 are excepted. However, M-5 and M-6 achieve complete mixing at the expense of a high pressure drops. M5 and M-6 show approximately 7.5 times and 168 times higher pressure drops, respectively, as compared to M-10 at *Re* = 60. Hence, M-10 shows approximately 7.5 times higher *MC* as compared to M-5 in this *Re* range. The worst mixing performance among the tested micromixers is obtained by M-8.

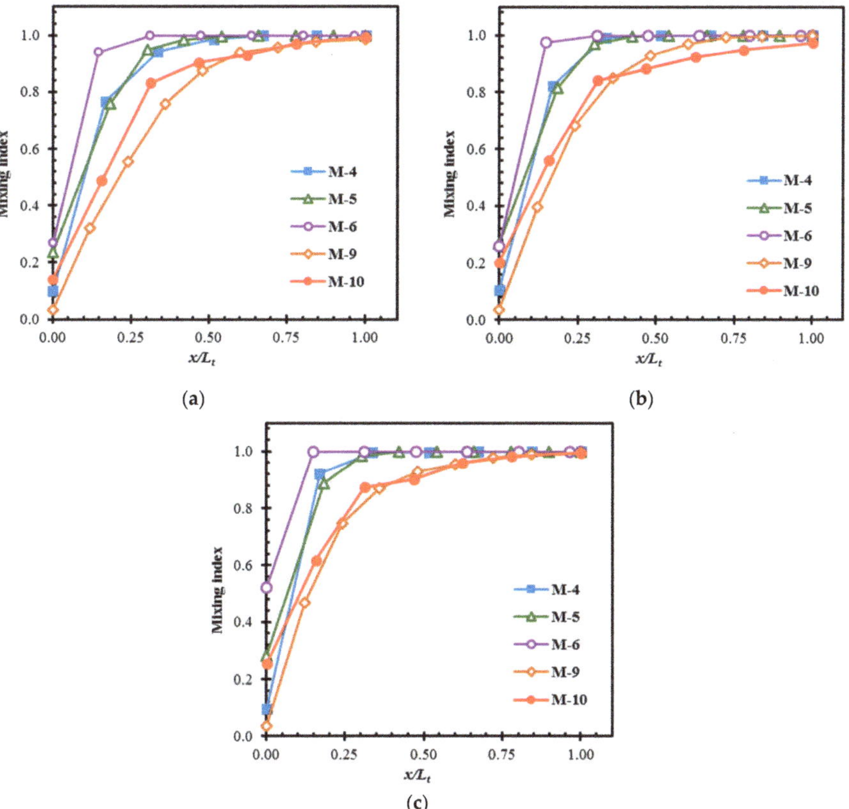

Figure 14. Developments of mixing along the length of the micromixers: (**a**) *Re* = 60, (**b**) *Re* = 80 and (**c**) *Re* = 120.

Figure 14 shows the developments of mixing in M-4, M-5, M-6, M-9, and M-10 at different Reynolds numbers. The rates of development of mixing near the inlets of M-4, M-5, and M-6 are found to be higher than those of M-9 and M-10 in this *Re* range. M-4 shows the highest development rate, and M-5, and M-6 show almost the same developments of mixing in the whole high-*Re* range (*Re* = 60, 80, and 120). As Reynolds number increases, the development rates of these three micromixers generally increase. M-4 achieves almost complete mixing within 15–30% of the length, and M-5 and M-6 achieve it within 30–50% of the length in this *Re* range. We found that 50% of the length of M-5 and M-6 where complete mixing is achieved, still show approximately 3 times and 84 times higher pressure drops, respectively, compared to M-10 at *Re* = 60. M-10 shows approximately 3.4 times higher *MC* as compared to 50%-length of M-5 at this Reynolds number. M-10 shows slightly better mixing rate than M-9 at *Re* = 60. M-9 and M-10 show similar trends of the mixing development especially at *Re* = Re–120, where the mixing improves continuously until the exit is approached.

4.2.4. Comparison of Velocity and Concentration Fields between M-4 and M-10

For M-4 and M-10, which show the best mixing performances among the 2D and 3D micromixers, respectively, flow structures and concentration distributions were compared at different Reynolds numbers. Figure 15 shows the flow structures on a y-z plane at x/L_t = 0.16 (plane A2 marked in Figures 4 and 10) for different Reynolds numbers. At *Re* = 0.01, M-4 shows the flow moving parallel to the wall, and there is no flow disturbance. However, at *Re* = 40, two symmetrical counter-rotating vortices are seen. These vortices get stronger with the increase in Reynolds number, as shown at *Re* = 120 with the additional vortices. Contrarily, velocity vectors in M-10 show a saddle-shaped flow structure even at the lowest Reynold number (*Re* = 0.01). This is the reason for the high mixing performance of M-10 in the low-*Re* range, which is discussed in Section 4.2.1. This result is in line with the observation in previous studies [84,87] that the saddle-shaped flow structure enhances mixing by stretching and folding of the fluid interfaces.

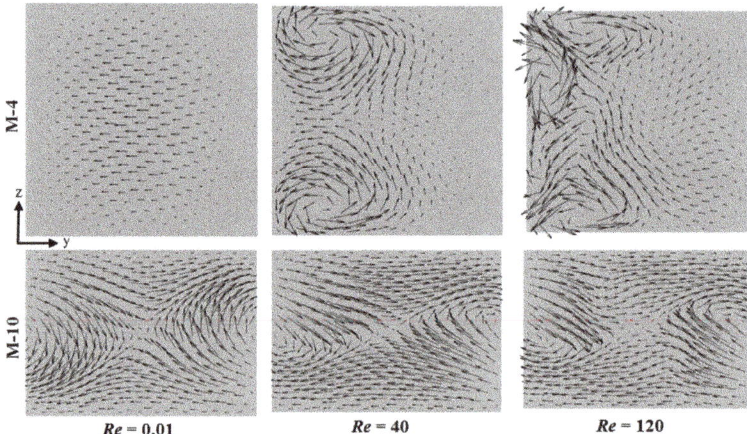

Figure 15. Velocity vectors in M-4 and M-10 at x/L_t = 0.16 (plane A2 in Figures 4 and 10) at different Reynolds numbers.

Figure 16 shows the dye concentrations on y-z planes of M-4 and M-10 for different Reynolds numbers. The progress of homogenization of the concentration in M-4 is much slower than that in M-10 at *Re* = 0.01. As shown in Figure 15, the fluids are mixed by pure diffusion in M-4, while M-10 shows a strong secondary flow structure even at this low Reynolds number, which is effective in promoting mixing. However, at high Reynolds numbers, the progress of mixing in M-4 is much faster than that in M-10 due to the occurrence of flow disturbances in the form of vortices (Figure 15) and high-inertia collisions of the streams on their recombinations as discussed in [77].

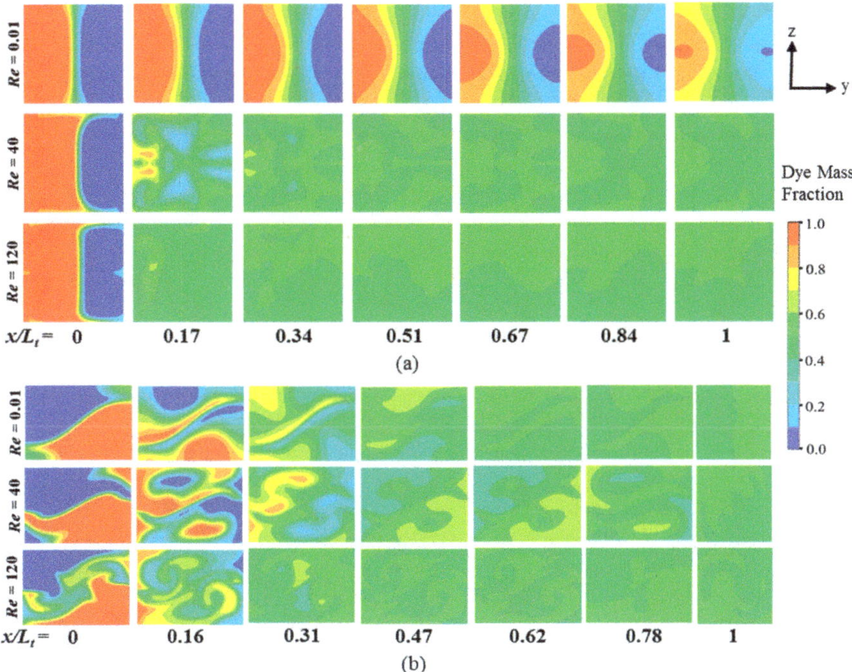

Figure 16. Dye concentration distributions at different Reynolds numbers: (**a**) M-4; and (**b**) M-10.

A selection procedure for micromixers is suggested as follows. Once the Reynolds number is fixed depending on the application, the micromixers showing preferred mixing index level should be selected considering ease of fabrication (2D or 3D). Pressure drop should also be considered as a vital factor in the case where the deformation of a sample will impact the outcome of the process, such as in biological applications. After narrowing down the micromixer types, mixing cost should be checked for the final selection.

5. Conclusions

Ten typical micromixers representing five different mixing mechanisms were analyzed quantitatively using Navier-Stokes equations under same working fluids, flow conditions, and axial channel length across a wide range of Reynolds number (Re = 0.01–120). In the results, M-9 and M-10 showed the best overall mixing indices among all the tested micromixers over the entire range of Re, while M-4 showed the best overall mixing indices among the 2D micromixers. However, M-3 showed the best overall MC values among the 2-D micromixers due to its incredibly low pressure drop regardless of Reynolds number. Compared to M-9 and M-10, M-4 showed far lower mixing indices in the low-Re range, but was represented among the similar or better mixing indices in the intermediate- and high-Re ranges. M-9 showed lower mixing performance than M-10 in the low-Re range, but showed higher mixing indices in the intermediate-Re range. M-10 showed the best overall MC values among the tested micromixers, which were slightly higher than those of M-9. The worst mixing performance was obtained by M-3 in the low-Re range, but by M-8 in the high-Re range. The stretching and folding of the fluid streams around the hyperbolic points created by the saddle-shaped flow structure in the crossing channel are more effective for mixing as compared to those generated by the grooves and barriers. Furthermore, the chaotic advection by crossing channels is generated over the entire range of Reynolds numbers, while for the other 2D and 3D channel designs, it is observed only at higher Reynolds numbers. In the crossing-channel micromixers, mixing does not depend upon inertial force, as indicated by good mixing even at low Reynolds numbers. Hence, these micromixers can also be applied to mix fluids having high viscosity.

Therefore, among the tested micromixers, the Tesla structure micromixer (M-4) is recommended in the intermediate- and high-*Re* ranges considering high mixing performance and easy fabrication due to the planar structure, unless pressure drop is critical. The 3D micromixers, M-9 and M-10 are recommended in the low-*Re* range considering their excellent mixing performance. But, the fabrication of their two-layer structures with the traditional photolithography process is a challenging task due to the misalignment issue of the top and bottom layers. The results of this comparative analysis are intended to provide guidance for the selection of effective micromixers under different flow conditions in various microfluidic applications.

Author Contributions: Conceptualization, W.R., S.H., and K.-Y.K.; numerical simulation, W.R. and S.H.; formal analysis, W.R.; investigation, W.R.; resources, K.-Y.K.; data curation, W.R.; writing—original draft preparation, W.R., S.H.; writing—review and editing, W.R. and K.-Y.K.; supervision, K.-Y.K.; project administration, W.R., S.H., and K.-Y.K.; funding acquisition, K.-Y.K. All authors have read and agreed to the published version of the manuscript.

Funding: This work was supported by the National Research Foundation of Korea (NRF) grant funded by the Korean government (MSIT) (No. 2019R1A2C1007657).

Conflicts of Interest: The authors declare there is no conflict of interest.

Nomenclature

c	mass fraction
D	diffusion coefficient (m^2/s)
L_i	length of initial part of main channel (μm)
L_e	exit channel length (μm)
L_t	total length of the micromixer (μm)
M	mixing index
M_0	mixing index at the exit
N	number of sampling points
P	pressure drop (Pa)
Pi	pitch length (μm)
Re	Reynolds number
SAR	split and recombine
U	average inlet velocity (m/s)
W	width of main channel (μm)
x, y, z	Cartesian coordinates
Greek Symbols	
μ	fluid dynamic viscosity ($kg{\cdot}m^{-1}{\cdot}s^{-1}$)
ν	fluid Kinematic viscosity (m^2/s)
ρ	fluid density (kg/m^3)
σ	standard deviation
Subscripts	
i	sampling point or fluid component
m	optimal mixing
max	maximum value
x	axial distance

References

1. Park, T.; Lee, S.; Seong, G.H.; Choo, J.; Lee, E.K.; Kim, Y.S.; Ji, W.H.; Hwang, S.Y.; Gweon, D.G.; Lee, S. Highly sensitive signal detection of duplex dye-labelled DNA oligonucleotides in a PDMS microfluidic chip: Confocal surface-enhanced Raman spectroscopic study. *Lab Chip* **2005**, *5*, 437–442. [CrossRef]
2. Rapp, B.E.; Gruhl, F.J.; Länge, K. Biosensors with label-free detection designed for diagnostic applications. *Anal. Bioanal. Chem.* **2010**, *398*, 2403–2412. [CrossRef]
3. Rahman, M.; Rebrov, E. Microreactors for gold nanoparticles synthesis: From faraday to flow. *Processes* **2014**, *2*, 466–493. [CrossRef]
4. Stone, H.A.; Stroock, A.D.; Ajdari, A. Engineering flows in small devices. *Annu. Rev. Fluid Mech.* **2004**, *36*, 381–411. [CrossRef]
5. Squires, T.M.; Quake, S.R. Microfluidics: Fluid physics at the nanoliter scale. *Rev. Mod. Phys.* **2005**, *77*, 977–1026. [CrossRef]
6. Hessel, V.; Löwe, H.; Schönfeld, F. Micromixers—A review on passive and active mixing principles. *Chem. Eng. Sci.* **2005**, *60*, 2479–2501. [CrossRef]
7. Nguyen, N.T. *Micromixers: Fundamentals, Design and Fabrication*; William Andrew Publishers: New York, NY, USA, 2011.
8. Ingham, C.J.; van Hylckama Vlieg, J.E.T. MEMS and the microbe. *Lab Chip* **2008**, *8*, 1604–1616. [CrossRef] [PubMed]
9. Schulte, T.H.; Bardell, R.L.; Weigl, B.H. Microfluidic technologies in clinical diagnostics. *Clin. Chim. Acta* **2002**, *321*, 1–10. [CrossRef]
10. Razzacki, S.Z.; Thwar, P.K.; Yang, M.; Ugaz, V.M.; Burns, M.A. Integrated microsystems for controlled drug delivery. *Adv. Drug Deliv. Rev.* **2004**, *56*, 185–198. [CrossRef] [PubMed]
11. Yang, A.S.; Chuang, F.C.; Chen, C.K.; Lee, M.H.; Chen, S.W.; Su, T.L.; Yang, Y.C. A high-performance micromixer using three-dimensional Tesla structures for bio-applications. *Chem. Eng. J.* **2015**, *263*, 444–451. [CrossRef]
12. Liu, R.H.; Lenigk, R.; Druyor-Sanchez, R.L.; Yang, J.; Grodzinski, P. Hybridization enhancement using cavitation microstreaming. *Anal. Chem.* **2003**, *75*, 1911–1917. [CrossRef] [PubMed]
13. Chen, Q.; Wu, J.; Zhang, Y.; Lin, J.M. Qualitative and quantitative analysis of tumor cell metabolism via stable isotope labeling assisted microfluidic chip electrospray ionization mass spectrometry. *Anal. Chem.* **2012**, *84*, 1695–1701. [CrossRef] [PubMed]
14. Westerhausen, C.; Schnitzler, L.G.; Wendel, D.; Krzysztoń, R.; Lächelt, U.; Wagner, E.; Rädler, J.O.; Wixforth, A. Controllable acoustic mixing of fluids in microchannels for the fabrication of therapeutic nanoparticles. *Micromachines* **2016**, *7*, 150. [CrossRef] [PubMed]
15. Huang, P.H.; Zhao, S.; Bachman, H.; Nama, N.; Li, Z.; Chen, C.; Yang, S.; Wu, M.; Zhang, S.P.; Huang, T.J. Acoustofluidic synthesis of particulate nanomaterials. *Adv. Sci.* **2019**, *6*, 1900913. [CrossRef]
16. Culbertson, C.T.; Mickleburgh, T.G.; Stewart-James, S.A.; Sellens, K.A.; Pressnall, M. Micro total analysis systems: Fundamental advances and biological applications. *Anal. Chem.* **2014**, *86*, 95–118. [CrossRef] [PubMed]
17. Bouvier, E.S.P.; Koza, S.M. Advances in size-exclusion separations of proteins and polymers by UHPLC. *TrAC Trends Anal. Chem.* **2014**, *63*, 85–94. [CrossRef]
18. Bachman, H.; Huang, P.H.; Zhao, S.; Yang, S.; Zhang, P.; Fu, H.; Huang, T.J. Acoustofluidic devices controlled by cell phones. *Lab Chip* **2018**, *18*, 433–441. [CrossRef]
19. Nama, N.; Huang, P.H.; Huang, T.J.; Costanzo, F. Investigation of micromixing by acoustically oscillated sharp-edges. *Biomicrofluidics* **2016**, *10*, 024124. [CrossRef]
20. Huang, P.H.; Chan, C.Y.; Li, P.; Nama, N.; Xie, Y.; Wei, C.H.; Chen, Y.; Ahmed, D.; Huang, T.J. A spatiotemporally controllable chemical gradient generator via acoustically oscillating sharp-edge structures. *Lab Chip* **2015**, *15*, 4166–4176. [CrossRef]
21. Huang, P.H.; Chan, C.Y.; Li, P.; Wang, Y.; Nama, N.; Bachman, H.; Huang, T.J. A sharp-edge-based acoustofluidic chemical signal generator. *Lab Chip* **2018**, *18*, 1411–1421. [CrossRef]
22. Harnett, C.K.; Templeton, J.; Dunphy-Guzman, K.A.; Senousy, Y.M.; Kanouff, M.P. Model based design of a microfluidic mixer driven by induced charge electroosmosis. *Lab Chip* **2008**, *8*, 565–572. [CrossRef] [PubMed]

23. Ahmed, D.; Mao, X.; Shi, J.; Juluri, B.K.; Huang, T.J. A millisecond micromixer via single-bubble-based acoustic streaming. *Lab Chip* **2009**, *9*, 2738. [CrossRef]
24. Nguyen, N.T.; Wu, Z. Micromixers—A review. *J. Micromech. Microeng.* **2005**, *15*, R1–R16. [CrossRef]
25. Ebadi, M.; Moshksayan, K.; Kashaninejad, N.; Saidi, M.S.; Nguyen, N.T. A tool for designing tree-like concentration gradient generators for lab-on-a-chip applications. *Chem. Eng. Sci.* **2020**, *212*, 115339. [CrossRef]
26. Rismanian, M.; Saidi, M.S.; Kashaninejad, N. A new non-dimensional parameter to obtain the minimum mixing length in tree-like concentration gradient generators. *Chem. Eng. Sci.* **2019**, *195*, 120–126. [CrossRef]
27. Aref, H. Stirring by chaotic advection. *J. Fluid Mech.* **1984**, *143*, 1–21. [CrossRef]
28. Ottino, J.M. *The Kinematics of Mixing: Stretching, Chaos, and Transport*; Cambridge Texts in Applied Mathematics; Cambridge University Press: Cambridge, Oxford, UK, 1989.
29. Stroock, A.D.; Dertinger, S.K.W.; Ajdari, A.; Mezic, I.; Stone, H.A.; Whitesides, G.M. Chaotic mixer for microchannels. *Science* **2002**, *295*, 647–651. [CrossRef]
30. Fu, L.M.; Fang, W.C.; Hou, H.H.; Wang, Y.N.; Hong, T.F. Rapid vortex microfluidic mixer utilizing double-heart chamber. *Chem. Eng. J.* **2014**, *249*, 246–251. [CrossRef]
31. Swickrath, M.J.; Burns, S.D.; Wnek, G.E. Modulating passive micromixing in 2-D microfluidic devices via discontinuities in surface energy. *Sens. Actuators B Chem.* **2009**, *140*, 656–662. [CrossRef]
32. Mengeaud, V.; Josserand, J.; Girault, H.H. Mixing processes in a zigzag microchannel: Finite element simulations and optical study. *Anal. Chem.* **2002**, *74*, 4279–4286. [CrossRef]
33. Wang, W.; Zhao, S.; Shao, T.; Jin, Y.; Cheng, Y. Visualization of micro-scale mixing in miscible liquids using? LIF technique and drug nano-particle preparation in T-shaped micro-channels. *Chem. Eng. J.* **2012**, *192*, 252–261. [CrossRef]
34. Roudgar, M.; Brunazzi, E.; Galletti, C.; Mauri, R. Numerical study of split T-micromixers. *Chem. Eng. Technol.* **2012**, *35*, 1291–1299. [CrossRef]
35. Hsieh, S.S.; Lin, J.W.; Chen, J.H. Mixing efficiency of Y-type micromixers with different angles. *Int. J. Heat Fluid Flow* **2013**, *44*, 130–139. [CrossRef]
36. Hengzi, W.; Pio, I.; Erol, H.; Syed, M.; Wang, H.; Iovenitti, P.; Harvey, E.; Masood, S. Optimizing layout of obstacles for enhanced mixing in microchannels. *Smart Mater. Struct.* **2002**, *11*, 662.
37. Ait Mouheb, N.; Malsch, D.; Montillet, A.; Solliec, C.; Henkel, T. Numerical and experimental investigations of mixing in T-shaped and cross-shaped micromixers. *Chem. Eng. Sci.* **2012**, *68*, 278–289. [CrossRef]
38. Wang, L.; Liu, D.; Wang, X.; Han, X. Mixing enhancement of novel passive microfluidic mixers with cylindrical grooves. *Chem. Eng. Sci.* **2012**, *81*, 157–163. [CrossRef]
39. Vanka, S.P.; Luo, G.; Winkler, C.M. Numerical study of scalar mixing in curved channels at low Reynolds numbers. *AIChE J.* **2004**, *50*, 2359–2368. [CrossRef]
40. Hossain, S.; Ansari, M.A.; Kim, K.Y. Evaluation of the mixing performance of three passive micromixers. *Chem. Eng. J.* **2009**, *150*, 492–501. [CrossRef]
41. Alam, A.; Kim, K.Y. Analysis of mixing in a curved microchannel with rectangular grooves. *Chem. Eng. J.* **2012**, *181–182*, 708–716. [CrossRef]
42. Liu, R.H.; Stremler, M.A.; Sharp, K.V.; Olsen, M.G.; Santiago, J.G.; Adrian, R.J.; Aref, H.; Beebe, D.J. Passive mixing in a three-dimensional serpentine microchannel. *J. Microelectromechan. Syst.* **2000**, *9*, 190–197. [CrossRef]
43. Beebe, D.J.; Adrian, R.J.; Olsen, M.G.; Stremler, M.A.; Aref, H.; Jo, B.H. Passive mixing in microchannels: Fabrication and flow experiments. *Mecaniqe Ind.* **2001**, *2*, 343–348. [CrossRef]
44. Liu, Y.Z.; Kim, B.J.; Sung, H.J. Two-fluid mixing in a microchannel. *Int. J. Heat Fluid Flow* **2004**, *25*, 986–995. [CrossRef]
45. Ansari, M.A.; Kim, K.Y. Parametric study on mixing of two fluids in a three-dimensional serpentine microchannel. *Chem. Eng. J.* **2009**, *146*, 439–448. [CrossRef]
46. Lee, K.; Kim, C.; Shin, K.S.; Lee, J.W.; Ju, B.K.; Kim, T.S.; Lee, S.K.; Kang, J.Y. Fabrication of round channels using the surface tension of PDMS and its application to a 3D serpentine mixer. *J. Micromech. Microeng.* **2007**, *17*, 1533–1541. [CrossRef]
47. Lin, Y. Numerical characterization of simple three-dimensional chaotic micromixers. *Chem. Eng. J.* **2015**, *277*, 303–311. [CrossRef]

48. Lin, Y.; Yu, X.; Wang, Z.; Tu, S.T.; Wang, Z. Design and evaluation of an easily fabricated micromixer with three-dimensional periodic perturbation. *Chem. Eng. J.* **2011**, *171*, 291–300. [CrossRef]
49. Park, J.M.; Kwon, T.H. Numerical characterization of three-dimensional serpentine micromixers. *AIChE J.* **2008**, *54*, 1999–2008. [CrossRef]
50. Cortes-Quiroz, C.A.; Azarbadegan, A.; Zangeneh, M.; Goto, A. Analysis and multi-criteria design optimization of geometric characteristics of grooved micromixer. *Chem. Eng. J.* **2010**, *160*, 852–864. [CrossRef]
51. Du, Y.; Zhang, Z.; Yim, C.; Lin, M.; Cao, X. Evaluation of floor-grooved micromixers using concentration-channel length profiles. *Micromachines* **2010**, *1*, 19–33. [CrossRef]
52. Tóth, E.L.; Holczer, E.G.; Iván, K.; Fürjes, P. Optimized simulation and validation of particle advection in asymmetric staggered herringbone type micromixers. *Micromachines* **2015**, *6*, 136–150. [CrossRef]
53. Sarkar, A.; Narváez, A.; Harting, J. Quantification of the performance of chaotic micromixers on the basis of finite time Lyapunov exponents. *Microfluid. Nanofluidics* **2012**, *13*, 19–27. [CrossRef]
54. Hossain, S.; Husain, A.; Kim, K.Y. Shape optimization of a micromixer with staggered-herringbone grooves patterned on opposite walls. *Chem. Eng. J.* **2010**, *162*, 730–737. [CrossRef]
55. Afzal, A.; Kim, K.Y. Three-objective optimization of a staggered herringbone micromixer. *Sens. Actuators B Chem.* **2014**, *192*, 350–360. [CrossRef]
56. Lin, D.; He, F.; Liao, Y.; Lin, J.; Liu, C.; Song, J.; Cheng, Y. Three-dimensional staggered herringbone mixer fabricated by femtosecond laser direct writing. *J. Opt.* **2013**, *15*, 025601. [CrossRef]
57. Ansari, M.A.; Kim, K.Y. Shape optimization of a micromixer with staggered herringbone groove. *Chem. Eng. Sci.* **2007**, *62*, 6687–6695. [CrossRef]
58. Kim, D.S.; Lee, S.W.; Kwon, T.H.; Lee, S.S. A barrier embedded chaotic micromixer. *J. Micromech. Microeng.* **2004**, *14*, 798–805. [CrossRef]
59. Hossain, S.; Kim, K.Y. Mixing analysis in a three-dimensional serpentine split-and-recombine micromixer. *Chem. Eng. Res. Des.* **2015**, *100*, 95–103. [CrossRef]
60. Ohkawa, K.; Nakamoto, T.; Izuka, Y.; Hirata, Y.; Inoue, Y. Flow and mixing characteristics of? Type plate static mixer with splitting and inverse recombination. *Chem. Eng. Res. Des.* **2008**, *86*, 1447–1453. [CrossRef]
61. Lee, S.W.; Lee, S.S. Rotation effect in split and recombination micromixing. *Sens. Actuators B Chem.* **2008**, *129*, 364–371. [CrossRef]
62. Hardt, S.; Pennemann, H.; Schönfeld, F. Theoretical and experimental characterization of a low-Reynolds number split-and-recombine mixer. *Microfluid. Nanofluidics* **2006**, *2*, 237–248. [CrossRef]
63. Lee, S.W.; Kim, D.S.; Lee, S.S.; Kwon, T.H. A split and recombination micromixer fabricated in a PDMS three-dimensional structure. *J. Micromech. Microeng.* **2006**, *16*, 1067–1072. [CrossRef]
64. Viktorov, V.; Nimafar, M. A novel generation of 3D SAR-based passive micromixer: Efficient mixing and low pressure drop at a low Reynolds number. *J. Micromech. Microeng.* **2013**, *23*, 055023. [CrossRef]
65. Nimafar, M.; Viktorov, V.; Martinelli, M. Experimental comparative mixing performance of passive micromixers with H-shaped sub-channels. *Chem. Eng. Sci.* **2012**, *76*, 37–44. [CrossRef]
66. Nimafar, M.; Viktorov, V.; Martinelli, M. Experimental investigation of split and recombination micromixer in confront with basic T- and O- type micromixers. *Int. J. Mech. Appl.* **2012**, *2*, 61–69. [CrossRef]
67. Carrier, O.; Funfschilling, D.; Debas, H.; Poncin, S.; Löb, P.; Li, H.Z. Pressure drop in a split-and-recombine caterpillar micromixer in case of newtonian and non-newtonian fluids. *AIChE J.* **2013**, *59*, 2679–2685. [CrossRef]
68. Li, J.; Xia, G.; Li, Y. Numerical and experimental analyses of planar asymmetric split-and-recombine micromixer with dislocation sub-channels. *J. Chem. Technol. Biotechnol.* **2013**, *88*, 1757–1765. [CrossRef]
69. Ansari, M.A.; Kim, K.Y. Mixing performance of unbalanced split and recombine micomixers with circular and rhombic sub-channels. *Chem. Eng. J.* **2010**, *162*, 760–767. [CrossRef]
70. Xia, G.; Li, J.; Tian, X.; Zhou, M. Analysis of flow and mixing characteristics of planar asymmetric split-and-recombine (P-SAR) micromixers with fan-shaped cavities. *Ind. Eng. Chem. Res.* **2012**, *51*, 7816–7827. [CrossRef]
71. Pennella, F.; Rossi, M.; Ripandelli, S.; Rasponi, M.; Mastrangelo, F.; Deriu, M.A.; Ridolfi, L.; Kähler, C.J.; Morbiducci, U. Numerical and experimental characterization of a novel modular passive micromixer. *Biomed. Microdevices* **2012**, *14*, 849–862. [CrossRef]
72. Hong, C.C.; Choi, J.W.; Ahn, C.H. A novel in-plane passive microfluidic mixer with modified Tesla structures. *Lab Chip* **2004**, *4*, 109–113. [CrossRef]

73. Chung, C.K.; Chang, C.K.; Lai, C.C. Simulation and fabrication of a branch-channel rhombic micromixer for low pressure drop and short mixing length. *Microsyst. Technol.* **2014**, *20*, 1981–1986. [CrossRef]
74. Xia, G.D.; Li, Y.F.; Wang, J.; Zhai, Y.L. Numerical and experimental analyses of planar micromixer with gaps and baffles based on field synergy principle. *Int. Commun. Heat Mass Transf.* **2016**, *71*, 188–196. [CrossRef]
75. Hossain, S.; Kim, K.Y. Mixing analysis of passive micromixer with unbalanced three-split rhombic sub-channels. *Micromachines* **2014**, *5*, 913–928. [CrossRef]
76. Ansari, M.A.; Kim, K.Y.; Anwar, K.; Kim, S.M. A novel passive micromixer based on unbalanced splits and collisions of fluid streams. *J. Micromech. Microeng.* **2010**, *20*, 055007. [CrossRef]
77. Hossain, S.; Ansari, M.A.; Husain, A.; Kim, K.Y. Analysis and optimization of a micromixer with a modified Tesla structure. *Chem. Eng. J.* **2010**, *158*, 305–314. [CrossRef]
78. Sheu, T.S.; Chen, S.J.; Chen, J.J. Mixing of a split and recombine micromixer with tapered curved microchannels. *Chem. Eng. Sci.* **2012**, *71*, 321–332. [CrossRef]
79. Afzal, A.; Kim, K.Y. Passive split and recombination micromixer with convergent-divergent walls. *Chem. Eng. J.* **2012**, *203*, 182–192. [CrossRef]
80. Kim, D.S.; Lee, S.H.; Kwon, T.H.; Ahn, C.H. A serpentine laminating micromixer combining splitting/recombination and advection. *Lab Chip* **2005**, *5*, 739–747. [CrossRef]
81. Park, J.M.; Kim, D.S.; Kang, T.G.; Kwon, T.H. Improved serpentine laminating micromixer with enhanced local advection. *Microfluid. Nanofluidics* **2008**, *4*, 513–523. [CrossRef]
82. Feng, X.; Ren, Y.; Jiang, H. An effective splitting-and-recombination micromixer with self-rotated contact surface for wide Reynolds number range applications. *Biomicrofluidics* **2013**, *7*, 1–10. [CrossRef]
83. Lim, T.W.; Son, Y.; Jeong, Y.J.; Yang, D.Y.; Kong, H.J.; Lee, K.S.; Kim, D.P. Three-dimensionally crossing manifold micro-mixer for fast mixing in a short channel length. *Lab Chip* **2011**, *11*, 100–103. [CrossRef] [PubMed]
84. Xia, H.M.; Wan, S.Y.M.; Shu, C.; Chew, Y.T. Chaotic micromixers using two-layer crossing channels to exhibit fast mixing at low Reynolds numbers. *Lab Chip* **2005**, *5*, 748–755. [CrossRef] [PubMed]
85. Li, L.; Chen, Q.D.; Tsai, C.T. Three dimensional triangle chaotic micromixer. *Adv. Mater. Res.* **2014**, *875–877*, 1189–1193. [CrossRef]
86. Fang, W.F.; Yang, J.T. A novel microreactor with 3D rotating flow to boost fluid reaction and mixing of viscous fluids. *Sens. Actuators B Chem.* **2009**, *140*, 629–642. [CrossRef]
87. Hossain, S.; Lee, I.; Kim, S.M.; Kim, K.Y. A micromixer with two-layer serpentine crossing channels having excellent mixing performance at low Reynolds numbers. *Chem. Eng. J.* **2017**, *327*, 268–277. [CrossRef]
88. Raza, W.; Hossain, S.; Kim, K.Y. Effective mixing in a short serpentine split-and-recombination micromixer. *Sens. Actuators B Chem.* **2018**, *258*, 381–392. [CrossRef]
89. Hossain, S.; Kim, K.Y. Parametric investigation on mixing in a micromixer with two-layer crossing channels. *Springerplus* **2016**, *5*, 794. [CrossRef]
90. Lee, C.Y.; Fu, L.M. Recent advances and applications of micromixers. *Sens. Actuators B Chem.* **2018**, *259*, 677–702. [CrossRef]
91. Cai, G.; Xue, L.; Zhang, H.; Lin, J. A review on micromixers. *Micromachines* **2017**, *8*, 274. [CrossRef]
92. Jeong, G.S.; Chung, S.; Kim, C.B.; Lee, S.H. Applications of micromixing technology. *Analyst* **2010**, *135*, 460–473. [CrossRef]
93. Capretto, L.; Cheng, W.; Hill, M.; Zhang, X. Micromixing within microfluidic devices. *Top. Curr. Chem.* **2011**, *304*, 27–68. [PubMed]
94. Lee, C.Y.; Wang, W.T.; Liu, C.C.; Fu, L.M. Passive mixers in microfluidic systems: A review. *Chem. Eng. J.* **2016**, *288*, 146–160. [CrossRef]
95. Pennella, F.; Mastrangelo, F.; Gallo, D.; Massai, D.; Deriu, M.A.; Labate, G.F.D.; Bignardi, C.; Montevecchi, F.; Morbiducci, U. A Survey of microchannel geometries for mixing of species in biomicrofluidics. *Single Two Phase Flows Chem. Biomed. Eng.* **2012**, 548–578.
96. Falk, L.; Commenge, J.M. Performance comparison of micromixers. *Chem. Eng. Sci.* **2010**, *65*, 405–411. [CrossRef]
97. Viktorov, V.; Mahmud, M.; Visconte, C. Comparative analysis of passive micromixers at a wide range of reynolds numbers. *Micromachines* **2015**, *6*, 1166–1179. [CrossRef]
98. Bošković, D.; Loebbecke, S.; Gross, G.A.; Koehler, J.M. Residence time distribution studies in microfluidic mixing structures. *Chem. Eng. Technol.* **2011**, *34*, 361–370. [CrossRef]

99. ANSYS. *Solver Theory Guide, CFX-15.0*; ANSYS Inc.: Canonsburg, PA, USA, 2013.
100. Chung, C.K.; Shih, T.R. A rhombic micromixer with asymmetrical flow for enhancing mixing. *J. Micromech. Microeng.* **2007**, *17*, 2495–2504. [CrossRef]

Article

Toward the Next Generation of Passive Micromixers: A Novel 3-D Design Approach

Mahmut Burak Okuducu [1,*] and Mustafa M. Aral [2]

1 School of Civil and Environmental Engineering, Georgia Institute of Technology, Atlanta, GA 30332, USA
2 Design and Simulation Technologies Inc., Istanbul 34860, Turkey; mmaral@live.com
* Correspondence: mbokuducu@gatech.edu

Abstract: Passive micromixers are miniaturized instruments that are used to mix fluids in microfluidic systems. In microchannels, combination of laminar flows and small diffusion constants of mixing liquids produce a difficult mixing environment. In particular, in very low Reynolds number flows, e.g., Re < 10, diffusive mixing cannot be promoted unless a large interfacial area is formed between the fluids to be mixed. Therefore, the mixing distance increases substantially due to a slow diffusion process that governs fluid mixing. In this article, a novel 3-D passive micromixer design is developed to improve fluid mixing over a short distance. Computational Fluid Dynamics (CFD) simulations are used to investigate the performance of the micromixer numerically. The circular-shaped fluid overlapping (CSFO) micromixer design proposed is examined in several fluid flow, diffusivity, and injection conditions. The outcomes show that the CSFO geometry develops a large interfacial area between the fluid bodies. Thus, fluid mixing is accelerated in vertical and/or horizontal directions depending on the injection type applied. For the smallest molecular diffusion constant tested, the CSFO micromixer design provides more than 90% mixing efficiency in a distance between 260 and 470 μm. The maximum pressure drop in the micromixer is found to be less than 1.4 kPa in the highest flow conditioned examined.

Keywords: micromixer; diffusive mixing; passive mixing; fluid overlapping; sequential injection; segmentation; concentric flow; CFD

Citation: Okuducu, M.B.; Aral, M.M. Toward the Next Generation of Passive Micromixers: A Novel 3-D Design Approach. *Micromachines* **2021**, *12*, 372. https://doi.org/10.3390/mi12040372

Academic Editors: Takasi Nisisako and Kwang-Yong Kim

Received: 25 January 2021
Accepted: 28 March 2021
Published: 30 March 2021

1. Introduction

Over the past few decades, improvements in microfabrication technology [1] and successful implementation of complex processes at microscales [2] have led the development of microfluidic systems. Micro total analysis systems (μTAS) or lab-on-a-chip (LOC) devices [3,4] are commonly referred to describe a centimeter-size compact unit in which physical, chemical, and biological processes take place in a microchannel network. Micromixers are typically one of the major operational sub-units of microfluidic schemes [5] and are employed to mix fluids using active or passive [6–9] mixing principles. In active mixing, extra modules are required to generate external disturbance forces on the flow domain (e.g., electrical, thermal, magnetic, acoustic, and pressure [10–12]) which develops a complexity in terms of fabrication and integration of these components with other microchip elements [11,13,14]. On the contrary, despite offering a relatively poor mixing performance, passive micromixers are simple devices with notable structural and operational advantages. In passive mixing approach, fluid mixing is carried out by exploiting the fluid flow energy within the micromixer and two well-known mixing phenomena: advection and molecular diffusion [4]. As such, additional module and power source requirements are eliminated, and mixing performance is mainly governed by micromixer geometry. Therefore, passive micromixers are typically easy to fabricate, provide simple and stable operating conditions, and offer better integrability [1,6,15], which make these devices prevalent to mix fluids at microscales.

In passive micromixers, fluid mixing arises as a challenging work due to advection-dominant transport developed in microchannels. Strictly laminar fluid flow that is usually Reynolds (Re) number << 100 [16,17] and very low molecular diffusion coefficients (e.g., typically in the range of 10^{-9}–10^{-11} m^2/s [12]) fundamentally create tough mixing conditions. These difficulties are generally dealt with generating a so-called chaotic fluid motion in special micromixer designs that is typically achieved when Re > 10–20. Thus, contact surface area between fluids is enhanced, which in turn expedites diffusive mixing. Typical design and fluid mixing examples may be seen in References [18–21]. At very low Re numbers (e.g., Re < 5–10); however, the effective use of advection is inhibited drastically due to unidirectional fluid flow developed in microchannels. In this case, mixing length increases significantly since mixing process mainly depends on molecular diffusion which is a slow process. Therefore, a long mixing channel is required to obtain an adequate mixing efficiency (e.g., 80% [22]). Increase in mixing length induces three major problems in microfluidic systems as follows: the integration of the micromixer into a microfluidic scheme, high energy requirement, and a long mixing time.

To date, several passive micromixer geometries have been devised to improve the degree of fluid mixing over a short distance. Although majority of these efforts enhance fluid mixing by developing secondary flows when Re > 10–20, usually a considerable mixing length increase is observed at very low flow conditions. For instance, Gidde et al. [8] studied a planar passive micromixer with circular and square mixing chambers. The authors performed a wide range of Re number scenarios between 0.1 and 75 and showed that fluid mixing is promoted based on the development of chaotic advection. In both designs, mixing efficiency was improved noticeably when the flowrate increased, and the highest values were achieved at Re = 75. The distance that is required to yield 80% mixing efficiency is reported as approximately 3500, 9000, 11,000, and 5000 μm for Re = 0.1, 1, 5, and 75, respectively. A similar behavior is also observed in spiral, omega, and interlocking semi-circular shape passive micromixer designs in [23]. While all the three designs perform well beyond Re = 1, mixing efficiencies follow a decreasing trend with rising flowrates between Re = 0.01 and 1. In the article [23], this is related to the short residence time of fluids and inadequate Dean vortices developed in curved microchannels. Although the overall channel length is around 22,000 μm in the spiral micromixer design, mixing efficiencies vary between 65–70% for the Re number range of 0.01–1. Bhagat et al. [24] examined the effects of several obstruction geometries in the mixing channel of a Y-shape micromixer. While obstructed micromixer provides almost a complete fluid mixing at Re = 0.01 (measured at ~5000 μm axial distance), roughly 20% and 40% efficiency drop is observed at Re = 0.1 and 1, respectively. Nonetheless, further increase of flowrate after Re = 1, which is defined as an inflection point in the study, increases the mixing performance by means of chaotic advection formed in the mixing channel. In our previous work [18], a similar inflection point was seen at Re = 5 after which the effective use of advection was accelerated in convex semi-circular ridge (CSCR) micromixer design. Although the CSCR micromixer geometry developed a complex flow profile even at very low flow conditions (e.g., Re $\leq$ 5), and hence enlarged the contact surface between fluids, short residence time of fluids reduced the diffusive activity across the interfacial area formed. Thus, a diminishing mixing efficiency profile was observed between Re = 0.1 and 5.

In several other studies, researchers focused on improving fluid mixing specifically at very low Re numbers. Fang et al. [25] studied a T-shape passive micromixer with several baffle-embedded mixing units along a mixing channel. Numerical and experimental results showed that the design proposed increased mixing performance by inducing chaotic advection in mixing units at Re = 0.29. Based on the simulation and experimental outcomes, it was reported that 28-period mixing unit is required (i.e., ~1.7 cm axial distance) to mix fluids completely. Ortega-Casanova and Lai [26] surveyed the effects of multiple inlets on mixing. The authors applied alternative inlet combinations on a passive micromixer design which has a single mixing unit similar to the one used in the study above [25]. Numerical outcomes show that increasing the number of inlets is more effective for the most challenging flow and transport conditions simulated in the study. When the effects of seven– and two–inlet configurations are compared, seven–inlet configuration increases mixing efficiency approximately 5.5 and 3.5 times for Re = 0.29 and 0.1 flow conditions, respectively. Mixing efficiencies quantified at the exit of the seven–inlet design (i.e., ~5000 µm axial distance) were reported as around 90% and 80% in Re = 0.1 and 0.29 flow cases, respectively. Lin et al. [21] proposed a 3-D circular passive micromixer design to generate vortex effects under low flow conditions. Using eight equally spaced inlets, a rotational fluid motion is created in the circular microchamber beyond a critical Re number that is 2.32. While the mixing efficiency of the micromixer is around 50% and 74% at Re = 0.5 and 2.32, respectively, more than 90% mixing values are obtained when Re $\geq$ 4 (measured at a ~1000 µm distance). Although vortex formation accelerated diffusive interaction between fluid bodies, high mixing efficiencies could also be reached as a result of using a relatively high molecular diffusion constant in the test cases, which is on the order of 10^{-7} m^2/s. Goullet et al. [27] surveyed fluid mixing in a classical T-shape micromixer under sinusoidal fluid injection conditions at Re = 0.3. Results indicate that sinusoidal injection of fluids forms sequential fluid segments in the mixing channel, which in turn enhance contact surface and diffusive mixing. The degree of mixing was reported as 38.4, 70.3, and 86.5% (measured at 500 µm axial distance) for the injection frequencies of 1.25, 5, and 20 Hz, respectively.

As discussed above, mixing performance can be boosted in virtue of complex flow patterns which are mostly generated at relatively high Re numbers. When, however, very low Re conditions are examined, a substantial decrease in mixing performance is seen because of yielding a small contact surface between fluids. This is mainly caused by ineffective flow manipulations in microchannels. Although some injection and design strategies help to improve mixing at low Re numbers, overall micromixer length can still rise to the centimeter level which is not desired as noted earlier. In the present paper, we propose a novel fluid overlapping passive micromixer design to surpass the small interfacial area restrictions at very low fluid flow conditions. Unlike the conventional micromixer configurations, where the effective use of advection process is prioritized to enlarge contact surface, the novel design proposed enables forming a predefined interfacial area between fluid bodies in a compact geometry. Therefore, a rapid inter-diffusion between fluids is ensured, and mixing distance is decreased significantly.

2. Governing Equations and Mathematical Methods

Computational Fluid Dynamics (CFD) simulations are performed to numerically describe the fluid flow and scalar transport in the micromixer. In all scenarios tested, it is assumed that the system is isothermal and gravitational effects are negligible. Fluids to be mixed are of constant density and viscosity, miscible, and non-reactive. Based on the assumptions above, flow field is simulated using incompressible Navier-Stokes (NS) and continuity equations as presented in Equations (1) and (2), respectively. Advection–diffusion (AD) equation is employed to resolve the scalar transport domain in the micromixer as given in Equation (3).

$$\rho\left[\frac{\partial \mathbf{u}}{\partial t} + \mathbf{u}\cdot\nabla\mathbf{u}\right] = -\nabla p + \mu\nabla^2\mathbf{u} \quad (1)$$

$$\nabla\cdot\mathbf{u} = 0 \quad (2)$$

$$\frac{\partial c}{\partial t} + \mathbf{u}\cdot\nabla c = D\nabla^2 c \quad (3)$$

where μ is the viscosity (Pa·s), p is pressure (Pa), ρ is the density (kg/m^3), $\mathbf{u}$ = (u_x, u_y, u_z) is the velocity vector (m/s), c is scalar concentration, and D is molecular diffusion coefficient (m^2/s).

Equations (1)–(3) are solved using OpenFOAM [28] (v5.0, OpenFOAM Foundation, OpenCFD Ltd., Bracknell, UK, 2015) CFD package, in which governing equations are discretized based on finite volume method (FVM). While *simpleFOAM* and *scalarTransportFOAM* solvers are employed to simulate steady-state incompressible fluid flow and passive scalar transport, respectively, a modified form *icoFOAM* solver is exploited in time-dependent solutions of coupled flow and scalar transport equations. Solvers are arranged to treat the advection and diffusion terms in the governing equations with second-order accurate upwind [29] and central difference numerical schemes, respectively. In the *simpleFOAM* solver, SIMPLEC (semi-implicit method for pressure linked equations-consistent) [30] algorithm is used to solve pressure-velocity coupling. To ensure stability in steady-state simulations, under-relaxation technique [31] is used with a factor of 0.9. Steady-state solutions are accepted to be converged when residuals reach 1×10^{-6} threshold as typically practiced in several numerical micromixer studies [18,32,33]. In time-dependent simulations, temporal terms are discretized using Crank-Nicolson scheme with a blending coefficient of 0.9.

Dimensionless Reynolds (Re) and Peclet (Pe) numbers are used to determine the characteristics of fluid flow (e.g., laminar or turbulent) and scalar transport (e.g., advection– or diffusion–dominant) in the micromixer, respectively. Re and Pe numbers are computed in the exit channel of the micromixer as defined in Reference [34]. In the present study, flowrate-averaged mixing approach [33] is employed to quantify the degree of mixing on a given plane as formulized in Equation (4). Unmixed and fully mixed conditions are specified in a mixing index (MI) scale between 0 and 1, respectively. It should be noted that computing a mixing efficiency only relying on the scalar concentration distribution on a certain cross-section may provide imprecise mixing outcomes in cases where the distribution of scalar concentration is non-uniform in a flow profile. More information on this point can be found in References [18,33].

$$MI = 1 - \sqrt{\frac{\sigma^2}{\sigma^2_{\text{max}}}},\ \sigma^2 = \frac{\int_A (c - \bar{c})^2 \cdot u dA}{\int_A u dA},\ \bar{c} = \frac{\int_A c \cdot u dA}{\int_A u dA} \tag{4}$$

where A is area, u is velocity, σ and σ^2 are variance and the maximum variance, respectively, c is concentration, and $\bar{c}$ is the average concentration. In grid independence tests, Equation (5) is used to measure numerical discrepancy between a mesh density and the finest mesh for a given parameter.

$$\Delta D_{M\text{-}F} = \left| \frac{P_M - P_F}{P_F} \right| \times 100 \tag{5}$$

where P_M and P_F denote the parameter values, obtained from the numerical solutions of a certain mesh density and the finest mesh, respectively. $\Delta D_{M\text{-}F}$ shows the difference, as a percentage, between a mesh level and the finest mesh with respect to the parameter employed. In this research, pressure drop (Δp) and mixing efficiency (MI) parameters are used to assess numerical errors in fluid flow and scalar transport simulations, respectively.

3. Micromixer Design and Simulation Setup

In this study, a 3-D circular-shaped fluid overlapping (CSFO) passive micromixer is developed. As shown in Figure 1, the micromixer geometry consists of three main branches that are inlet channel, mixing units, and exit channel. The dimensions of the circular inlet and exit channels are equal with a length (l_i and l_e) and cross-section area (A_c) of 200 µm and 2×10^4 µm^2, respectively. In the CSFO design, five identical mixing units are used to observe the effect of fluid overlapping approach in a wide range of flow conditions. The height (h_u) and radius (r_u) of a single mixing unit are 60 µm and 300 µm, respectively. Each mixing unit is divided equally in the z-direction with a solid, impermeable, and thin-plate disk element which is coaxial with the mixing unit and has a radius (r_d) of 270 µm. It is assumed that the existence of physical joining parts between a disk element and mixing unit will affect the overlapping flow pattern trivially. Therefore, for the sake of designing convenience in the present study, these parts are excluded in the CSFO geometry. In physical applications of the CSFO design, the disk elements can be attached to mixing units from various points as indicated by the line arrows I, II, and

III in Figure 1. Other than that, the mixing units are linked to each other via cylindrical extensions, of which height (h_c) is 10 µm and radius (r_c) is equal to that of inlet and outlet channels. Nonetheless, it should be noted that the purpose of including these connection parts is only to be able to measure mixing performance at the outlet of the mixing chambers. In a physical design, these extensions can be omitted since the short residence time of fluids in these sections will contribute to diffusive mixing negligibly. The dimensions of the CSFO micromixer were selected consistent with the 3-D passive micromixer designs studied in the literature. The 3-D CSFO passive micromixer proposed can be fabricated using the current microfabrication technology. Multi-layer fabrication methods [35,36] can be used in physical construction of the CSFO design. Example micromixer studies may be seen in References [21,37–39] for detailed fabrication process of 3-D geometries at microscales.

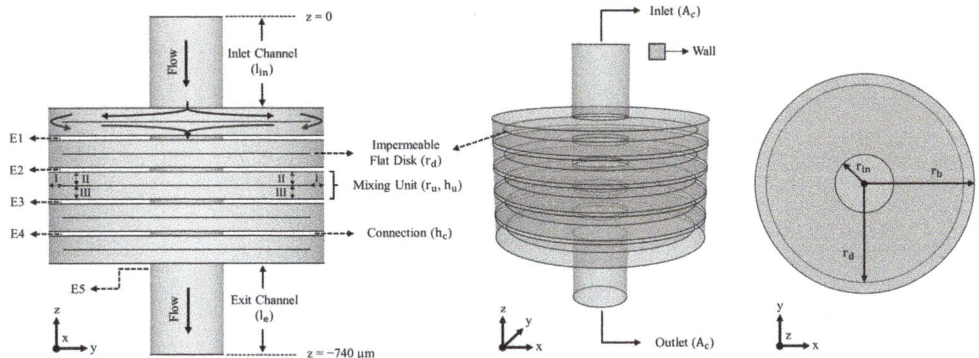

Figure 1. Different views of the 3-D CSFO micromixer geometry. E1, E2, E3, E4, and E5 are the locations 5 µm after the exit of each mixing box.

In the CSFO micromixer, nested-type inlets are used to create overlapping flow profile throughout the disk surfaces in mixing units. The core and outer segments of inlet surfaces are used to inject fluids as depicted in Figure 2. Please note that these segments have an equal surface area in all injection types applied, and these surfaces are further split equally in injection B. The development of different injection patterns in both circular and rectangular geometries can be seen from Figures A1 and A2 in Appendix A. In the present study, both symmetrical and alternating injection patterns (see Figures 2 and A2) are applied over the inlet boundary.

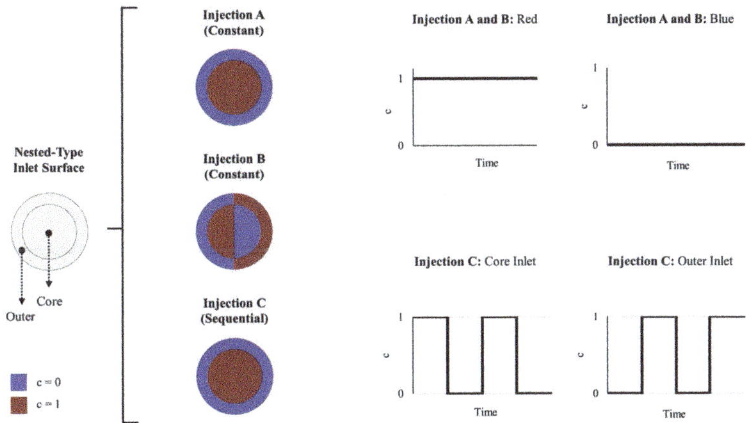

Figure 2. Fluid injection types applied over the inlet boundary of the CSFO micromixer design.

Micromixer performance is examined extensively establishing several molecular diffusion constants, i.e., $D_1 = 3.0 \times 10^{-10}$ (crystal violet dye in water [40]), $D_2 = 1.5 \times 10^{-9}$ (fluorescein solution in water [41]) and $D_3 = 3.0 \times 10^{-9}$ m^2/s (self-diffusion coefficient of water [42]), in a broad range of flow conditions that are Re = 0.1, 0.5, 1, 5, and 10. In all mixing scenarios, equal amount of fluid is injected from each inlet segment in injection A, B, and C. The physical properties of mixing fluids are chosen close to that of water at 20 °C [43,44] with a density (ρ) and viscosity (μ) of 10^3 kg/m^3 and 10^{-3} Pa·s, respectively. To solve the governing flow equations, the following boundary conditions are prescribed in numerical simulations: a uniform velocity profile at the inlets, zero-gauge pressure at the outlet, and zero fluid velocity, i.e., no-slip condition, on solid surfaces. In scalar transport simulations, the gradient of scalar concentration is set to zero at the outlet and solid boundaries to prevent the scalar undergo diffusion over these surfaces. To investigate fluid mixing in the micromixer, relative scalar concentrations, 0 and 1, are imposed on the inlet surface as described schematically in Figure 2. In injection A and B cases, fluid injection is constant over time, and therefore numerical simulations are conducted in two steps as follows. First, a steady-state flow domain is obtained from the simultaneous solution of Equations (1) and (2). Second, a steady-state passive scalar transport simulation is performed by solving the AD equation with the stationary flow domain obtained. In injection C, fluids are injected over the core and outer inlet regions as a square wave with the same injection frequency (f). Thus, a time-dependent simulation is carried out to solve coupled fluid flow and scalar transport equations. In transient simulations, overall simulation times were chosen long enough—for a given flow condition, at least three times of the theoretical fluid mean residence time in the micromixer—to observe the complete development of fluid mixing in the micromixer. In the rest of the paper, CSFO–A, –B and –C notations are used to describe the CSFO micromixer configurations with fluid injection modes A, B and C, respectively. Before the numerical simulations of the test scenarios, the CFD code was validated against the experimental data of two different T-shaped passive micromixer studies in References [45,46]. The validation outcomes, given in Figure A3 in Appendix A, indicate that numerical simulation results are in a good agreement with the experimental data reported in both studies. Thus, the numerical method presented above can be used to predict the mixing performance of the CSFO micromixer proposed.

4. Mesh Study

Numerical solution of advection-dominant transport systems is prone to create high amount of numerical (or false) diffusion errors [34,47]. This is due to inaccurate approximation of steep scalar gradients that inherently develop at high Pe transport conditions. In FVM, maintaining an orthogonality between flow and grid boundaries is pivotal to minimize numerical diffusion. For further information about managing numerical errors in high Pe transport systems, see References [34,47–49]. In this research, hexahedron elements are used to discretize the computational domain in numerical simulations. A systematic mesh study is performed by determining four different grid levels in the computational domain of the CSFO micromixer. Total element numbers in L1, L2, L3, and L4 grid levels are 3.90×10^6, 2.45×10^6, 1.58×10^6, and 1.05×10^6, respectively. Numerical simulations are carried out for the highest Pe number scenario examined in the CSFO–B micromixer configuration (i.e., $Pe = 3.33 \times 10^4$ when Re = 10 and $D = D_1 = 3.0 \times 10^{-10}$ m^2/s). Mesh study outcomes are presented in Figure A4a–c in Appendix A.

Figure A4a shows that there is a good agreement between the finest and coarser mesh levels when the pressure drop is used to quantify the relative numerical errors in numerical solutions. The maximum variation is calculated as 2.1% between L1 and L4 meshes which evidently indicates that even the coarsest grid level, L4, can provide quite accurate results in fluid flow simulations. The same agreement between different mesh level solutions is also seen in Figure A4b, which shows the distribution of velocity along the diameter of outlet plane. On the contrary, when outlet mixing efficiency is employed in error analysis, numerical solutions exhibit a high divergence as indicated by the rising trendline in Figure A4a. In fact, such a discrepancy between the two trendlines occurs due to quite different transport conditions in fluid flow and scalar transport simulations. While a mild Re number (Re = 10) in the former offers a better control of numerical errors even in relatively coarse grids, the latter is carried out at a very high Pe number (Pe = 3.33×10^4). Hence, much smaller mesh elements are required to approximate sharp scalar gradients accurately. Accordingly, mesh study outcomes need to be evaluated in refence to scalar transport simulations to employ a suitable mesh density in the simulations. For the MI parameter given in Figure A4a, the differences in L1–L4, L1–L3, and L1–L2 comparisons are measured to be nearly 28, 13, and 2.7%, respectively. The lessening percentages indicate that false diffusion generation is suppressed noticeably with increasing mesh densities. The convergent trend of mesh refinement can also be seen in Figure A4c, which shows the development mixing efficiency along the CSFO micromixer for all mesh levels tested. Considering the small variation, i.e., 2.7%, against a large mesh density difference, i.e., 1.45×10^6 elements, between L1–L2 mesh levels, L2 mesh level is determined to conduct numerical simulations. Furthermore, this selection is also validated by estimating an effective diffusivity coefficient from the scalar transport solution of L2 mesh level as proposed in [48]. The ratio of effective diffusivity coefficient to molecular diffusion constant was found to be 1.112 which is quite close to 1. This implies that the molecular diffusion constant simulated is mostly recovered from the numerical solution and the amount of numerical diffusion errors is trivial. Therefore, the use of L2 mesh density provides mostly physical mixing outcomes even in the worst-case scenario. Please note that in other mixing scenarios established in the present paper, numerical solutions will generate much less numerical diffusion due to diminishing magnitude of Pe number in mild transport conditions.

5. Results

5.1. Fluid Mixing in the CSFO–A and CSFO–B Micromixer Configurations

At small Re numbers, ineffective manipulation of fluid bodies cause yielding a small contact area between fluid bodies, which in turn limits mixing by diffusion. To overcome this problem and enlarge the interfacial area between mixing fluids, a typical approach is to create several laminations in microchannels [38]. In this method, the main flows are divided into numerous sub-streams or layers of fluid sections which are aligned in microchannels to be in serial or parallel flow regions. In laminating micromixers [37,50–52], the overall contact surface is proportional to the number of different fluid segments generated in the micromixer. Although diffusive mixing is promoted over the interfacial area shared by the fluid segments, usually a complex channel network is required to align fluids in microchannels. In the CSFO micromixer design proposed, the enhancement of contact area is ensured without generating multiple flow sectors in the flow domain. Instead, entire fluid bodies are overlapped and stretched in compact mixing units. As can be seen from Figure A5 in Appendix A, which shows the flow pathlines and 3-D flow domain in the CSFO micromixer, fluids that are injected from core and outer inlet segments flow concentrically through the inlet channel and are stretched over the disk surface. During the fluid flow in the CSFO micromixer, the injected fluids occupy different volumes of the flow domain. As the core flow (shown in red in Figure A5) follows a path around disk elements at the central region of the micromixer, the outer flow (shown in blue in Figure A5) develops between the core flow and micromixer walls. Therefore, a quite large interfacial area is yielded between the two fluid bodies due to the encapsulation of the core flow by the outer flow across the CSFO micromixer domain. The development of contact surface in both upper and lower compartments of a single mixing unit is shown schematically in Figure 3.

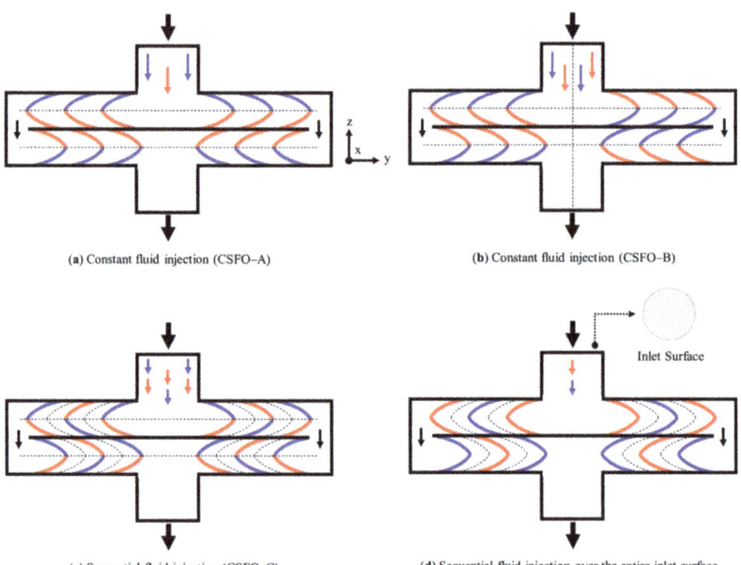

Figure 3. The distribution of injected fluids in each mixing unit of the micromixer configurations (**a**) CSFO–A; (**b**) CSFO–B; and (**c**) CSFO–C. (**d**) The distribution of fluids in a single mixing unit when the entire inlet surface is used to inject fluids sequentially. The dashed lines and curves show the contact surfaces formed between different fluids. Red and blue colors represent the mixing fluids injected from inlet surface(s). Black arrows show flow direction.

As shown in Figure 3 the overlapping (or stratified) fluid pattern expands throughout the disk surface in the upper volume of mixing chamber and flows to the lower volume through the gap between the mixing chamber and the disk element. In the lower section, the above streams are converged at the exit of the cylindrical box and transferred to the next mixing unit. In all design configurations, the same flow cycle is repeated until the fluids are conveyed to the main exit channel of the micromixer. While both CSFO–A and CSFO–B configurations develop a contact surface on the horizontal plane, CSFO–B micromixer also forms an interface in the vertical direction due to alternating fluid injection imposed on the core and outer inlet segments. The horizontal and vertical contact areas formed between mixing fluids are represented by the dashed lines in Figure 3. Meanwhile, it should be noted that the total area of the gap region is approximately 2.7 times higher than that of exit cross-sections. Thus, the fluid flow is not restricted in the gap region and the residence time of fluid particles in a single mixing unit is controlled by the area of the exit cross-section. In the present CSFO micromixer design, the surface area of inlet, outlet, and exit planes are kept equal as noted earlier.

When the mixing performance of micromixers are measured, the outcomes evidently show that diffusive mixing—in the vertical direction—is activated across the large interfacial areas formed. Figure A7 in Appendix A shows the development of fluid mixing along the CSFO–A and CSFO–B micromixers for all mixing conditions tested. Firstly, regarding the results in Figure A7, it can be said that the vertical contact surface formed in the CSFO–B micromixer affects the mixing performance trivially. The MI values of both configurations indicates that even the maximum difference is less than 5%. This is because the degree of mixing is mainly controlled by the horizontal surface areas developed in the upper and lower sections of mixing boxes. The contribution of the additional interface to the diffusive mixing—in the horizontal direction—is more visible at low flow conditions, whereas this effect vanishes by lessening residence time of fluid particles at higher Re numbers. In the lowest flow condition (Re = 0.1), almost a complete fluid mixing (MI > 94%)

is observed at the exit of the first mixing unit. Moreover, although it is not reflected in the plots, the distribution of scalar concentration in simulation results showed that the biggest portion of the mixing takes place only in the upper section of the first mixing box. In all molecular diffusion scenarios, more than 90% mixing efficiency is yielded in a distance less than 260 μm in the main streamwise direction. At Re = 0.5, while at least two mixing units are required to provide more than 90% mixing value when the smallest diffusion constant is used, this is not the case in higher diffusivity conditions. In D_2 and D_3 mixing scenarios, more than 94% mixing efficiency is obtained at the exit of the first mixing unit of the CSFO–A micromixer. As can be pursued from the change of the trendlines in increasing Re numbers, reducing contact time between fluid bodies suppresses the inter–diffusion continually. Hence more mixing units are required to enhance the degree of mixing in the micromixer. Notably, the combination of low contact times with small diffusion coefficients develops the most challenging mixing conditions in the micromixer. At Re = 1, while D_2 and D_3 diffusivities can still be tolerated against the fluid residence time reduced, the use of the smallest diffusion constant becomes difficult. In that case, the mixing distance increases to 400 μm (E3) and 470 μm (E4) to obtain nearly 86% and 93% MI, respectively. In all mixing scenarios tested, the lowest mixing efficiencies are obtained for the two highest flow conditions of the D_1 case as expected. The MI values at the exit of the micromixers are found to be nearly 65% and 54% for Re = 5 and 10, respectively. As mentioned previously, although the contact surface area is increased substantially in CSFO geometry, the development of mixing by diffusion is prohibited by rising flowrates. Nonetheless, the MI values are still promising for the higher molecular diffusion constants, D_2 and D_3, as shown in Figure A7. At Re = 5, while D_2 scenario provides more than 86% mixing efficiency at the exit of the third mixing unit (E3), only two units are required to reach a MI value of nearly 91% (E2) in D_3 case. For the same diffusivity scenarios, D_2 and D_3, the highest flow condition, Re = 10, yields 84% and 92% fluid mixing at E5 and E4 exits, respectively. For all mixing conditions examined, the distribution of scalar concentrations on the outlets of CSFO–A and CSFO–B micromixers are presented in Figure 4. In addition, for the smallest diffusion constant, D_1, the development fluid mixing on different cross-sections along the CSFO–A configuration can be seen from Figure A6 in Appendix A.

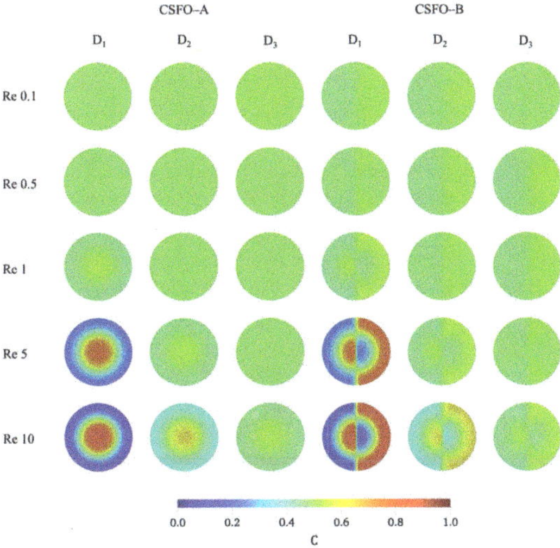

Figure 4. The distribution of scalar concentrations on the exit cross-section of the CSFO–A (first three columns on the left) and CSFO–B (last three columns on the right) micromixer configurations.

5.2. Fluid Mixing in the CSFO–C Micromixer Configuration

As discussed above, when the CSFO micromixer is operated with constant fluid injection, large contact surfaces are yielded between fluid bodies in the horizontal directions. However, in the CSFO design, the overall interfacial area can be enhanced further if the fluids are injected sequentially over the core and outer inlet segments as described in Figure 2. Sequential or pulse injection of fluids can be achieved by manipulating micropumps as described and used in References [53–56]. In the case of sequential injection in CSFO–C configuration, the development of additional contact areas between consecutive fluid pairs is enabled as shown schematically in Figure 3c (see the dashed curves). In mixing units, these new interfaces move dynamically by expanding and shrinking in the upper and lower mixing sections, respectively, which creates a wave pattern throughout the disk surfaces. Therefore, in each half volume of the mixing units, diffusive mixing is also promoted in the horizontal directions. It should be noted that unlike the CSFO–A and CSFO–B configurations, where entire fluid bodies are overlapped on the horizontal plane, in the CSFO–C micromixer, different fluid segments develop the overlapped fluid structure due to wave pattern in the horizontal direction. The mixing performance of the CSFO–C configuration is investigated in various injection frequencies—between 10 and 250 Hertz (Hz) depending on the flow condition—for the most challenging mixing scenarios (i.e., $D = D_1$ and Re = 1, 5, and 10). The evolution of mixing efficiencies is observed with respect to time at the exit of each mixing unit. The results are plotted in Figure A8 in Appendix A. Please note that in Figure A8, f = 0 Hz plots show time-dependent numerical solutions of CSFO–A micromixer, in which fluid injection is constant over time as described in Section 3 and Figure 2. These solutions are used to compare the relative effects of constant and sequential fluid injections in the CSFO design.

Figure A8 evidently shows that the formation of additional contact surface areas accelerated diffusive mixing substantially. At Re = 1, even the lowest injection frequency, f = 10 Hz, is adequate to reduce the mixing distance (MI > 90%) to the exit of the second mixing unit (E2). However, further increase of the injection frequency contributes to the overall mixing efficiency slightly. When the mixing outcomes are compared with that of CSFO–A micromixer, CSFO–C configuration (f = 10 Hz) provides a rapid fluid mixing over a very short distance. To reach a MI value around 85%, the time and distance required are "240 millisecond (ms) and 330 μm" and "640 ms and 400 μm" in CSFO–C and CSFO–A configurations, respectively. Therefore, the use of sequential injection reduces mixing time and distance by the factors of 2.7 and 1.2, respectively. Much higher improvements in mixing values are seen in Re = 5 and 10 flow conditions as indicated by the rising trendlines in Figure A8. Meanwhile, it needs to be explained that before reaching their steady values, the mixing efficiencies follow a declining and rising trend after a sharp increase at early stages. The spikes in the trendlines are observed at the exit of each mixing unit in both CSFO–A and CSFO–C micromixer configurations. These peak points essentially occur due to the following reason explained. At the beginning of the fluid flow in the inlet channel, the formation parabolic flow profile yields a relatively high contact area and diffusive mixing starts developing on this surface. During the fluid flow, the diffusive mixing on the parabolic front travels in the micromixer and leaves the micromixer in the end. Therefore, the peak mixing efficiency, which is generated at the very early stage, is observed at the exit of the mixing units. After the peak values of MI, the declining and rising trends show the actual development of mixing efficiency in the micromixers.

At Re = 5, as constant fluid injection (f = 0 Hz) can only offer a MI value around 63% (t = 280 ms) at E5 location, more than 85% MI (t = 120 ms) is obtained at the exit of the second mixing unit (E2) by the use of sequential fluid injection in the CSFO geometry (f = 25 Hz). For higher injection frequencies, f = 50 and 100 Hz, the degree of mixing rises to 95% (t = 160 ms) and 98% (t = 180 ms) levels at the same location (E2), respectively. Unlike the Re = 1 flow condition, the effect of injection frequency is more visible at Re = 5. While f = 25 Hz case provides nearly 65% (t = 140 ms) mixing efficiency at the exit of the first mixing box (E1), the MI values reach 81% (t = 120 ms) and 90% (t = 100 ms) levels in f = 50 and 100 Hz scenarios, respectively. In the highest flow scenario, Re = 10, while nearly 53% MI (t = 140 ms) can be measured at the last exit location (E5) of the CSFO–A micromixer, CSFO–C configuration provides more than 61% MI (t = 70 ms) at the exit of the first mixing unit (E1) for the lowest injection frequency tested (f = 50 Hz). When the injection frequency is set to f = 100 and 250 Hz, the MI value reaches 77% (t = 80 ms) and 88% (t = 100 ms) on the same exit location (E1), respectively. As the best-case scenarios at Re = 5 (f = 25 Hz) and Re = 10 (f = 50 Hz), CSFO–C configuration develops approximately 85% (t = 120 ms) and 83% (t = 70 ms) mixing efficiencies in a distance less than 330 and 400 μm, respectively. When these mixing figures are compared with the outputs of the CSFO–A configuration, mixing conditions are improved significantly in terms of efficiency, distance, and time. Notably, such an improvement could be achieved by means of the extra contact areas formed between consecutive fluid segments during the sequential injection.

6. Discussion

The CSFO micromixer and nested-type inlets developed in this study offer a novel design approach to mix fluids at microscales. Unlike the conventional micromixer designs, where the enhancement of interfacial area strongly depends on the effective manipulation of fluid flow in microchannels, the CSFO geometry inherently develops a large contact area without requiring a complex flow formation in the micromixer domain. Therefore, better operating conditions are obtained. As can be seen from Figure A4d, the CSFO design improves fluid mixing under reasonable pressure drop conditions. Even the highest flow condition, Re = 10, yields a pressure drop value of less than 1.4 kPa, which is quite acceptable compared to that of reported in the literature [57,58]. The pressure values in Figure A4d can be decreased further when the number of mixing units are reduced in the design. When the mixing performance of the CSFO micromixer is compared with other studies in the literature, a substantial amount of mixing efficiency is achieved over a very short distance as presented in Table 1.

Table 1. The comparison of the CSFO–A configuration with the micromixers reported in the literature in terms of mixing performance in low flow conditions (Re < 10).

Micromixer	Re	Mixing Efficiency (%)	Mixing Length (μm)	Reference
Crossing Channels	0.1	88	6400	[57]
Multi-Inlet	0.1–0.29	90–80	5000	[26]
Serpentine	0.2	100	7500	[59]
Baffled	0.29	52	7200	[25]
T-Shaped (f = 20 Hz)	0.3	86.5	500	[27]
T-Shaped (split inlet)	0.5	42	2000	[18]
Vortex	0.5	50	1000	[21]
Rhombic	1	55	6000	[58]
Obstructed Channels	1	55	1180	[60]
CSFO–A (D_1)	0.1	94	260	Present study
	0.5	94	400	
	1	91	470	

It should also be noted that the use of nested-type inlets is not only limited to the CSFO micromixer, but also can be used in any type of active or passive micromixer designs. Concentric flows that are developed in nested-type inlets basically provide two main advantages. First, when the fluids are injected concentrically, the deformation of fluid bodies in the micromixer becomes relatively much easy compared to the conventional fluid injections in separate channels. For instance, in split-and-recombination (SAR) micromixers [43,61,62], several mixing units are required to increase the distribution of inlet streams in sub-channels. When, however, fluids are injected concentrically, the distribution ratio of different fluids in the sub-channels is increased, and hence the number of mixing units required can be reduced. Second, the nested-type inlets inherently create a contact area between the two fluids being injected. Thus, fluid mixing is initiated at the beginning of the inlet channel before fluids reach to the micromixer. The test simulations, which we do not report here, showed that the use of concentric flows in circular or rectangular channels improves diffusive mixing significantly when Re ≤ 0.1. Therefore, in extremely slow flow conditions, only a straight or curved channel with a nested-type inlet can be used as a micromixer.

The CSFO micromixer can also function without employing the nested-type inlets when the fluid injection is sequential. In such a case, the entire inlet surface is used to feed the micromixer with different fluids sequentially. However, in this condition, the interfacial area on the horizontal plane is not formed and the overall contact surface is developed by the wave pattern as described schematically in Figure A4d. Besides this function, the CSFO geometry can be modified to be operated at much higher flow conditions by generating chaotic advection in the micromixer. For this purpose, the disk elements can be redesigned with alternative grooves or obstacles to create complex flow patterns in the mixing units.

In addition to the circular micromixer design, the fluid overlapping mixing approach can also applied in rectangular or polygonal (e.g., pentagon, hexagon, etc.) geometries. However, when a rectangular geometry is used, a non-uniform velocity distribution can develop on the rectangular plane that divides mixing box volume equally. As displayed in Figure 5 which shows the flow pathlines and the distribution of flow vectors in single-mixing-box circular and square designs (Re = 10), the two geometries render varying flow profiles. In contrast to smooth flow distribution in the circular design, the fluid flow is dominated at the center of the horizontal directions in the square geometry, which creates dead flow zones at the corner regions of the square box (see the dashed red lines). Although that variation in the flow structure does not affect the development of the fluid overlapping pattern in the mixing box, the diffusive interaction is diminished. That is due to the yield of a relatively a smaller contact area and increased flow velocity in central directions, which reduces contact time. Regarding the outcomes in Figure 5, the circular geometry appears to be an optimal shape for the fluid overlapping mixing approach.

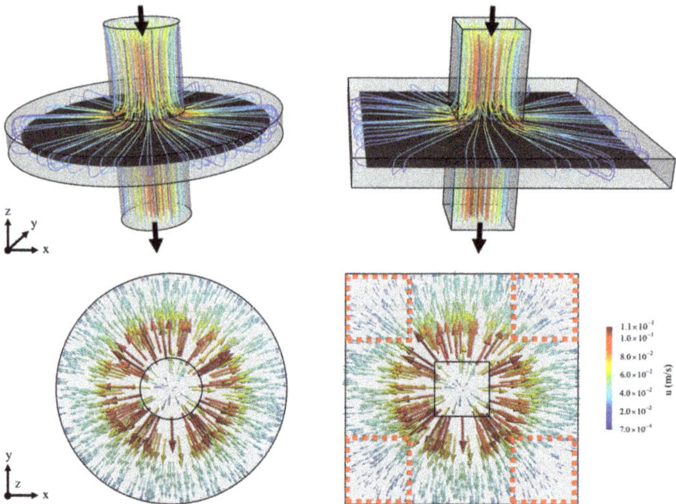

Figure 5. The distribution of flow vectors and flow pathlines in single–mixing–box circular (**left**) and square (**right**) fluid overlapping design configurations (Re = 10). Black arrows show flow direction.

Consequently, considering the plain design structure and high mixing performance, the CSFO passive micromixer can be integrated with microfluidic systems or used as a stand-alone device to mix fluids at microscales.

7. Conclusions

In this research, fluid overlapping mixing approach and nested-type inlets are introduced for passive micromixers. A 3-D circular-shaped passive micromixer design is developed to enhance fluid mixing particularly at low flow conditions that is Re < 10. The mixing performance of the CSFO micromixer is examined numerically in various fluid flow and molecular diffusion conditions. The effects of alternative design configurations and injection strategies are tested. Numerical simulation results indicate that the CSFO design creates a large contact surface between mixing fluids in both upper and lower volumes of each mixing unit. In the case of constant fluid injection, the overlapping fluid pattern develops an interfacial area throughout the disk elements on the horizontal plane. However, when the fluids are injected sequentially, additional contact areas are formed between consecutive fluids. While symmetrical and alternating fluid feeding types provide almost identical results in the constant injection scenarios, the mixing effect of injection frequency is increased with rising Re numbers in the sequential injection cases. In both injection conditions, high mixing efficiency values could be achieved with a reasonable pressure drop in the CSFO micromixer. The maximum pressure drop is found to be less than 1.4 kPa at Re = 10. For the smallest diffusion coefficient and constant fluid injection, more than 90% mixing efficiency is quantified in a distance of 260, 400, and 470 μm for Re = 0.1, 0.5, and 1 flow scenarios, respectively. The mixing distances are reduced further even in high flow conditions when fluids are injected sequentially. When the mixing outcomes are compared with that of reported in the literature, the CSFO design offers a high amount of fluid mixing over a very short distance. Therefore, the CSFO micromixer can be employed in next generation microfluidic systems, where short mixing distances will be required, to mix fluids at microscales.

Author Contributions: Both authors contributed to the development of micromixer design, analysis and discussion of the results, and the preparation of the paper. M.B.O. contributed to setup and completion of the numerical simulations. M.M.A. edited the paper and supervised the overall research. Both authors have read and agreed to the published version of the manuscript.

Funding: This research received no external funding.

Institutional Review Board Statement: Not applicable.

Informed Consent Statement: Not applicable.

Data Availability Statement: Data is contained within the article.

Acknowledgments: This research did not receive any grants from funding agencies in the public, commercial, or not-for-profit sectors.

Conflicts of Interest: The authors declare no conflict of interest.

Appendix A

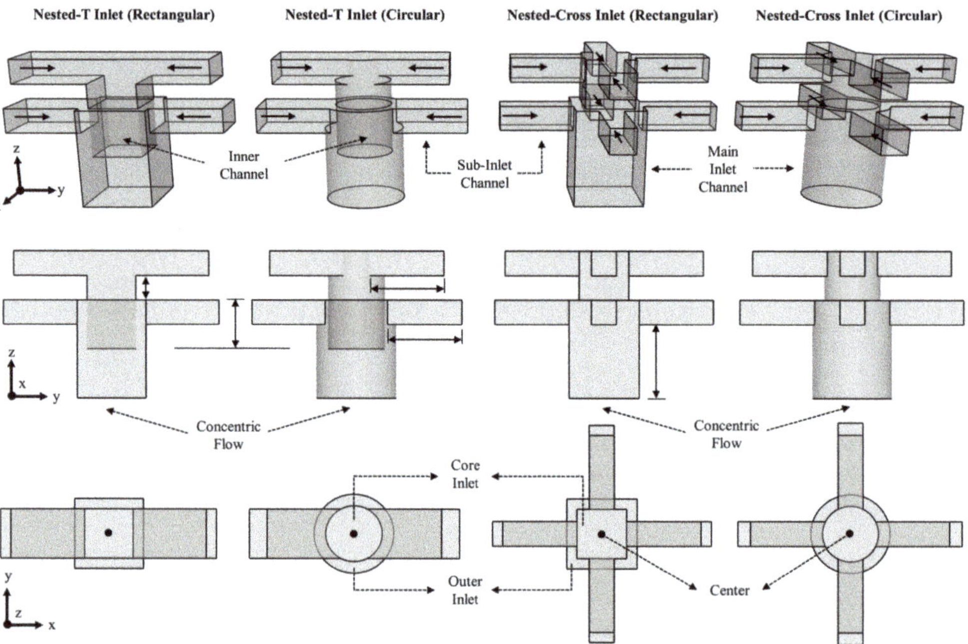

Figure A1. Nested-T (**left two columns**) and nested-cross (**right two columns**) inlet structures to create concentric fluid flow in circular and rectangular microchannels. 3-D view (**top**), yz-plane (**middle**), and xy-plane (**bottom**).

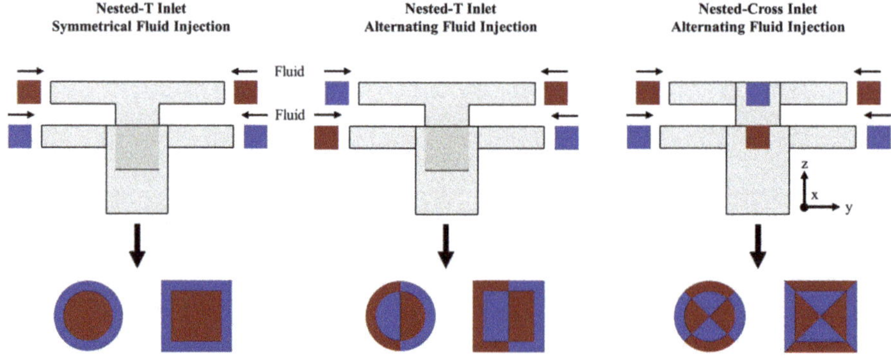

Figure A2. Formation of concentric flows with symmetrical and alternating fluid injections over the nested-t and nested-cross inlet structures. Circular and rectangular planes show the distribution of fluids in concentric flows. Red and blue colors represent different fluids.

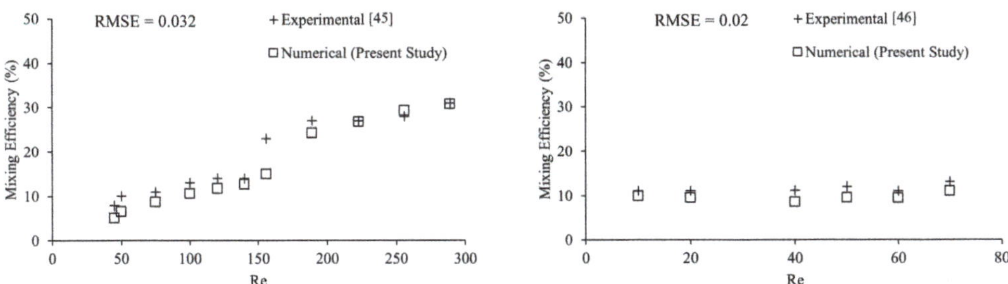

Figure A3. CFD code validation results. Numerical simulation outcomes (present study) against experimental data of two different T-shaped passive micromixers. Root mean square error (RMSE) is used to measure the fit of the numerical simulation results to the experimental data.

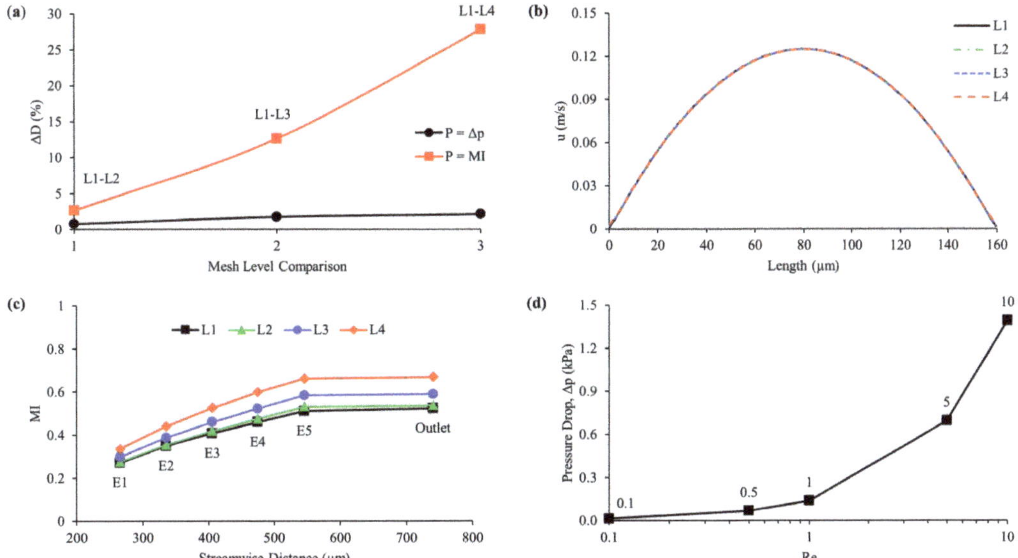

Figure A4. Mesh study outcomes and the change of pressure drop in the CSFO micromixer: (**a**) difference, as a percentage, between L1 and L2, L3, L4 mesh densities with respect to Δp and MI parameters; (**b**) velocity distribution on the diameter of outlet cross-section obtained from L1, L2, L3, and L4 mesh level solutions; (**c**) development of mixing efficiency along the CSFO micromixer in L1, L2, L3, and L4 mesh solutions. MI values are computed on E1, E2, E3, E4, and E5 cross-sections which are normal to the z-direction; (**d**) Δp vs. Re number in the CSFO micromixer.

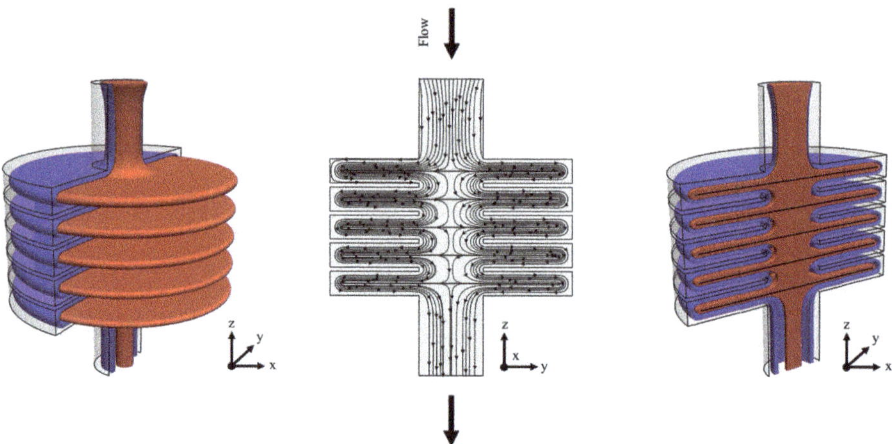

Figure A5. 3-D flow domain and flow pathlines in the CSFO micromixer. Fluid overlapping flow profile (**left and right**) and flow pathlines between inlet and outlet on the central plane of the CSFO design (**center**). Red and blue colors show the fluids injected from core and outer inlets, respectively.

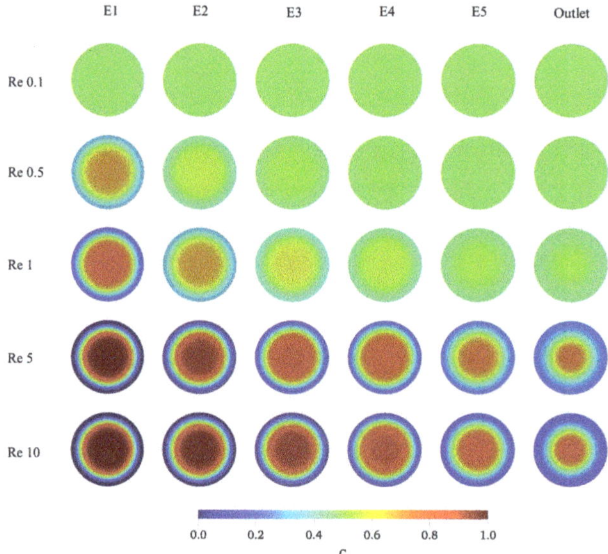

Figure A6. The development of fluid mixing along the CSFO–A micromixer configuration for all flow scenarios of D_1 diffusion constant. Circular planes show the distribution of scalar concentration on the exit cross-section of mixing units (E1, E2, E3, E4, and E5) and outlet of the CSFO micromixer. E1, E2, E3, E4, E5, and outlet planes are normal to the z-direction.

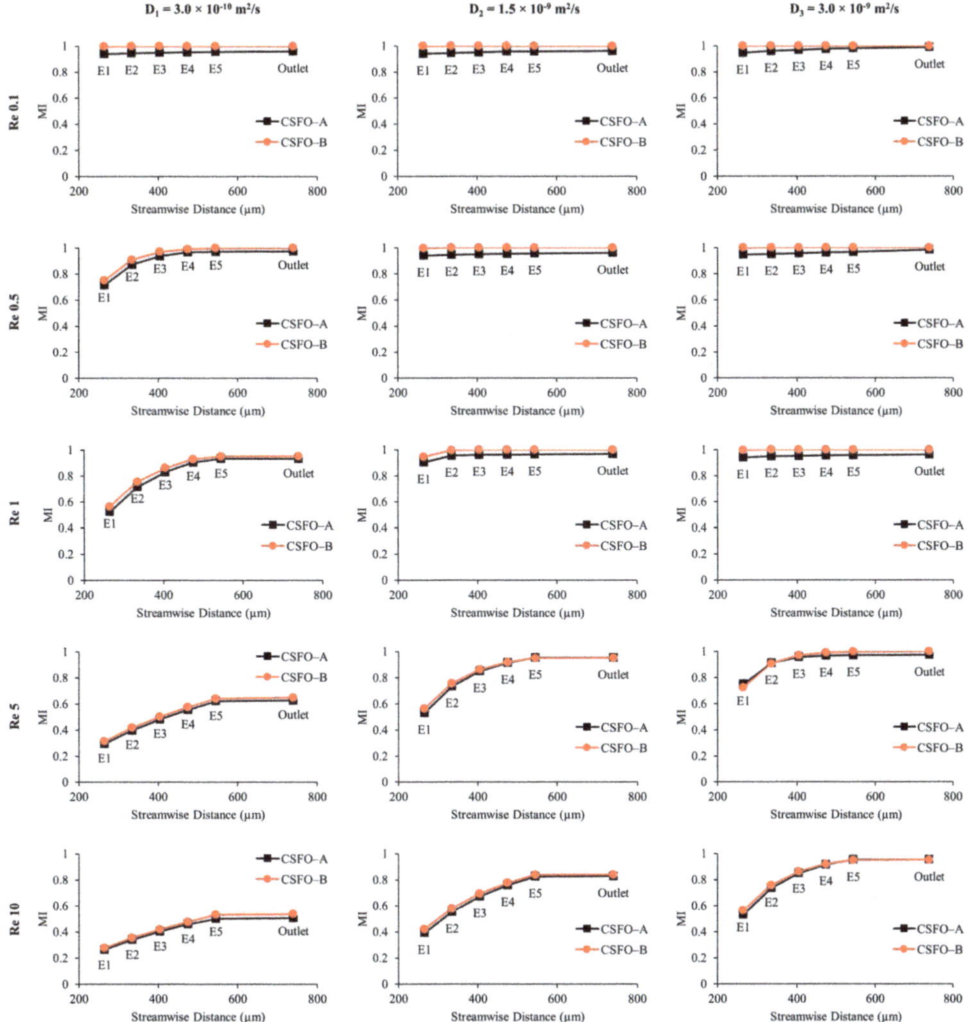

Figure A7. The development of MI along the CSFO–A and CSFO–B micromixer configurations for all flow conditions (i.e., Re = 0.1, 0.5, 1, 5, and 10) and molecular diffusion constants (D_1, D_2, and D_3). MI values are calculated on the E1, E2, E3, E4, and E5 cross-sections which are normal to the z-direction.

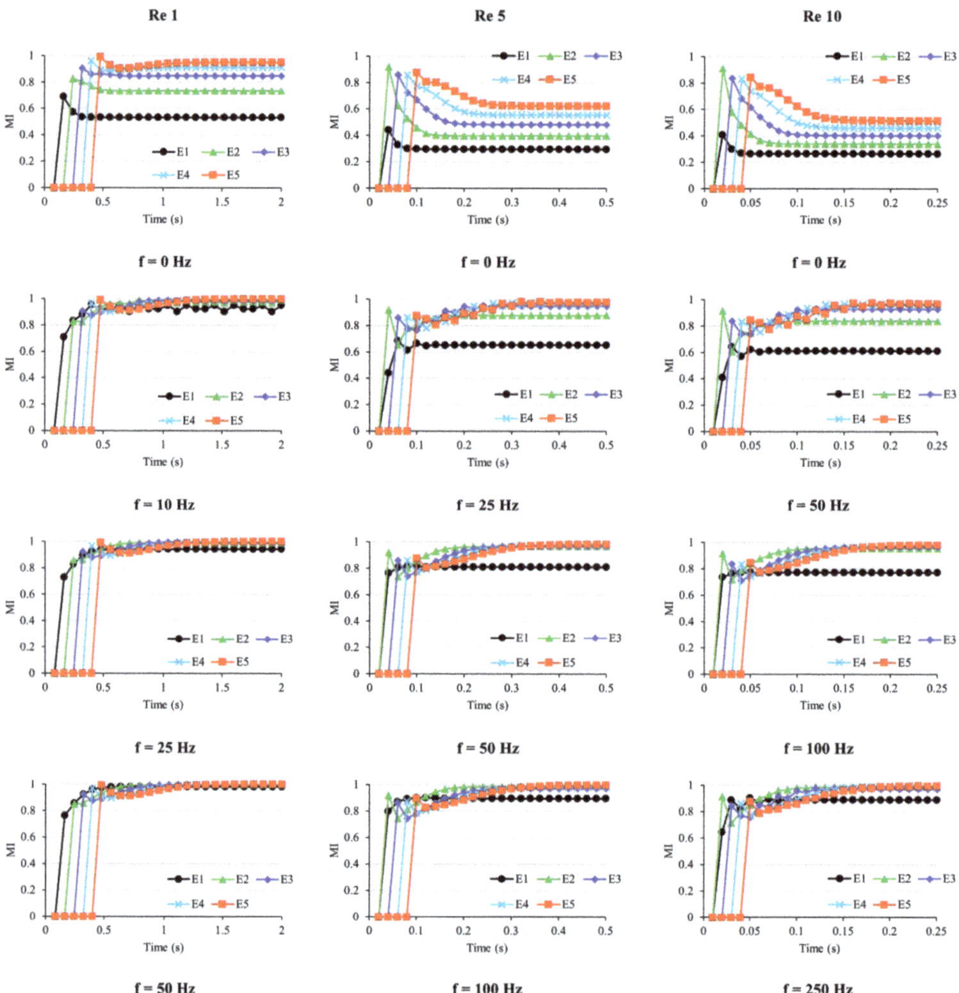

Figure A8. The development of fluid mixing with respect to time at the exit of mixing units in the CSFO–C micromixer configuration when Re = 1, 5, and 10 and D = D_1. f = 0 Hz plots (**first row**) show time-dependent solutions of CSFO–A micromixer. MI values are calculated on the E1, E2, E3, E4, and E5 cross-sections which are normal to the z-direction.

References

1. Mansur, E.A.; Ye, M.; Wang, Y.; Dai, Y. A state-of-the-art review of mixing in microfluidic mixers. *Chin. J. Chem. Eng.* **2008**, *16*, 503–516. [CrossRef]
2. Streets, A.M.; Huang, Y. Chip in a lab: Microfluidics for next generation life science research. *Biomicrofluidics* **2013**, *7*, 011302. [CrossRef] [PubMed]
3. Park, J.M.; Seo, K.D.; Kwon, T.H. A chaotic micromixer using obstruction-pairs. *J. Micromechanics Microengineering* **2009**, *20*, 015023. [CrossRef]
4. Reyes, D.R.; Iossifidis, D.; Auroux, P.-A.; Manz, A. Micro total analysis systems. 1. Introduction, theory, and technology. *Anal. Chem.* **2002**, *74*, 2623–2636. [CrossRef]
5. Gambhire, S.; Patel, N.; Gambhire, G.; Kale, S. A review on different micromixers and its micromixing within microchannel. *Int. J. Curr. Eng. Technol.* **2016**, *4*, 409–413.
6. Nguyen, N.-T.; Wu, Z. Micromixers—A review. *J. Micromechanics Microengineering* **2005**, *15*, R1–R16. [CrossRef]
7. Suh, Y.K.; Kang, S. A review on mixing in microfluidics. *Micromachines* **2010**, *1*, 82–111. [CrossRef]

8. Gidde, R.R.; Pawar, P.M.; Ronge, B.P.; Misal, N.D.; Kapurkar, R.B.; Parkhe, A.K. Evaluation of the mixing performance in a planar passive micromixer with circular and square mixing chambers. *Microsyst. Technol.* **2017**, *24*, 2599–2610. [CrossRef]
9. Lee, C.Y.; Chang, C.L.; Wang, Y.N.; Fu, L.M. Microfluidic mixing: A review. *Int. J. Mol. Sci.* **2011**, *12*, 3263–3287. [CrossRef]
10. Kumar, V.; Paraschivoiu, M.; Nigam, K.D.P. Single-phase fluid flow and mixing in microchannels. *Chem. Eng. Sci.* **2011**, *66*, 1329–1373. [CrossRef]
11. Cai, G.; Xue, L.; Zhang, H.; Lin, J. A review on micromixers. *Micromachines* **2017**, *8*, 274. [CrossRef]
12. Nguyen, N.-T. Chapter 1-introduction. In *Micromixers*, 2nd ed.; Nguyen, N.-T., Ed.; William Andrew Publishing: Oxford, UK, 2012; pp. 1–8.
13. Chung, Y.-C.; Hsu, Y.-L.; Jen, C.-P.; Lu, M.-C.; Lin, Y.-C. Design of passive mixers utilizing microfluidic self-circulation in the mixing chamber. *Lab A Chip* **2004**, *4*, 70–77. [CrossRef]
14. Alam, A.; Afzal, A.; Kim, K.-Y. Mixing performance of a planar micromixer with circular obstructions in a curved microchannel. *Chem. Eng. Res. Des.* **2014**, *92*, 423–434. [CrossRef]
15. Capretto, L.; Cheng, W.; Hill, M.; Zhang, X. Micromixing within microfluidic devices. *Top Curr. Chem.* **2011**, *304*, 27–68.
16. Le The, H.; Le Thanh, H.; Dong, T.; Ta, B.Q.; Tran-Minh, N.; Karlsen, F. An effective passive micromixer with shifted trapezoidal blades using wide reynolds number range. *Chem. Eng. Res. Des.* **2015**, *93*, 1–11. [CrossRef]
17. Bhopte, S.; Sammakia, B.; Murray, B. Numerical study of a novel passive micromixer design. In Proceedings of the 2010 12th IEEE Intersociety Conference on Thermal and Thermomechanical Phenomena in Electronic Systems, Las Vegas, NV, USA, 2–5 June 2010; pp. 1–10.
18. Okuducu, M.B.; Aral, M.M. Novel 3-d t-shaped passive micromixer design with helicoidal flows. *Processes* **2019**, *7*, 637. [CrossRef]
19. Chen, X.; Li, T. A novel passive micromixer designed by applying an optimization algorithm to the zigzag microchannel. *Chem. Eng. J.* **2017**, *313*, 1406–1414. [CrossRef]
20. Hossain, S.; Ansari, M.A.; Kim, K.-Y. Evaluation of the mixing performance of three passive micromixers. *Chem. Eng. J.* **2009**, *150*, 492–501. [CrossRef]
21. Lin, C.-H.; Tsai, C.-H.; Fu, L.-M. A rapid three-dimensional vortex micromixer utilizing self-rotation effects under low reynolds number conditions. *J. Micromechanics Microengineering* **2005**, *15*, 935–943. [CrossRef]
22. Tran-Minh, N.; Dong, T.; Karlsen, F. An efficient passive planar micromixer with ellipse-like micropillars for continuous mixing of human blood. *Comput. Methods Programs Biomed.* **2014**, *117*, 20–29. [CrossRef]
23. Al-Halhouli, A.a.; Alshare, A.; Mohsen, M.; Matar, M.; Dietzel, A.; Büttgenbach, S. Passive micromixers with interlocking semi-circle and omega-shaped modules: Experiments and simulations. *Micromachines* **2015**, *6*, 953–968. [CrossRef]
24. Bhagat, A.A.S.; Peterson, E.T.K.; Papautsky, I. A passive planar micromixer with obstructions for mixing at low reynolds numbers. *J. Micromechanics Microengineering* **2007**, *17*, 1017–1024. [CrossRef]
25. Fang, Y.; Ye, Y.; Shen, R.; Zhu, P.; Guo, R.; Hu, Y.; Wu, L. Mixing enhancement by simple periodic geometric features in microchannels. *Chem. Eng. J.* **2012**, *187*, 306–310. [CrossRef]
26. Ortega-Casanova, J.; Lai, C.H. Cfd study about the effect of using multiple inlets on the efficiency of a micromixer. Assessment of the optimal inlet configuration working as a microreactor. *Chem. Eng. Process. Process Intensif.* **2018**, *125*, 163–172. [CrossRef]
27. Goullet, A.; Glasgow, I.; Aubry, N. Dynamics of microfluidic mixing using time pulsing. *Discret. Contin. Dyn. Syst.* **2005**, *2005*, 327–336.
28. OpenFOAM. *The Openfoam Foundation, Opencfd Ltd., v5*; OpenFOAM: Bracknell, UK, 2015.
29. Warming, R.F.; Beam, R.M. Upwind second-order difference schemes and applications in aerodynamic flows. *AIAA J.* **1976**, *14*, 1241–1249. [CrossRef]
30. Wang, L.; Yang, J.-T.; Lyu, P.-C. An overlapping crisscross micromixer. *Chem. Eng. Sci.* **2007**, *62*, 711–720. [CrossRef]
31. Moukalled, F.; Mangani, L.; Darwish, M. *The Finite Volume Method in Computational Fluid Dynamics: An Advanced Introduction with Openfoam and Matlab*; Springer Publishing Company Switzerland, Incorporated: Geneva, Switzerland, 2015; p. 791.
32. Ansari, M.A.; Kim, K.Y.; Kim, S.M. Numerical and experimental study on mixing performances of simple and vortex micro t-mixers. *Micromachines (Basel)* **2018**, *9*, 204. [CrossRef]
33. Roudgar, M.; Brunazzi, E.; Galletti, C.; Mauri, R. Numerical study of split t-micromixers. *Chem. Eng. Technol.* **2012**, *35*, 1291–1299. [CrossRef]
34. Okuducu, M.B.; Aral, M.M. Performance analysis and numerical evaluation of mixing in 3-d t-shape passive micromixers. *Micromachines* **2018**, *9*, 210. [CrossRef]
35. Zhang, M.; Wu, J.; Wang, L.; Xiao, K.; Wen, W. A simple method for fabricating multi-layer pdms structures for 3d microfluidic chips. *Lab A Chip* **2010**, *10*, 1199–1203. [CrossRef]
36. Chen, X. Fabrication and performance evaluation of two multi-layer passive micromixers. *Sens. Rev.* **2018**, *38*, 321–325. [CrossRef]
37. Branebjerg, J.; Gravesen, P.; Krog, J.P.; Nielsen, C.R. Fast mixing by lamination. In Proceedings of the Ninth International Workshop on Micro Electromechanical Systems, San Diego, CA, USA, 11–15 February 1996; pp. 441–446.
38. Gray, B.L.; Jaeggi, D.; Mourlas, N.J.; van Drieënhuizen, B.P.; Williams, K.R.; Maluf, N.I.; Kovacs, G.T.A. Novel interconnection technologies for integrated microfluidic systems1paper presented as part of the ssaw-98 workshop.1. *Sens. Actuators A Phys.* **1999**, *77*, 57–65. [CrossRef]
39. Yang, J.; Qi, L.; Chen, Y.; Ma, H. Design and fabrication of a three dimensional spiral micromixer. *Chin. J. Chem.* **2013**, *31*, 209–214. [CrossRef]

40. Galletti, C.; Roudgar, M.; Brunazzi, E.; Mauri, R. Effect of inlet conditions on the engulfment pattern in a t-shaped micro-mixer. *Chem. Eng. J.* **2012**, *185–186*, 300–313. [CrossRef]
41. Wu, Z.; Nguyen, N.-T.; Huang, X. Nonlinear diffusive mixing in microchannels: Theory and experiments. *J. Micromech. Microeng.* **2004**, *14*, 604–611. [CrossRef]
42. Holz, M.; Heil, S.; Sacco, A. Temperature-dependent self-diffusion coefficients of water and six selected molecular liquids for calibration in accurate 1h nmr pfg measurements. *Phys. Chem. Chem. Phys.* **2000**, *2*, 4740–4742. [CrossRef]
43. Ansari, M.A.; Kim, K.-Y.; Anwar, K.; Kim, S.M. A novel passive micromixer based on unbalanced splits and collisions of fluid streams. *J. Micromechanics Microengineering* **2010**, *20*, 055007. [CrossRef]
44. Izadpanah, E.; Hekmat, M.H.; Azimi, H.; Hoseini, H.; Babaie Rabiee, M. Numerical simulation of mixing process in t-shaped and dt-shaped micromixers. *Chem. Eng. Commun.* **2018**, *205*, 363–371. [CrossRef]
45. Silva, J.P.; dos Santos, A.; Semiao, V. Experimental characterization of pulsed newtonian fluid flows inside t-shaped micromixers with variable inlets widths. *Exp. Therm. Fluid Sci.* **2017**, *89*, 249–258. [CrossRef]
46. Ansari, M.A.; Kim, K.-Y.; Anwar, K.; Kim, S.M. Vortex micro t-mixer with non-aligned inputs. *Chem. Eng. J.* **2012**, *181–182*, 846–850. [CrossRef]
47. Okuducu, M.B.; Aral, M.M. Computational evaluation of mixing performance in 3-d swirl-generating passive micromixers. *Processes* **2019**, *7*, 121. [CrossRef]
48. Liu, M. Computational study of convective–diffusive mixing in a microchannel mixer. *Chem. Eng. Sci.* **2011**, *66*, 2211–2223. [CrossRef]
49. Bailey, R.T. Managing false diffusion during second-order upwind simulations of liquid micromixing. *Int. J. Numer. Methods Fluids* **2017**, *83*, 940–959. [CrossRef]
50. Tofteberg, T.; Skolimowski, M.; Andreassen, E.; Geschke, O. A novel passive micromixer: Lamination in a planar channel system. *Microfluid. Nanofluidics* **2009**, *8*, 209–215. [CrossRef]
51. Hardt, S.; Schönfeld, F. Laminar mixing in different interdigital micromixers: Ii. Numerical simulations. *AIChE J.* **2003**, *49*, 578–584. [CrossRef]
52. Sabry, M.N.; El-Emam, S.H.; Mansour, M.H.; Shouman, M.A. Development of an efficient uniflow comb micromixer for biodiesel production at low reynolds number. *Chem. Eng. Process. Process Intensif.* **2018**, *128*, 162–172. [CrossRef]
53. Fujii, T.; Sando, Y.; Higashino, K.; Fujii, Y. A plug and play microfluidic device. *Lab Chip* **2003**, *3*, 193–197. [CrossRef]
54. Nguyen, N.-T.; Huang, X. Mixing in microchannels based on hydrodynamic focusing and time-interleaved segmentation: Modelling and experiment. *Lab on a chip* **2005**, *5*, 1320–1326. [CrossRef]
55. Nguyen, N.-T.; Huang, X. An analytical model for mixing based on time-interleaved sequential segmentation. *Microfluid. Nanofluidics* **2005**, *1*, 373–375. [CrossRef]
56. Glasgow, I.; Lieber, S.; Aubry, N. Parameters influencing pulsed flow mixing in microchannels. *Anal. Chem.* **2004**, *76*, 4825–4832. [CrossRef]
57. Alam, A.; Kim, K.-Y. Mixing performance of a planar micromixer with circular chambers and crossing constriction channels. *Sens. Actuators B Chem.* **2013**, *176*, 639–652. [CrossRef]
58. Chung, C.K.; Shih, T.R. A rhombic micromixer with asymmetrical flow for enhancing mixing. *J. Micromechanics Microengineering* **2007**, *17*, 2495–2504. [CrossRef]
59. Hossain, S.; Lee, I.; Kim, S.M.; Kim, K.-Y. A micromixer with two-layer serpentine crossing channels having excellent mixing performance at low reynolds numbers. *Chem. Eng. J.* **2017**, *327*, 268–277. [CrossRef]
60. Sadegh Cheri, M.; Latifi, H.; Salehi Moghaddam, M.; Shahraki, H. Simulation and experimental investigation of planar micromixers with short-mixing-length. *Chem. Eng. J.* **2013**, *234*, 247–255. [CrossRef]
61. Ansari, M.A.; Kim, K.-Y. Mixing performance of unbalanced split and recombine micomixers with circular and rhombic sub-channels. *Chem. Eng. J.* **2010**, *162*, 760–767. [CrossRef]
62. Raza, W.; Hossain, S.; Kim, K.-Y. Effective mixing in a short serpentine split-and-recombination micromixer. *Sens. Actuators B Chem.* **2018**, *258*, 381–392. [CrossRef]

Article

Kinematic Measurements of Novel Chaotic Micromixers to Enhance Mixing Performances at Low Reynolds Numbers: Comparative Study

Toufik Tayeb Naas [1,†], Shakhawat Hossain [2,†], Muhammad Aslam [3], Arifur Rahman [4], A. S. M. Hoque [2], Kwang-Yong Kim [5,*] and S. M. Riazul Islam [6,*]

1 Gas Turbine Joint Research Team, University of Djelfa, 17000 Djelfa, Algeria; toufiknaas@gmail.com
2 Department of Industrial and Production Engineering, Jashore University of Science and Technology, Jashore 7408, Bangladesh; shakhawat.ipe@just.edu.bd (S.H.); asm.hoque2@mail.dcu.ie (A.S.M.H.)
3 Department of Chemical Engineering, COMSATS University Islamabad (CUI), Lahore Campus, Defense Road, Off Raiwind Road, Lahore 45550, Pakistan; maslam@cuilahore.edu.pk
4 Department of Mechanical Engineering, Bangabandhu Textile Engineering College, Kalihati, Tangail 1970, Bangladesh; arif02me@gmail.com
5 Department of Mechanical Engineering, Inha University, 100 Inha-Ro, Michuhol-Gu, Incheon 22212, Korea
6 Department of Computer Science and Engineering, Sejong University, Seoul 05006, Korea
* Correspondence: kykim@inha.ac.kr (K.-Y.K.); riaz@sejong.ac.kr (S.M.R.I.); Tel.: +82-32-872-3096 (K.-Y.K.); Fax: +82-32-868-1716 (K.-Y.K.); +82-02-3408-2969 (S.M.R.I.)
† These authors contributed equally to this work. Co-first author (T.T.N. and S.H.).

Citation: Naas, T.T.; Hossain, S.; Aslam, M.; Rahman, A.; Hoque, A.S.M.; Kim, K.-Y.; Islam, S.M.R. Kinematic Measurements of Novel Chaotic Micromixers to Enhance Mixing Performances at Low Reynolds Numbers: Comparative Study. *Micromachines* **2021**, *12*, 364. https://doi.org/10.3390/mi12040364

Academic Editor: Chiara Galletti

Received: 14 February 2021
Accepted: 23 March 2021
Published: 28 March 2021

Abstract: In this work, a comparative investigation of chaotic flow behavior inside multi-layer crossing channels was numerically carried out to select suitable micromixers. New micromixers were proposed and compared with an efficient passive mixer called a Two-Layer Crossing Channel Micromixer (TLCCM), which was investigated recently. The computational evaluation was a concern to the mixing enhancement and kinematic measurements, such as vorticity, deformation, stretching, and folding rates for various low Reynolds number regimes. The 3D continuity, momentum, and species transport equations were solved by a Fluent ANSYS CFD code. For various cases of fluid regimes (0.1 to 25 values of Reynolds number), the new configuration displayed a mixing enhancement of 40%–60% relative to that obtained in the older TLCCM in terms of kinematic measurement, which was studied recently. The results revealed that all proposed micromixers have a strong secondary flow, which significantly enhances the fluid kinematic performances at low Reynolds numbers. The visualization of mass fraction and path-lines presents that the TLCCM configuration is inefficient at low Reynolds numbers, while the new designs exhibit rapid mixing with lower pressure losses. Thus, it can be used to enhance the homogenization in several microfluidic systems.

Keywords: TLCCM configuration; mixing rate; kinematics; deformation; vorticity; stretching; folding

1. Introduction

Mass transfer induced by different fluid concentrations is generally applied in kinematic processes that involve the dynamic transport of chemical species within physical technology. The mixing fluids process of micromixers occurs in multiple applications of different mechanisms and industrial applications that have much utility in bioengineering fields, biomedical treatments, and chemical reactions [1–3], etc.

The role of mass transfer is a subject of interest for many researchers who have given solutions to enhance the kinematic behavior and fluid mixing performance while reducing pressure drops. However, due to the small size of micromixers, it is difficult to enhance the molecular distribution operation during fluid flow, while the improvement in dynamics and kinematic behaviors is limited. The generation of secondary flows through chaotic advection is a suitable technique to enhance mixing efficiency using passive micromixers [4,5].

To understand the behavior of mass fraction inside T-type micromixers, many researchers have tried to analyze solutions within split and recombined micromixers [6] or modified curved T-type mixers [4]. The impacts of various flow regimes on velocity contours, initial sensitivity, mixing capacity, pressure drop, energy performances, and entropy generation rates were investigated numerically and experimentally. Their results showed that for low flow regimes, the molecular diffusion dominated for mixing fluid; therefore, a high mixing performance occurred for all configurations. In addition, the strong secondary flows in the proposed configuration enhanced the entropy generation. In real-time, Huanhuan et al. [7] investigated fluid mixing via a serpentine micromixer based on an elliptical groove design. The results indicated a significantly enhanced mixing performance (90%). Numerical works were performed to characterize the dynamic flow of Newtonian fluids for complex micromixers, such as zigzag, square-wave [8], and two-layer crossing channel [9,10]. The authors argued that the best mixing rate was obtained for a chaotic configuration with low cost. For low Reynolds numbers, Koudari et al. [11] investigated the quality of mixing flow of Newtonian fluids within a two-layer micromixer, which was proposed recently by Shakhawat et al. [12]. The authors modified the crossing zone to reduce the number of units and enhance the mixing performance through a parametric study. The authors also solved the species transport equations using CFD software. Arshad and Kim [10] found that a convergent-divergent micromixer with a T-junction described the most efficient coupling with pulsatile flow. The mass transfer showed that the interfacial area rises and separates to produce discrete puffs of flows for enhanced mixing efficiency. Chih et al. [13] proposed a new procedure to enhance fluid mixing in T-shaped micromixers by adding vortex-inducing obstacles. They present the particle-tracking flow with an estimation of diffusion prototypes to determine the mass distribution within the micromixer. An experimental study was carried out to fabricate the fused silica microfluidic mixer via employing low viscosity, by Dena et al. [14]. They illustrated the utilization of a newly fabricated mixer to examine the folding of the Pin1 WW region at a low time range.

Recently, a new technique to create a strong vortex inside the fluid flow by using a direct current (DC) electric field was investigated [15,16]. Newtonian fluid flow in T-channel [15] and corrugated wall [16] configurations were suggested to improve the mixing process. They found that the fluid homogenization rate enhances more than 200% by adding a direct current (DC) electric field inside the configurations. To characterize mixing efficiency with secondary flows, researchers use a different parameter focused on both fluid homogenization and fluid kinematic behavior for active [17] and passive mixers [18–20] called Local Physical Process (LPP) or kinematic measurements. They analyzed the behavior of the fluid in motion by calculating elongation, vorticity, helicity, stretching, and folding rates for different flow regimes.

In this work, novel multi-layer micromixers have been proposed to offer excellent kinematic performances and mixing enhancement with low-pressure drops, compared with a recent two-layer micromixer created by Kouadri et al. [11]. Kinematic measurements of new chaotic micromixers at low Reynolds numbers are investigated in detail.

2. Geometries Description and Boundary Conditions

New chaotic multi-layer micromixers alongside the relevant boundary conditions are proposed, namely, Two-Layer Crossing Modified geometry (TLC-M1, TLC-M2, TLC-M3, and TLC-M4) compared with the Two-Layer Crossing Channel Micromixer (TLCCM), which was used recently by Kouadri et al. [11] and Shakhawat et al. [12], in terms of kinematic behavior and mixing process. This micromixer is formed of two twisted channels of the same dimensions; the upper and lower channels are arranged with a periodic chamber. The serial mixing parts occur in reconstructed models of various grooves. Each mixer is created with an unfolded length equal to 7.5 mm, the same width and depth of 1.5 mm, and a hydraulic diameter of 1.5 mm. Figure 1 displays a 3D view design of the proposed micromixers. Two water liquids are imposed at the inlet sections with different colors and uniform velocity. The outsides are supposed to be adiabatic, and other flow boundaries include no-slip. The pressure outlet condition is considered at the outflow section.

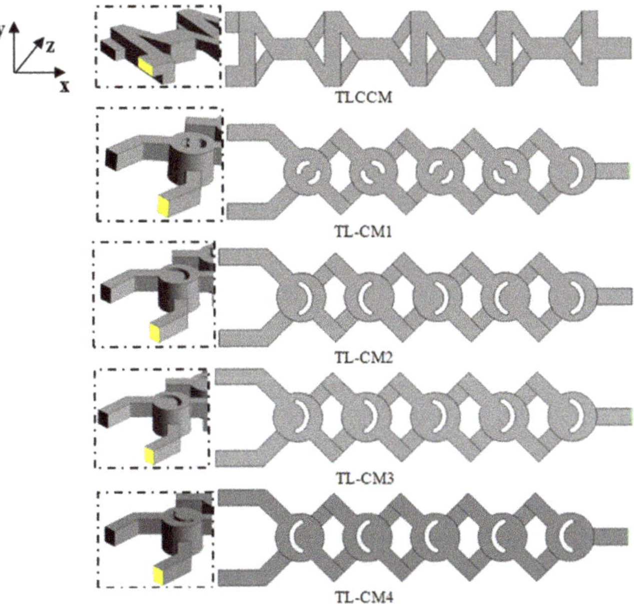

Figure 1. Schematic representation of the micromixers with geometric parameters.

3. Governing Equations and Numerical Methodology

Continuity, momentum, and species transport equations of incompressible steady flows are presented by the following equations:

- Continuity equation:

$$\nabla \vec{V} = 0 \tag{1}$$

- Momentum equation:

$$\left(\vec{V}\nabla\right)\vec{V} = -\frac{1}{\rho}\nabla P + \nu\nabla^2\vec{V} \tag{2}$$

where P, V, and ρ represent the static pressure, velocity, and density of the fluid, respectively. For Newtonian fluid flow, the species transport equation is expressed as follows:

$$\vec{V}\nabla = D\nabla^2 C_i \tag{3}$$

Uniform velocities were imposed at the inlets with a no-slip boundary condition applied on the other parts of the walls, and zero static pressure at the outlet section. Additionally, in one of the fluid entries, the mass fraction equals 1, and the other equals zero.

All governing equations in the proposed micromixers were solved in a laminar regime by using ANSYS Fluent™ 16 (Ansys, Canonsburg, PL, USA) CFD software [21], which is essentially based on the finite volumes method. The SIMPLEC scheme was chosen for velocity coupling and pressure. A second-order upwind scheme was selected to solve the concentration and momentum equations. The computations were ensured and simulated to be converged at 10^{-7} of root mean square (RMS) residual values. Pure water liquids were used as a working fluid, with a fluid density of 1000 kg/m^3 and the diffusion coefficient equal to 1×10^{-11} m^2/s.

The Reynolds number is defined as follows:

$$Re = \frac{\rho v D_h}{\mu} \tag{4}$$

4. Mixing Rate

The mixing rate (M_r) of two fluids, given in the following equation, is an efficient parameter for quantifying scalar mixing:

$$M_r = 1 - \frac{\sqrt{\frac{1}{N}\sum_{i=1}^{N}(C_i - \overline{C})^2}}{\sigma_0} \tag{5}$$

where N is the number of nodes on one section, C_i is the mass fraction at the sampling point i, and $\overline{C}$ is the most favorable of the mass fraction of C_i. σ_0 is the standard deviation at the inlet flow section. The values of M_r range from zero for the no mixture case, to 1 for optimum mixing.

5. Kinematic Behavior Process of the Chaotic Flows

The velocity field is strongly dependent on the components of the velocity gradient $(\partial Ui)/(\partial xj)$. Therefore, these components contribute to the kinematic flow of the fluid, such as vortex velocity, strain rate, rotation speed, and vortex stretch and compression [18]. It should be noted that these characteristics are almost negligible for a very slow regime (Re < 1), as the fluid flow is laminar and settling.

5.1. Stretching and Folding Process

The transport equation of the vorticity is written and mathematically articulated as [19,20]:

$$\partial\vec{\Omega}/\partial t + \vec{V}.\overrightarrow{\overline{\nabla}}\vec{\Omega} = \vec{\Omega}.\overrightarrow{\overline{\nabla}}\vec{V} + \nu\Delta\vec{\Omega} \tag{6}$$

The expression $\vec{\Omega}.\overrightarrow{\overline{\nabla}}\vec{V}$ provokes the development of vortex formations of different measurements by creating the compression and stretching vortex; see Figure 2. These phenomena work simultaneously on the dimensions of the vortex. At a given moment, the stretching operation develops the length of the vortex and reduces its cross-section, while the folding operation decreases the length of the vortex and augments its cross-section. These phenomena are generated as a result of the conservation of mass and kinematic moment.

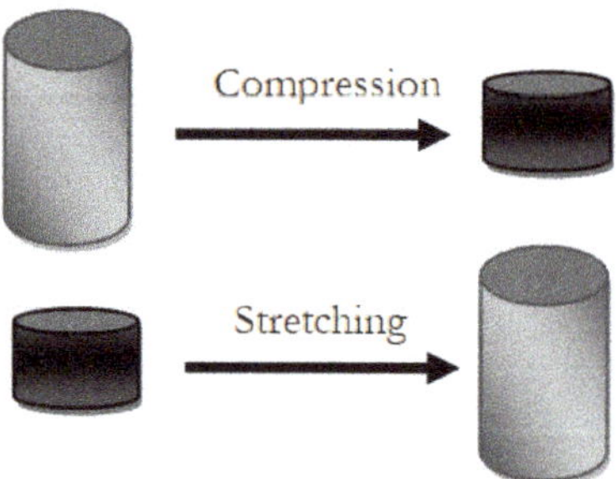

Figure 2. The stretching and compression processes.

Stretching leads to the divergence of neighboring points; bending leads to the mixing of distant points. These operations damage the energetic boundary layers and limit their recovery. As the boundary layer is a barrier upon parietal heat transfer, its destruction considerably improves heat and mass transfer, especially on multi-layer micromixers [18].

These processes enhance the contact surface between the fluids to be mixed, even in the presence of the interfacial barrier as a surface tension [17,18]. To define these behaviors, the coefficients of stretching and folding of the vortex α were investigated:

$$A = \frac{\vec{\Omega}.\overline{\overline{D}}.\vec{\Omega}}{\Omega^2} \tag{7}$$

where $\vec{\Omega}$ and $\overline{\overline{D}}$ are the vorticity vector and the deformation tensor, respectively. Where $A > 0$, the vortex stretching predominates on vortex folding [19]. A− provides the division mean of the negative values of the compression parameter. A+ denotes the division average of the positive values of the stretching parameter.

5.2. Deformation and Vorticity Intensity

The vorticity process achieves good macroscopic homogenization, while the deformation process achieves good mixing quality. For this purpose, chaotic micromixers can be a possible solution to increase both deformation and vorticity rates. These parameters are given in the following [18,19]:

$$D = \left[2\left(\frac{\partial u}{\partial x}\right)^2 + 2\left(\frac{\partial v}{\partial y}\right)^2 + 2\left(\frac{\partial w}{\partial z}\right)^2 + \left(\frac{\partial u}{\partial y} + \frac{\partial v}{\partial x}\right)^2 + \left(\frac{\partial u}{\partial z} + \frac{\partial w}{\partial x}\right)^2 + \left(\frac{\partial v}{\partial z} + \frac{\partial w}{\partial y}\right)^2\right]^{\frac{1}{2}} \tag{8}$$

$$D_{\text{mean}} = \frac{1}{-}\int Sd \tag{9}$$

$$\Omega = \frac{1}{2}\left[\left(\frac{\partial w}{\partial y} - \frac{\partial v}{\partial z}\right)^2 + \left(\frac{\partial u}{\partial z} - \frac{\partial w}{\partial x}\right)^2 + \left(\frac{\partial v}{\partial x} - \frac{\partial u}{\partial y}\right)^2\right]^{\frac{1}{2}} \tag{10}$$

$$\Omega_{\text{mean}} = \frac{1}{-}\int \Omega d \tag{11}$$

where $\mho$ presents the total fluid volume of the micromixer.

6. Numerical Validation

A numerical steady state of mixing flows of water fluid inside multi-layer micromixers was solved to validate the CFD accuracy with those received by Jibo et al. [22]. The results show the mixing rates as a function of various Reynolds numbers at the outlet section; see Figure 3. The computational comparison revealed satisfactory agreements among the results of Jibo et al. [22], where the relative error with the numerical results is less than 1%.

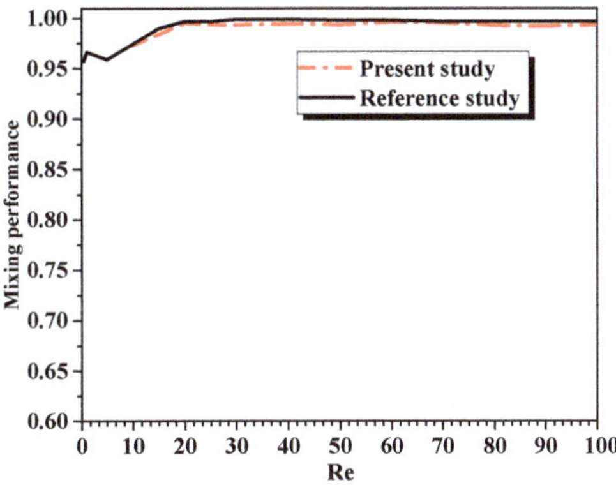

Figure 3. Comparison of current computational results for mixing rates at the outlet flow section with results of Jibo et al. [22] for different Reynolds numbers.

7. Results and Discussion

Behaviors of kinematic mixing flows for water fluid were investigated in detail for four new micromixers: TL-CM 1, TL-CM 2, TL-CM 3, and TL-CM 4, which were compared with the reference TLCCM configuration used by Kouadri et al. [11]. Local Physical Processes, such as deformation, vorticity, rotation rate, stretching and folding rates, were investigated as a function of a low Reynolds number ranging between 0.1 and 25.

The flow visualization of the mass fraction between black and yellow water fluids is presented in Figure 4 for all proposed configurations. Different regimes (Re = 5, 15, and 25) were implemented within the fluid pattern to understand the development of visual mixing inside the new configuration and compared to the reference geometry. We remark that the mixing augmented with the Reynolds number for all micromixers, and there was no completely black region in the second units for all new micromixers compared to the TLCCM configuration, as can be seen in the red circle. Therefore, the new micromixers have a quicker mixing performance than TLCCM.

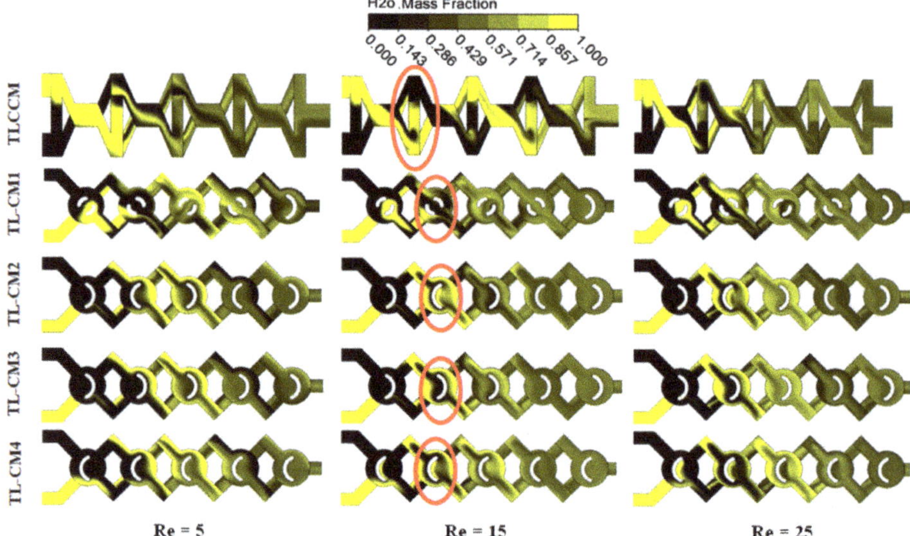

Figure 4. Qualitative representation of mass fraction contours for different Reynolds numbers at the horizontal middle section of each micromixer.

Figure 5 shows the evaluation of the mixing efficiency of the micromixers for various Reynolds numbers. As the flow homogenization was performed only by molecular diffusion for a very low Reynolds number (Re $\leq$ 5), the modified configurations were not effective in improving the mixing. When the Reynolds number increased, the fluid homogenization was more effective, and the mixing intensity developed quickly, so the most select mixing state was reached for TL-CM1, TL-CM2, TL-CM3, and TL-CM4 configurations. For the resulting plot, we found that there is scarcely a difference between the mixing rates obtained for various regimes.

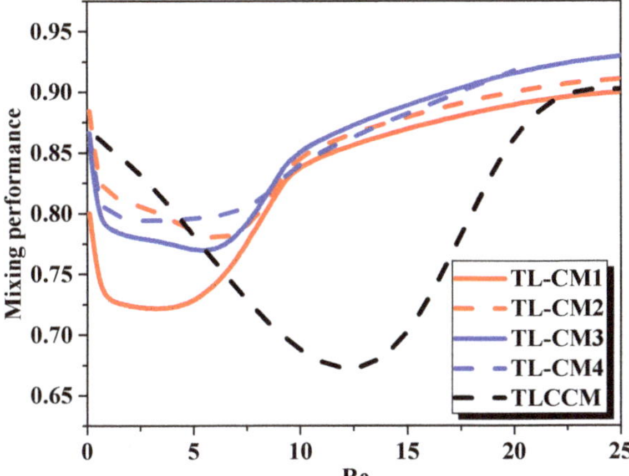

Figure 5. Development of mixing performance for different Reynolds numbers with various micromixers.

Moreover, it was observed that TL-CM4 showed a 30% enhancement of the mixing rate at the outflow with 48% lower pressure losses, as compared to the TLCCM [11] configuration at Re = 10, as shown in Figures 5 and 6. The simulation demonstrated that the TL-CM4 had the greatest rates of mixing compared to the other micromixers, due to the chaotic advection impact, which confirms the fluid homogenization.

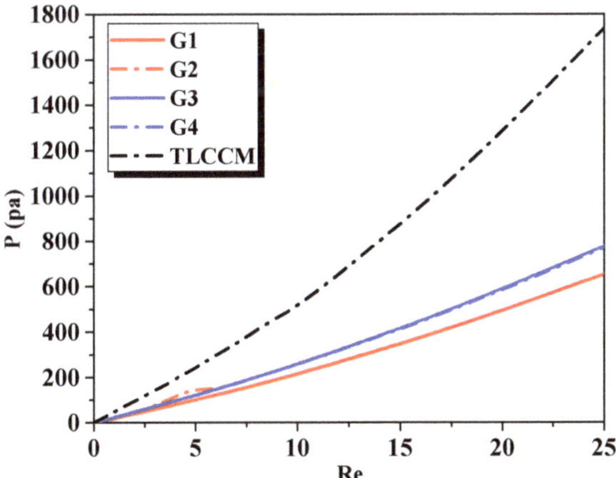

Figure 6. Pressure drops as a function of Reynolds number for different micromixers.

Figure 6 presents the evolutions of the pressure losses for different geometries with a low Reynolds number ranging from 0.1 to 25. The pressure tended towards asymptotic values for the new micromixer configurations. A large number of geometrical perturbations in TLLCM restricted the establishment of the boundary layer. This phenomenon increased the pressure losses, especially at TLCCM.

Figure 7 shows the secondary flow vectors of fluid particles at the middle regions (Re = 10) for all proposed geometries. The new configurations made a recirculation zone recorded by the sudden change in direction, due to the particular geometrical feature of the chaotic micromixer. Consequently, the main path of the vectors was distorted within 90°. Meanwhile, the vectors were observed to pass over each other, causing the fluid flow to be homogenized in the tangential direction; furthermore, significant secondary flows were created, which can contribute an extra advantage to kinematic performance and mixing fluids.

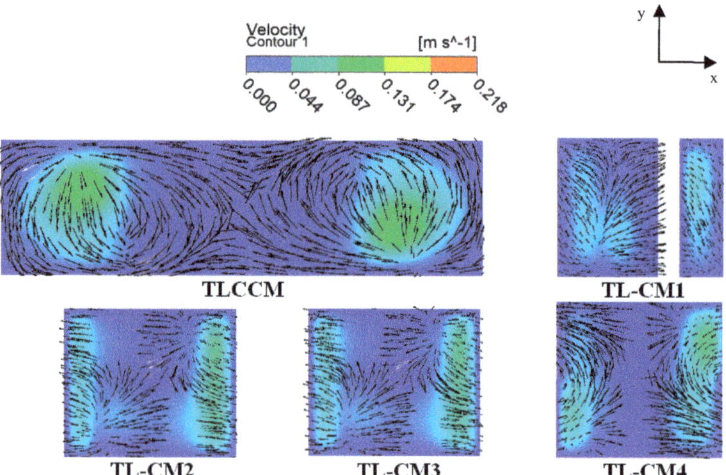

Figure 7. Qualitative representation of velocity vector plots and velocity contours on x-y planes of the middle third chamber with a distance of 3.25 mm from the inlet flow.

The effect of Reynolds numbers on the vortex intensity of the fluid within five cases of mixers is presented in Figure 8. As the Reynolds number augmented, for all micromixers, it was obvious that the flows were strong, which led to high kinematic energy. Therefore, the vorticity and secondary flow developed quickly with the Reynolds number. For a given value of Re, the dynamic flow is, likewise, powerful for the agitation of a chaotic flow within TL-CM2.

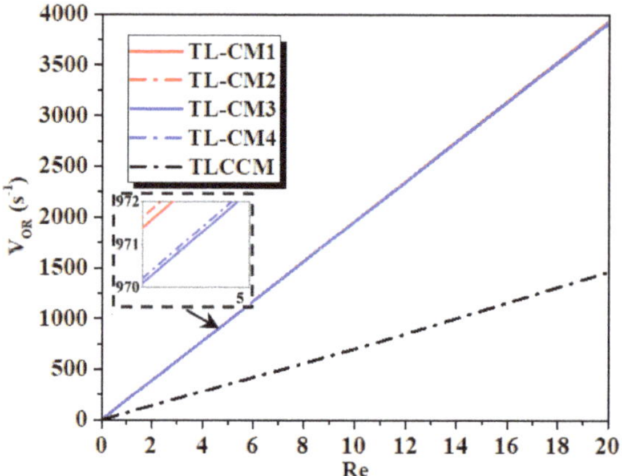

Figure 8. Evaluation of vortex intensity for various Reynolds numbers with different micromixers.

Equation (8) estimated the deformation intensity and the impact of the Reynolds number for different micromixers; see Figure 9. Due to the chaotic advection, the transversal movements of the particle fluid augmented, and the axial dispersion reduced the subsequently improved fluid homogenization. It was noted that the deformation rate increased with the growth of the Reynolds number. However, the permanence of the great chaotic flows in the novel shapes was compared to those revealed in the state of TLCCM. For low regimes, the evolutions were very close to each other. Consequently, the deformation rate had a lower effect on the secondary flow, and the flow passed easily within the micromixers, and thus, a vortex was not formed.

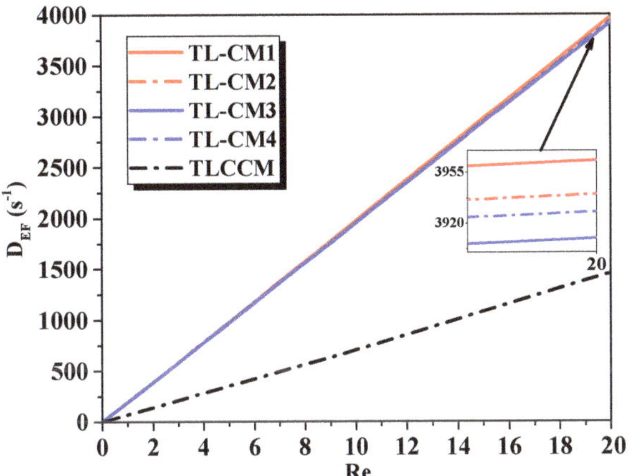

Figure 9. Evaluation of deformation intensity for various Reynolds numbers with different micromixers.

Figure 10presents the developments of a compression parameter (A−) and a vortex stretching coefficient (A+) as a function of the Reynolds number for the proposed micromixers. The configurations of low Re had qualitatively the same behavior in terms of stretching and folding intensity, due to the fluid being more viscous and the secondary flows not yet being active. When the Reynolds number exceeded the value of 10, the difference enhancement was obvious for both the stretching and folding processes; as it can be seen that, when the Reynolds number increased, the intensities of the compression and stretching became larger inside the TL-CM3 and TL-CM4 geometries, respectively, compared to the others, and the fluid flow became more sheared and agitated. Moreover, the flow in TLCCM displayed low rates of stretching compared to the other micromixers.

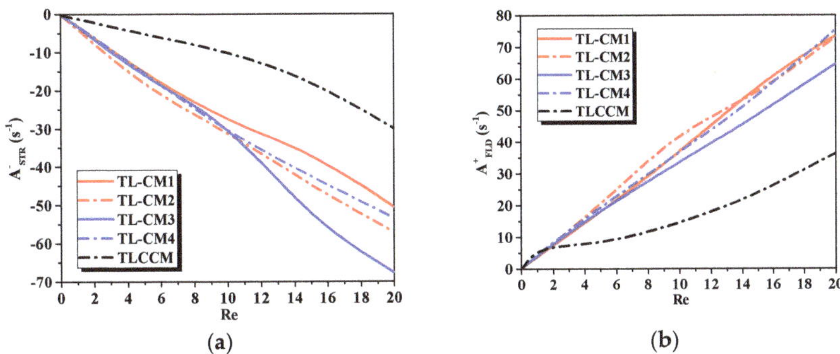

Figure 10. Evaluation of (**a**) stretching and (**b**) folding intensity for various Reynolds numbers with different micromixers.

To understand the structure of the fluid flow within the micromixers, a numerical visualization of streamlines with mass fraction is presented in Figure 11. Each configuration has an important role to enhance homogenization with improved kinematic behavior. As can be seen in Figure 11, TLCCM had a single strong vortex region inside each unit which developed the mixing rate of the geometry inwardly.

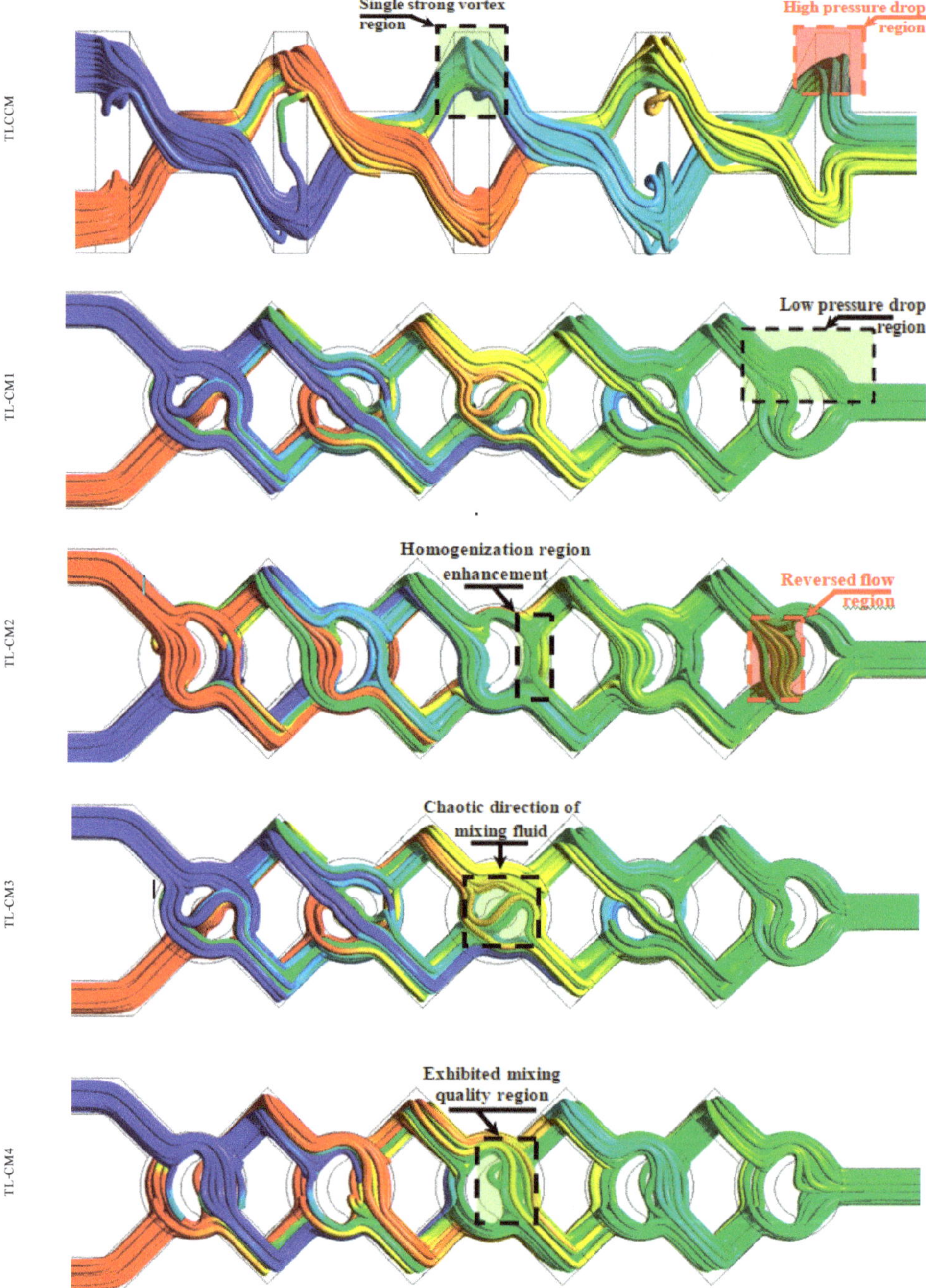

Figure 11. Streamlines of the mass fraction at Re = 10.

However, it had a high pressure loss near the outlet part. In TL-CM2, TL-CM3, and TL-CM4 configurations, the mixing was more dynamic and chaotic because of the kinematic behavior and the existence of the intense recirculation zones in the flow, which strongly affected the hydrodynamic mixing. In addition, it was seen that the nature of the trajectories inside the selected new micromixers created a reversed flow pattern, which can enhance the kinematic characteristics and ensure high-quality mixing.

8. Conclusions

We examined the effective kinematic performances and homogenization fluid quality for low Reynolds numbers with four different multi-layer micromixers based on the chaotic mixing behavior. According to this work, the following remarks can be given:

- The kinematic behavior was influenced by the Reynolds number for all proposed micromixers.
- The new design contributes an enhancement advantage to kinematic measurements, especially for the folding and stretching processes.
- Strong secondary flows were created inside the new micromixer (TL-CM 3), which enhanced the mixing quality compared to the other geometries.
- As the Reynolds number increased, the flow visualization revealed that the vortex created in each micromixer has more vigorous intensity.
- TLCCM exhibited low vortex intensity of stretching and folding compared to the preferable micromixers.
- TLLCM has many geometrical perturbations and chaotic flow that increase the pressure losses within fluid flow.
- Higher rates of Reynolds number have more effects that increase both deformation and vorticity rates.
- As a consequence, for low Reynolds numbers, mixing and kinematic performances for the TL-CM 3 configuration are more important compared to the other configurations.
- The proposed micromixers might be integrated with micrototal analysis systems and LOC systems to facilitate the study of reaction kinetics, enhanced reaction selectivity, and dilution of the fluid sample.

Author Contributions: T.T.N., methodology, numerical simulation, and writing original draft; S.H., conceptualization, methodology, data analysis, and writing original draft. M.A., writing—review and editing. A.R., literature survey and writing original draft. A.S.M.H., writing—review and editing; K.-Y.K., writing—review and editing, funding, and supervision. S.M.R.I., writing—review and editing. All authors have read and agreed to the published version of the manuscript.

Funding: This work was supported by the National Research Foundation of Korea (NRF) grant funded by the Korean government (MSIT) (No. 2019R1A2C1007657). The authors gratefully acknowledge this support.

Conflicts of Interest: The authors declare there is no conflict of interest.

References

1. Lee, C.Y.; Fu, L.M. Recent advances and applications of micromixers. *Sens. Actuators* **2018**, *259*, 677–702. [CrossRef]
2. Lee, C.Y.; Wang, W.T.; Liu, C.C.; Fu, L.M. Passive mixers in microfluidic systems: A review. *Chem. Eng. J.* **2016**, *288*, 146–160. [CrossRef]
3. Kim, K.-Y.; Wasim, R.; Shakhawat, H. A Review of Passive Micromixers with a Comparative Analysis. *Micromachines* **2020**, *11*, 455.
4. Mashaei, P.R.; Asiaei, S.; Hosseinalipour, S.M. Mixing efficiency enhancement by a modified curved micromixer; a numerical study. *Chem. Eng. Process. Process Intensification* **2020**, *154*, 1–57. [CrossRef]
5. Yao, C.; Chen, X.; Liu, S. Numerical and experimental investigations of novel passive micromixers with fractal-like tree structures. *Chem. Phys. Lett.* **2020**, *747*, 30–37.
6. Huang, C.-Y.; Hu, Y.-H.; Wan, S.-A.; Nagai, H. Application of pressure-sensitive paint for the characterization of mixing with various gases in T-type micromixers. *Int. J. Heat Mass Transf.* **2020**, *156*, 17–31. [CrossRef]
7. Huanhuan, S.; Kaixuan, N.; Bo, D.; Lemeng, C.; Fengxiao, G.; Mingyang, M.; Mengqiu, L.; Zhengchun, L. Mixing Enhancement via a Serpentine Micromixer for Real-time Activation of Carboxy. *Chem. Eng. J.* **2020**, *392*, 123642.
8. Shakhawat, H.; Ansar, M.A.; Kim, K.-Y. Evaluation of the mixing performance of three passive micromixers. *Chem. Eng. J.* **2009**, *150*, 492–501.

9. Ahmed, F.; Shakhawat, H.; Hyunil, R.; Mubashshir, A.A.; Muhammad, S.I.K.; Kim, K.-Y.; Tae-Joon, J.; Sun, M.K. Numerical and Experimental Study on Mixing in Chaotic Micromixers with Crossing Structures. *Chem. Eng. Technol.* **2020**, *43*, 1866–1875.
10. Arshad, A.; Kim, K.-Y. *Analysis and Design Optimization of Micromixers*; Springer: Singapore, 2021; pp. 2191–5318.
11. Kouadri, A.; Douroum, E.; Khelladi, S. Parametric study of the Crossing elongation effect on the mixing performances using short Two-Layer Crossing Channels Micromixer (TLCCM) geometry. *Chem. Eng. Res. Des.* **2020**, *158*, 33–43.
12. Shakhwat, H.; Lee, I.; Kim, S.M.; Kim, K.-Y. A micromixer with two-layer serpentine crossing channels having excellent mixing performance at low Reynolds numbers. *Chem. Eng. J.* **2017**, *327*, 268–277.
13. Chih-Yang, W.; Bing-Hao, L. Numerical Study of T-Shaped Micromixers with Vortex-Inducing Obstacles in the Inlet Channels. *Micromachines* **2020**, *11*, 12.
14. Dena, I.; Trieu, N.; Lisa, J.L. Complete Procedure for Fabrication of a Fused Silica Ultrarapid Microfluidic Mixer Used in Biophysical Measurements. *Micromachines* **2017**, *8*, 16.
15. Chengfa, W. Liquid Mixing Based on Electrokinetic Vortices Generated in a T-Type Microchannel. *Micromachines* **2021**, *12*, 2.
16. Soroush, N.; Saman, R. A new design of induced-charge electrokinetic micromixer with corrugated walls and conductive plate installation. *Int. Commun. Heat Mass Transf.* **2020**, *114*, 104564.
17. Telha, M.; Bachiri, M.; Lasbet, Y.; Naas, T.T. Effect of Temporal Modulation on the Local Kinematic Process of Two-Dimensional Chaotic Flow: A Numerical Analysis. *J. Appl. Fluid Mech.* **2021**, *14*, 187–199.
18. Naas, T.T.; Lasbet, Y.; Aidaoui, L.; Ahmed, L.B.; Khaled, L. High performance in terms of thermal mixing of non-Newtonian fluids using open chaotic flow: Numerical investigations. *Therm. Sci. Eng. Progress* **2020**, *16*, 100454. [CrossRef]
19. Galanti, B.; Gibbonyzk, J.D.; Heritagez, M. Vorticity alignment results for the three-dimensional Euler and Navier–Stokes equations. *Nonlinearity* **1997**, *10*, 1675–1694. [CrossRef]
20. Khakhar, D.V.; Ottino, J.M. Deformation and breakup of slender drops in linear flows. *J. Fluid Mech.* **1986**, *166*, 265–285. [CrossRef]
21. ANSYS, Inc. *ANSYS Fluent User's Guide, Release 16*; ANSYS, Inc.: Canonsburg, PL, USA, 2016.
22. Jibo, W.; Guojun, L.; Xinbo, L.; Fang, H.; Xiang, M. A Micromixer with Two-Layer Crossing Microchannels Based on PMMA Bonding Process. *Int. J. Chem. React. Eng.* **2019**, *17*, 8.

 *micromachines*

Article

Enhancement of Mixing Performance of Two-Layer Crossing Micromixer through Surrogate-Based Optimization

Shakhawat Hossain [1,*], Nass Toufiq Tayeb [2], Farzana Islam [3], Mosab Kaseem [3], P.D.H. Bui [4], M.M.K. Bhuiya [5], Muhammad Aslam [6] and Kwang-Yong Kim [7,*]

1 Department of Industrial and Production Engineering, Jashore University of Science and Technology, Jashore 7408, Bangladesh
2 Gas Turbine Joint Research Team, University of Djelfa, 17000 Djelfa, Algeria; toufiknaas@gmail.com
3 Department of Nanotechnology and Advanced Materials Engineering, Sejong University, Seoul 05006, Korea; farzanaislam2003@gmail.com (F.I.); mosabkaseem@sejong.ac.kr (M.K.)
4 Department of Mechanical Engineering, University of Tulsa, Tulsa, OK 74104, USA; phuc-bui@utulsa.edu
5 Department of Mechanical Engineering, Chittagong University of Engineering & Technology (CUET), Chittagong 4349, Bangladesh; mkamalcuet@gmail.com
6 Department of Chemical Engineering, Lahore Campus, COMSATS University Islamabad (CUI), Lahore 53720, Pakistan; maslam@cuilahore.edu.pk
7 Department of Mechanical Engineering, Inha University, 100 Inha-Ro, Michuhol-Gu, Incheon 22212, Korea
* Correspondence: shakhawat.ipe@just.edu.bd (S.H.); kykim@inha.ac.kr (K.-Y.K.); Tel.: +880-8810308-526191 (S.H.); +82-32-872-3096 (K.-Y.K.); Fax: +82-32-868-1716 (K.-Y.K.)

Abstract: Optimum configuration of a micromixer with two-layer crossing microstructure was performed using mixing analysis, surrogate modeling, along with an optimization algorithm. Mixing performance was used to determine the optimum designs at Reynolds number 40. A surrogate modeling method based on a radial basis neural network (RBNN) was used to approximate the value of the objective function. The optimization study was carried out with three design variables; viz., the ratio of the main channel thickness to the pitch length (H/PI), the ratio of the thickness of the diagonal channel to the pitch length (W/PI), and the ratio of the depth of the channel to the pitch length (d/PI). Through a primary parametric study, the design space was constrained. The design points surrounded by the design constraints were chosen using a well-known technique called Latin hypercube sampling (LHS). The optimal design confirmed a 32.0% enhancement of the mixing index as compared to the reference design.

Keywords: Navier–Stokes equations; mixing index; passive micromixers; optimization; RBNN

Citation: Hossain, S.; Tayeb, N.T.; Islam, F.; Kaseem, M.; Bui, P.D.H.; Bhuiya, M.M.K.; Aslam, M.; Kim, K.-Y. Enhancement of Mixing Performance of Two-Layer Crossing Micromixer through Surrogate-Based Optimization. *Micromachines* **2021**, *12*, 211. https://doi.org/10.3390/mi12020211

Academic Editor: Yangcheng Lu
Received: 14 January 2021
Accepted: 13 February 2021
Published: 19 February 2021

1. Introduction

The recent advancements in miniaturized research in biochemistry and biomedicine demand development in the area of microfluidic systems. Micromixers are one of the fundamental components of micro total analysis systems (μ-TAS) and microfluidic systems [1,2]. A microchannel has geometrical dimensions typically in the order of a few microns. The flow inside such a microchannel is characterized by a small value of Reynolds numbers (Re), where the inertia forces of a fluid are much lower than the viscous forces. Micromixing is largely controlled by molecular diffusion, particularly at low Reynolds numbers. Most importantly, the fluid flow within the micro devices is a ubiquity of laminar.Apart from the various advantages of micro devices, such as the high surface-to-volume ratio, portability, and decreased analysis time, there are also challenges to achieving high mixing efficiency, as development of molecular diffusion is extremely sluggish [3].

On the basis of their working principles, micromixers have been classified into passive micromixers and active micromixers. Mixing performance is increased by employing external energy sources within an active micromixer. There are various methods available for flow manipulation to enhance mixing, such as pressure, acoustic, thermal, magnetic, electrokinetic, etc. In an active micromixer, higher mixing efficiency can be achieved in a very short mixing length. However, these micromixers require complex manufacturing methods, and their integration into a microfluidic system is also difficult. In a passive micromixer, the mixing is simply increased with the help of a modification in the geometrical structure of the microchannel. Recently, passive micromixers have been the superior preference for researchers over active micromixers because of their easier fabrication and their incorporation technique into microfluidic systems [4–7].

Commercial software using computational fluid dynamics (CFD) is one of the most consistent and popular tools for fluid flow structural analyses, as well as for evaluating the performance of microfluidic devices [8–11]. In recent times, to achieve efficient mixing, numerous micro devices have been proposed by researchers. In a laminar flow regime, the fluid flow is based on a chaotic mechanism that is stimulated by the cyclic disturbance of the flow, which can evidently progress the mixing performance [12,13]. The three-dimensional serpentine microchannel [14] was designed and fabricated to generate chaotic advection using the recurring behavior of stretching and folding phenomena of the fluid flow streams. However, the proposed micromixer can work well and produce chaotic advection with a reasonably higher Reynolds number (>25). A micromixer incorporated with oblique grooves at the bottom, proposed by Stroock et al. [15], could create three-dimensional twisting flows. The authors observed chaotic advection in the microchannel with herringbone grooves because of the sporadic velocity fields of the fluid streams. Chaotic microchannels with rectangular obstacles on the slanted grooves designed by Kim et al. [16], because of the regular disturbance of the flow streams over the slanted grooves, created a chaotic mixing mechanism and enhanced the performance of the device. Chaotic fluid flow mechanisms were observed by Wang and Yang [17] using an overlapping crisscross microchannel; the device could generate chaotic advections by enhancing and shrinking the streams of miscible fluid. A two-layer crossing microchannel (TLCCM) based on chaotic mixing behavior has been reported [18]; the study demonstrated that the micromixer showed high performance at Reynolds number (Re < 0.2).

For performance development of the available micromixers, the design optimization technique using the CFD tool has progressively and increasingly become the method preferred by researchers. Structural optimization of a micromixer with a slanted groove microchannel was performed [11]; using electroosmotic flow, the study demonstrated that the performance of the devices was remarkably improved and that performance depended on the depth and angle of the pattern grooves. For augmentation of the performance of a herringbone-grooves microchannel, a detailed optimization study based on the radial basis neural network (RBNN) technique was used by Ansari and Kim [19] using three different parameters. Lynn and Dandy [20] performed structural optimization of a micromixer using four different parameters; the performance of the micromixer was improved surprisingly by the investigated parameters. Hosasin et al. [21] optimized a modified Tesla structure using a weighted-average (WTA) surrogate model with two different objective functions to originate a single-objective optimization problem. A micromixer with a herringbone grooves device was optimized [22,23] using two functions: pressure losses and mixing performance. A micromixer with sigma structure [24] and convergent-divergent sinusoidal walls [25] was optimized using a multi-objective optimization technique. Objective function values were correlated using the concave shape of a Pareto-optimal font. Performance of the asymmetrical shape with a split-and-recombined [26] microchannel was improved through an optimization technique at Re = 20. The authors performed both single-objective and multi-objective optimizations using particle swarm and genetic algorithm optimization methods. The study concluded that improvement of the mixing effectiveness (by 58.9%) was achieved with the reference design using the single-objective optimization technique. The multi-objective optimization showed a 48.5% improvement of the mixing index and a 55% decrease of the pressure drop with the reference device.

Through a literature survey, we have established that a numerical optimization process using three-dimensional CFD has been a constructive tool for improvement of existing micromixers. This study performed single-objective optimizations for the further enhancement of micromixer performance in terms of mixing index, as proposed by Ahmed et al. [27]. The optimization study was carried out with three dimensionless design variables at Re = 40. The design points surrounded by the design constraints were chosen using the well-known Latin hypercube sampling (LHS) technique. The surrogate modeling method based on the RBNN was applied to approximate the objective function.

2. Design of the Proposed Microchannel

In our previous study [27], to inspect the mixing performance a micromixer including a two-layer (top layer and bottom layer) microchannel was proposed. The two inlet channels were joined with the main crossing channel at an angle of 90°, as shown in Figure 1a. The fluids were interrelated all the way through the vertical segments and at the center of every crossing segment. A consecutive structure of ten mixing units split and reconnected the fluid streams in a cyclic approach. The geometric dimensions of the projected micromixer were as follows: diagonal channel width (W), pitch length (Pl), thickness of the main channel (H), depth of the single channel (d), vertical segment (b), and number of mixing units were 0.15 mm, 0.64 mm, 1.07 mm, 0.15 mm, 0.3 mm, 0.15 mm, and 10, respectively. The dimensions of the inlet channel were 0.15 × 0.3 mm, and the outlet channel was 0.3 × 0.3 mm.

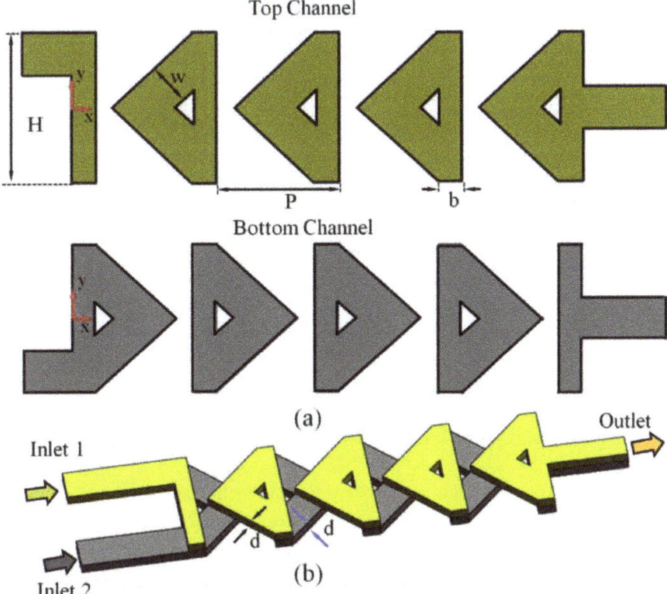

Figure 1. Two-layer crossing micromixer [27]. (**a**) Two-dimensional design of the top layer and bottom layer, respectively. (**b**) Proposed micromixer with three-dimensional image; both channels are interrelated at the center of the crossing-unit and the vertical sections.

3. Numerical Scheme

The present section formulates the numerical model used in this study for the proposed design of the microchannel. As discussed, the flow inside micro devices is characterized by the omnipresence of laminar flow; therefore, the flow was considered as laminar. In this study, numerical simulations were performed by assuming flow as laminar, steady, and incompressible. The detailed analysis of the mixing between the fluids was investigated by solving, continuity, mass diffusion, and three-dimensional Navier–Stokes equations. Under the considered assumptions, the equations are mathematically articulated as:

$$\vec{\nabla} \cdot \vec{V} = 0 \tag{1}$$

$$\left(\vec{V} \cdot \vec{\nabla}\right) \vec{V} = \nu \nabla^2 \vec{V} + \frac{1}{\rho} \nabla p \tag{2}$$

$$\left(\vec{V} \cdot \vec{\nabla}\right) c = D \, \nabla^2 c \tag{3}$$

where, V represents the velocity, ν represents the kinematic viscosity, p represents the pressure, ρ represents the density, D represents the diffusivity constant, and the mass fraction of species of the mixing fluids is denoted by c, respectively. The fluid flow inside the micromixer was defined as single-phase multi-component fluid flow. Therefore, apart from the transport equations, one other equation must be taken into account that depicts the effect of the variation [28] of the fluid properties along the flow:

$$\frac{\partial \rho_i}{\partial t} + \frac{\partial\left(\rho_i \vec{V}_j\right)}{\partial x_j} = -\frac{\partial\left[\rho_i\left(\vec{V}_{ij} - \vec{V}_j\right)\right]}{\partial x_j} + S_i \tag{4}$$

where, ρ_i represents the fluid density at sample point i within the mixture, $\vec{V}_j$ represents average velocity field, $\vec{V}_{ij}$ represents the average velocity of ith component within the mixture, and S_i represents the source term. For the entire domain, S_i must be equal to zero; therefore, when the above equation is applied and the results are summed for each of the components of the mixture, then the reduces to the following form:

$$\frac{\partial\left(\rho_i \vec{V}_j\right)}{\partial x_j} = 0 \tag{5}$$

where the equation is equivalent to the mass continuity equation of the fluid flow. Therefore, single velocity can be used for the computational analysis. In this study, water and the combination of dye–water was inserted through inlet 1 and inlet 2. It was assumed that temperature was constant at 20 °C. The dynamic viscosity (μ) and density (ρ) of the water were considered 8.84×10^{-4} kg/m s and 997 kg/m^3, respectively [29]. The diffusion coefficient for the dye–water mixture was constant throughout the numerical simulations at 1×10^{-11} m^2/s [18]. The numerical solution was executed by ANSYS CFX 15.0 [28] for principal equations. A high-density hexahedral mesh was created within the computational domain using ANSYS-ICEM CFD 15.0. The mesh density at the junction of the mixing units was kept high. The advection term was discretized using the high-resolution of the second-order approximation. The aspect ratio of cells was kept near to unity so that the numerical diffusion error was minimized and highly precise numerical results were obtained [30]. Additionally, for discretization of the convective parts, a high-resolution method [28] was used in the principal equations. Moreover, for pressure velocity pairing, the SIMPLEC algorithm [31] was used.

For numerical simulations, diverse boundary conditions were applied at the outlet, inlets, and side walls. Normal inlet velocity was calculated on the basis of the properties of water and specified as the inlet boundary condition at both inlets. Atmospheric pressure (zero) was applied at the outlet section, and the frictionless wall was specified. The convergence criteria of numerical solution were taken in terms of the root mean square value (10^{-6}) at each node of the computational domain.

4. Evaluation Parameter for the Performance of Micromixer

On the basis of variance of the dye mass fraction across a cross-sectional plane, the mixing efficiency was determined. The plane was taken at the end of the last mixing unit, and the mass fraction variation was calculated using following formula:

$$\sigma = \sqrt{\frac{1}{N}(c_i - \overline{c}_m)^2} \tag{6}$$

where, σ represents the standard variation across the sample plane, dye mass fraction at the ith sample point is denoted by c_i, c_m represents the mass fraction (mean) on the cross-sectional plane, and the number of sample points is denoted by N. Subsequently, the following mathematical expression was used to calculate the mixing index:

$$M = 1 - \sqrt{\frac{\sigma^2}{\sigma^2_{max}}} \tag{7}$$

where, σ^2 is the variation of the dye mass fraction, and the maximum variance value on the sample plane is denoted by the σ^2_{max}. Variance in the mass fraction is inversely related to the mixing index across the sample plane. For the best performance of a micromixer, the numerical value of the mixing index should be 1.0, which corresponds to the minimum variance of dye mass fraction.

5. Selection of Design Constraints and Objective Functions

The primary and very fundamental process during the optimization process is to select the appropriate design constraints that influence the objective function values. A detailed parametric investigation was carried out to select the sensitive parameters for structural optimization. Three dimensionless geometric parameters: the ratio of the main channel thickness to the pitch length (H/PI), the ratio of the thickness of the diagonal channel to the pitch length (W/PI), and the ratio of the depth of the channel to the pitch length (d/PI) were chosen to optimize the proposed micromixer. The design ranges of the parameters were restricted based on a preliminary study, shown in Table 1. For numerical analysis, the LHS method was applied to finalize the (twenty-eight) design points. A commonly used mixing index (F_{MI}) of the micromixer was considered for objective function values.

Table 1. Design parameters with their limits.

Design Variables	Ratio of The Main Channel Thickness to the Pitch Length (H/Pi)	Ratio of The Thickness of The Diagonal Channel to the Pitch Length (W/Pi)	Ratio of The Depth of The Channel to the Pitch Length (D/Pi)
Lower Limit	1.26	0.28	0.16
Upper Limit	1.89	0.57	0.31
Reference Design	1.67	0.47	0.24

6. Methodology of the Single-Objective Optimization

Figure 2 demonstrates the optimization procedure on the basis of surrogate modeling. Single-objective optimization was applied for the structural optimization. The optimization study, formulated as the maximization of the objective function value, can be mathematically articulated as:

$$\text{Min. } F(x) \text{ subjected to } x_l \leq x \leq x_u$$

where, x_u and x_l represent the upper and lower limits of the design variable x, respectively. In this study, the optimization procedure maintained the following steps; initially effective design parameters and proper design constraints were determined through the preliminary study to enhance the performance of the micromixer.

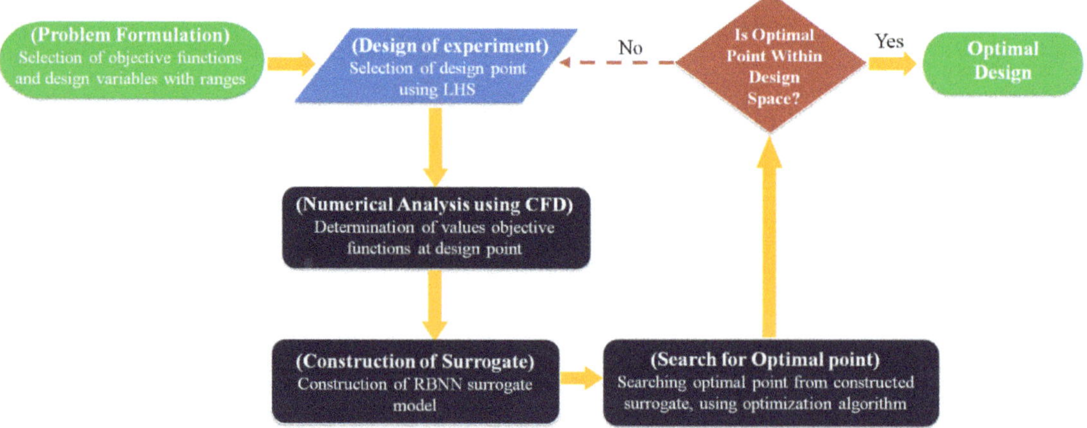

Figure 2. Flow process for single-objective optimization method. LHS = Latin hypercube sampling; CFD = computational fluid dynamics; RBNN = radial basis neural network.

To formulate the surrogate model, design of experiments (DOE) was applied to select the uniformly distributed design points within the design constraint. Two types of DOE methods are found in the literature: random design and orthogonal design. In the orthogonal design technique, model parameters are self-determining and represent that the factors are not related (experimentally) and can be varied separately. There are some limitations of orthogonal designs. Firstly, they are quite unconvincing when it comes to determining the important factor. As the fundamental function is determined, the probability of reproducing the design points is high, generally termed as the collapse problem [32]. Therefore, the method is inefficient in terms of the computational time period. To conquer the problem, a random design was applied to confirm the design points for the structural optimization. In a random design, design parameter values are determined on the basis of a random process [33]. As a result, there is no chance of reproducibility of the design points. Thus, each point provides unique information about the effect of another factor on the response. Accordingly, the method is very efficient in terms of computational time period [33].

In the present work, a random design based on LHS [34] was chosen to formulate the surrogate models with which to estimate the values of the objective function. Generally, the LHS method uses an m × n matrix, where sampling points are denoted by m and design parameters are denoted by n, respectively. In an LHS matrix, every n column contains the levels 1, 2, 3 tom, arbitrarily grouped to structure a Latin hypercube. Thus, the technique generates random sampling within the data range, which confirms every segment of the design constraint. To determine the optimum points within the design constraint a (GA) genetic algorithm [35] was considered for the search algorithm. For design points, the well-known MATLAB function (i.e., lhsdesign) was used. To exaggerate the minimum distance among neighboring design points, maxmin was used [36]. The values of objective functions at selected design points (twenty-eight) were calculated numerically.

Next, a surrogate was formulated depending on the (twenty-eight) objective function values. To determine the optimal points within the design constraint, the algorithms needed a large number of estimations of objective function. Therefore, for the sake of computational time saving, it was obligatory to formulate a surrogate model on the basis of distinct numerical analysis within the design constraints. In this study, to estimate the objective function values, the RBNN technique was used [37].

The RBNN is an artificial neural network (ANN) that considers the radial basis functions (RBF) as activation functions, constructed by three layers (i.e., hidden layer, input layer, and output layer), which contain linear or nonlinear neurons. Each activation function depends on the distance from the center vector to the input layer; thus, the function becomes symmetric along the center vector [38]. The hidden layer contains a set of functions of radial basis, which perform the same as activation functions [38]. The response differs within the distance between the center and the input. Furthermore, variation of coordinates was used to determine the distance between the center and the input. The radial basis model is able to reduce computational time as well as cost using its linear nature of the functions. Using the number of N basis functions of the model (linear), f (x) is mathematically expressed as:

$$f(x) = \sum_{j=1}^{N} w_j y_j \quad (8)$$

where weight is represented by w_j,and y_j is the basis function. Various techniques can be applied to select the functions. The function can be classified as nonlinear or linear. For the linear model, the parameters, including the basis function, are stable throughout the iteration progression. On the other hand, if the basis function differs through the iteration, it calls for a nonlinear model. The iteration procedure, corresponding to a search for the best plane within the multidimensional space, offers a suitable match to the learning data. The parameters of the surrogate model are termed as a user-defined error goal (EG) and the spread constant (SC) [38]. The finding of an appropriate value of EG and SC is very crucial. A large value of the EG will affect the precision of the model, whereas a small value of the EG will construct a model larger than the training experience of the network. Furthermore, estimation of the proper values of the SC is very crucial; if the SC value is large, the neuron may not react identically at each input; whereas for a small value of SC, the network would be highly sensitive. The appropriate EG value is resolute from the acceptable error of the mean input responses. The proposed modified RBNN function, called "newrb", accessible in MATLAB, was used in this study [39].

7. Results and Discussions

In any computational study to ensure a high-quality grid system, it is a predominant criterion to diminish the numerical inaccuracies generated during the discretization process. In this study a tetrahedral grid was built for the computational domain. To establish an optimum number of nodes as well as mesh size, an investigation of grid-dependency was executed. A grid system containing a number of nodes (five) starting from 0.5×10^6 to 1.9×10^6 were tested for mixing index evaluation down to the microchannel length at Reynolds numbers (Re = 40), as shown in Figure 3. To evaluate the mixing indices, perpendicular planes at six different positions along the axial direction were considered (Figure 3a). An almost identical development of the mixing performance is depicted for nodes 1.6×10^6 and 1.6×10^6, Re = 40. Additionally, tiny variation was found in the mixing development (0.27%) at the exit of the micromixer between the two grid systems, represented by the mesh element sizes 3.5 μm and 3.0 μm, respectively (Figure 3b). Therefore, from the tested results, the grid system representing nodes 1.6×10^6 (i.e., mesh element size 3.5 μm) was considered to be the most favorable grid system.

The numerical model was validated qualitatively and quantitatively with experimental findings [27], as demonstrated in Figure 4. The induced uncertainties during the experimental process were wall unevenness and dimensional variation (±5 μm) in fabrication. The phenomena play a vital role in the variations of mixing performance (Figure 4b). The numerical scheme was validated quantitatively and qualitatively with experimental results, shown in Figure 4a. The visual photograph of the dye mass fraction distribution of fluid mixing was confirmed with the numerical result on the x–y plane situated at the center of the top and bottom channel depth (Figure 4a). The following graph (4b) represents mixing indices at Re = 1, 15, 40, and 60, which were quantitatively evaluates with experimental data [27]. Numerical prediction values of the mixing indices were marginally varied with experimental data throughout the Reynolds number range. The variation between the experimental and numerical results occurred due to the microchannel fabrication procedure, including dimension variations (±5 μm), wall roughness, and experimental uncertainties such as focusing and evaluating the experimental images. Nevertheless, the quantitative and qualitative evaluation between the experimental and numerical results demonstrates satisfactory agreement.

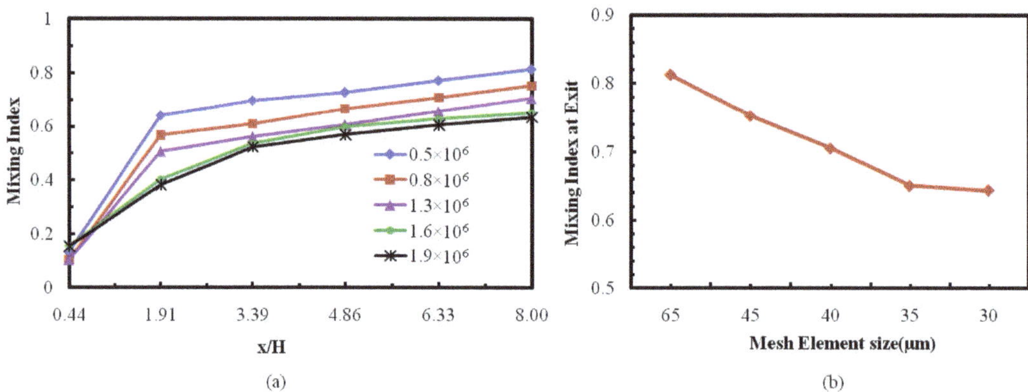

Figure 3. Grid dependency examination for reference design. (**a**) Mixing index with the channel length. (**b**) Mixing index at (x/H = 8.0) exit.

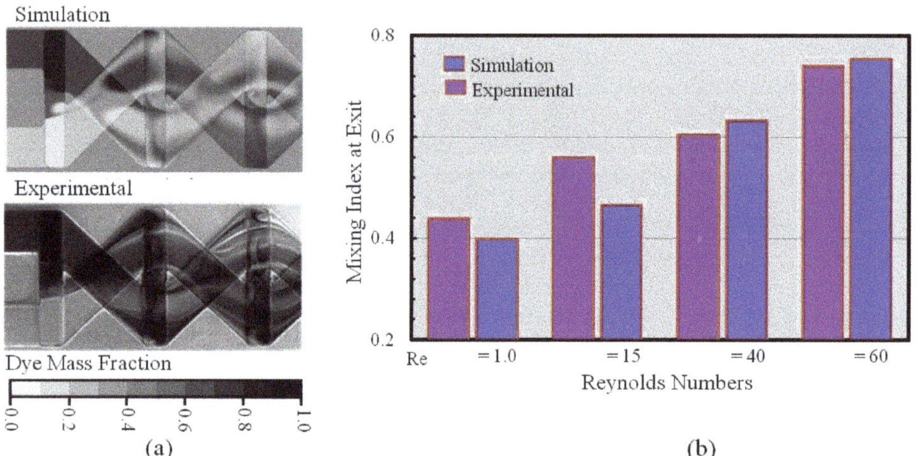

Figure 4. Confirmation of proposed model (numerical) evaluated by experimental findings [27]: (**a**) qualitative evaluation (Re = 60) and (**b**) quantitative evaluation at Re = 1, 15, 40, and 60.

Consequences of the Reynolds numbers on mixing performance were examined both quantitatively and qualitatively through a numerical procedure. To demonstrate the mixing developments, the mass fraction variations were captured on six cross-sectional planes (A_1–A_6) at the crossing nodes, shown in Figure 5a. At Re = 1.0 (Figure 5b), the interfacial area of the miscible sample is practically visible, has relatively fewer distortions, and is straight at each cross-sectional plane. Figure 5b shows two symmetric (at top and bottom) transverse fluid flow patterns. The figure demonstrates that the interfacial surface of the fluid becomes progressively wider as the flow proceeds and also with the Reynold numbers. The developments of mixing indices (Figure 5b) to the down way of the micromixer at four various Reynolds numbers (Re = 1, 20, 40, and 60) is presented. Six cross-sectional planes were selected for the estimation of the mixing index at the middle of the crossing structures.

Figure 5b shows that increasing phenomena of mixing performance significantly vary with the Reynolds numbers.

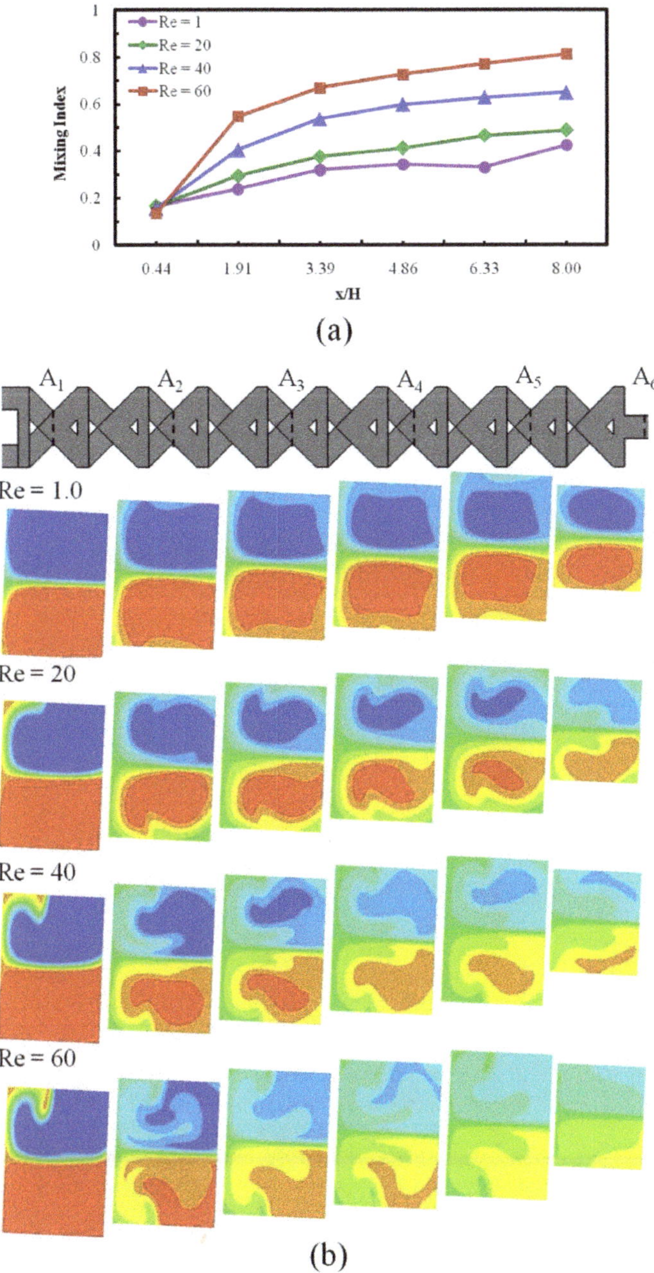

Figure 5. Mixing index development for reference design micromixer at different Reynolds numbers (Re = 1.0, 20, 40, and 60). (**a**) Improvement of mixing index along the microchannel. (**b**) Captured images of mass fraction distributions (dye) on cross-sectional planes.

Figure 6 illustrates the 3D streamlines of fluids, indicated with two different colors, originated from both inlets at Re = 1.0, 20, 40, and 60, which are plotted to scrutinize the fluid flow structure that enhances the mixing performance. Initially, the fluids start mixing at the middle of the first vertical section and enter to the main channel. Due to the 3D channel structure, the streams of fluid maintain their flow path after impact (Figure 1b). Thereafter, the fluid streams recombine at the first crossing node (A1 section). A fraction of swapping of streamlines happens at the crossing nodes, which produces an enlarging and shrinking of the fluids interface, and thus promotes chaotic advection. Therefore, the streams of fluids are gradually divided into many layers during a progression of ten mixing segments. The flow mechanism enlarges the interfacial section of fluids and diminishes the length of diffusion across the fluids' layers, and thus assists quicker diffusion and fast mixing. These phenomena increase with the Reynolds numbers, and thus enhance the mixing performance.

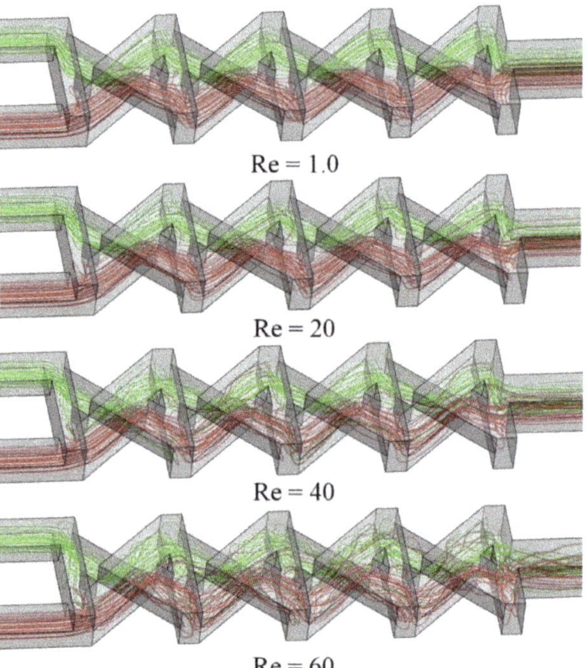

Figure 6. Streamlines originated from different inlets in the reference design micromixer for Re = 1.0, 20, 40, and 60.

Outcome of the structural design parameters (i.e., H/PI, W/PI, and d/PI) on the behavior of the mixing performance at Re = 1.0 and Re = 40 was performed as shown in Figure 7; three dimensionless parameters (i.e., H/PI, W/PI, and d/PI) listed in Table 1 were considered for numerical investigations. A number of mixing segments (ten) were kept constant; thus, the down way length of the micromixer remained the same during the investigations. At Re = 1.0, the mixing index was not significantly varied (Figure 7a) with the geometric parameters; thus, it is concluded that for this Reynolds number the mixing index generally depended on the molecular transmission of fluids rather than the geometric parameters. Mixing performance at Re = 40 (Figure 7b) is slightly varied with the value of W/PI. Parametric findings representing the variation of mixing index values were very much reactive to H/PI (42% deviation for $1.26 < H/PI < 1.89$) than to d/PI (31% deviation for $0.16 < d/PI < 0.31$) and W/PI (25% deviation for $0.28 < W/PI < 0.57$).

For predetermined down way length of the microchannel, the channel width (W) was proportionally varied along the values of channel width (H); thus, intensity of sample fluid velocity diminished as the H/PI value increased. Hence, lesser velocity corresponded with high residual duration of the fluid flow within the microchannel and weaker inertia force of fluids. The maximum mixing index value 0.72 was found at W/PI = 0.28 (minimum value), where the best matching of the fluids interfacial area and residential time was obtained. Thus, the higher value of W/PI caused a lower mixing index value at the exit.

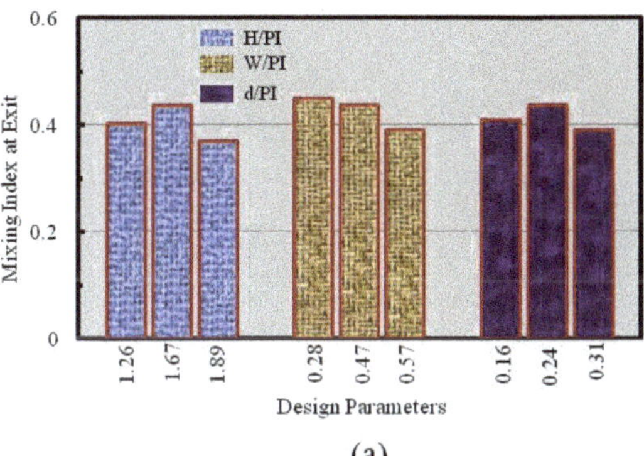

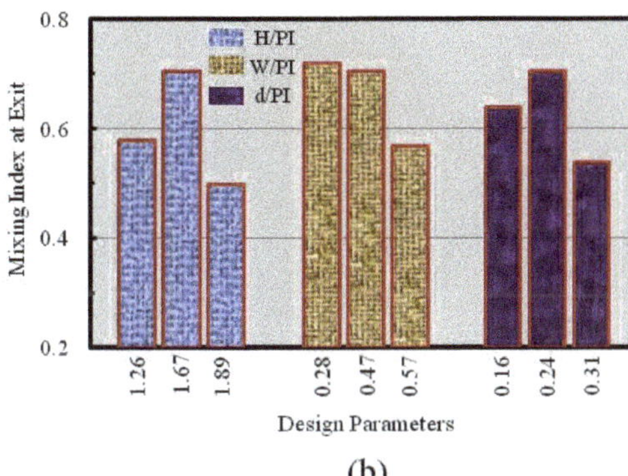

Figure 7. Effect of the geometrical parameters on the values of mixing performance with three dimensionless parameters at (**a**) Re = 1.0 and (**b**) Re = 40.

For qualitative evaluations of mixing performance values, mass fraction distributions were plotted (on x–y planes) at Re = 40, shown in Figure 8. The dye mass fraction distributions are plotted at the center of the top layer microchannel (as indicated by the dotted rectangular box) of the first mixing unit. With the increase of the W/PI values, the interface of the fluids becomes wider; additionally, mixing performance increases (Figure 7b). The inertia force of fluids and the values of W/PI are inversely proportional. The stronger inertia force of fluid (with lower values of W/PI) enhances the chaotic advection within the micromixer, and thus the mixing performance at the exit improves significantly.

In this study, the well-known surrogate models, RBNN, were usedto achieve the optimized micromixer.Table 2 shows the design variables values (H/PI = 1.73, W/PI = 0.42, and d/PI = 0.18) and objective function (MI = 0.86) for an optimum micromixer using the surrogate model. Predicted design variable values of the optimum micromixer were compared with the reference micromixer. Table 2 also represents the objective function values of the reference design (0.65) and predicted optimal design (0.86), representing a 32% relative increase in the mixing index through surrogate-based optimization. Considering the reference design, the optimum design was found at the lower value of d/PI (0.18). The calculated objective function value using the Navier–Stokes equation also compared with the predicted values, 0.86 and 0.81, respectively. This comparison signifies that the surrogate based optimization technique shows 6.2% deviation of objective function value from the optimum point.

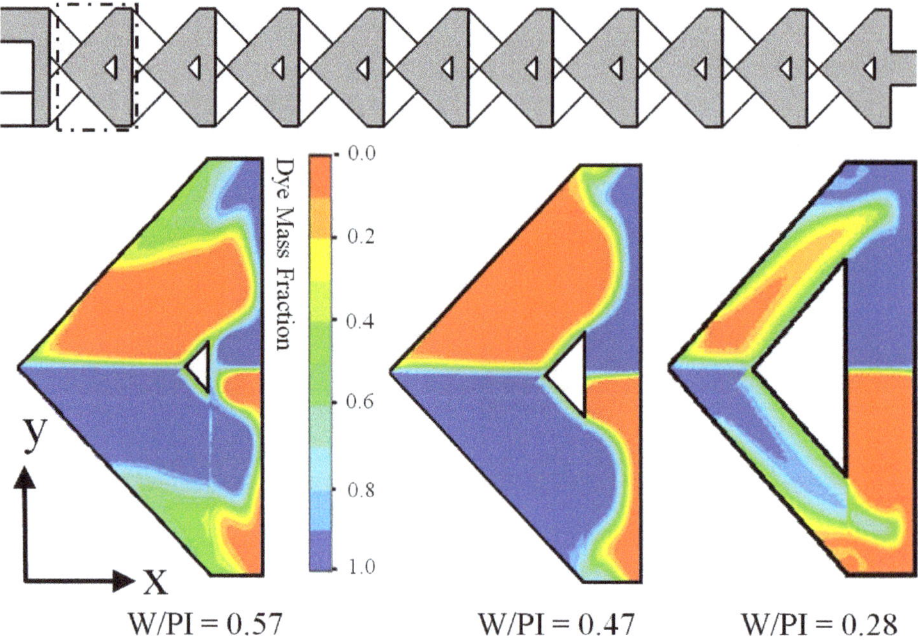

Figure 8. Dye mass fraction distributions at the middle of the top channel (designated by the dotted lines) of the first mixing unit for different values of W/PI = 0.57, 0.47, and 0.28 for reference design micromixers at Re = 40.

Table 2. Comparison of the objective function (MI)-oriented optimum geometry with the reference geometry.

Design Variables	H/PI	W/PI	d/PI	Objective Function (MI)	
				Predation by RBNN	Calculation by Navier–Stokes Analysis
Reference Design	1.67	0.47	0.24	-	0.65
Optimum Design	1.73	0.42	0.18	0.86	0.81

The developments of objective function (MI) value along the down way length of the micromixer for optimized geometry and reference design are represented at Reynolds number 40. Developing rates of the mixing index increase for both designs along the channel length, depicted in Figure 9. Figure 9 also represents the optimum design as having better mixing performance throughout the microchannel length, and the value of the mixing index-optimized design micromixer (x/H = 8.0) as having 1.4 times higher performance compared to the reference design.

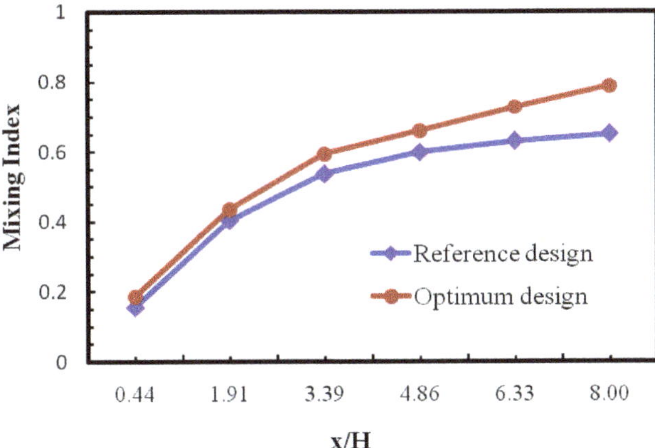

Figure 9. Improvement of objective function values along the channel length for reference design and optimum designs at Re = 15.

Figure 10a,b shows the velocity vectors plot and local vorticity variations on y–z planes of the reference and optimum designs, respectively. The cross-sectional plane was plotted (x/H = 8.0) at the end of the last mixing segment. A pair of counter-rotating vortices was observed in both cross-sectional planes. The optimized micromixer visualized a pair of small round-shaped (counter-rotating) vortices that filled the entire plane, while in the reference micromixer, two oval-shaped counter-rotating vortices shifted to the side wall; thus, velocity vectors became relatively weaker. On the other hand, the velocity vector for the optimum design micromixer indicated a strong transverse flow pattern, and velocity vectors were consistently spread all through the cross-sectional plane. The strongest transverse flow phenomena (Figure 10a) produced the difference in mixing performance for the optimized design micromixer. Figure 10b shows local vorticity distributions, plotted on the y–z plane for the reference and optimum design. The vorticity was designed by following formula:

$$\omega_x = \left(\frac{\partial v_z}{\partial y} - \frac{\partial v_y}{\partial z}\right) \tag{9}$$

where, ω_x represents vorticity along the x-direction, and w and v are the velocity components along z and y directions. As the vorticity plot signifies, as compared to the reference design, the strength of the vorticity is augmented in the optimum design and creates the potential difference in mixing performance at the exit. Normalized circulation development along the reference design and optimum design micromixers is represented in Figure 11. Circulation values signify the potentiality of vertical movement on the plotted plane, as shown in Figure 10b. The circulation (Ω_x) is articulated by incorporating the streamwise vorticity on y–z planes, mathematically represented as:

$$\Omega_x = \int_{A_{yz-plane}} \left(\frac{\partial v_z}{\partial y} - \frac{\partial v_y}{\partial z}\right) dydz \tag{10}$$

where, v_y represents the y direction velocity components and v_z represents the z direction velocity component. Along the channel length, the value of the circulation increases in both micromixers. Figure 11 illustrates that, as compared to the reference design, the optimized design has a higher circulation value, which enhances mixing performance. The dye mass fraction distributions at the middle of the top channel (designated by the dotted lines) of first mixing unit were plotted for the reference design and the optimum design micromixers, as shown in Figure 12. It is observed that the enhancement of the secondary flow induced in the optimized micromixer was superior compared to the reference design micromixer. Interface of sample fluids in the optimum design was wide and, accordingly, a quicker improvement of mixing performance along the microchannel length is observed (Figure 9). The result confirms that, for the optimum design micromixer, mixing performance is around 25% higher than the reference design micromixer.

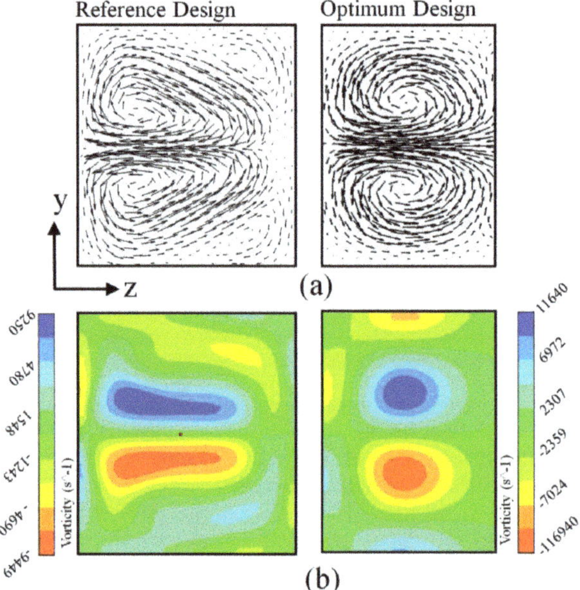

Figure 10. Qualitative comparison of reference and optimum designs at Re = 40. (**a**) Velocity vectors plot and (**b**) local vorticity distributions plot, on y–z planes (at x/H = 8.0).

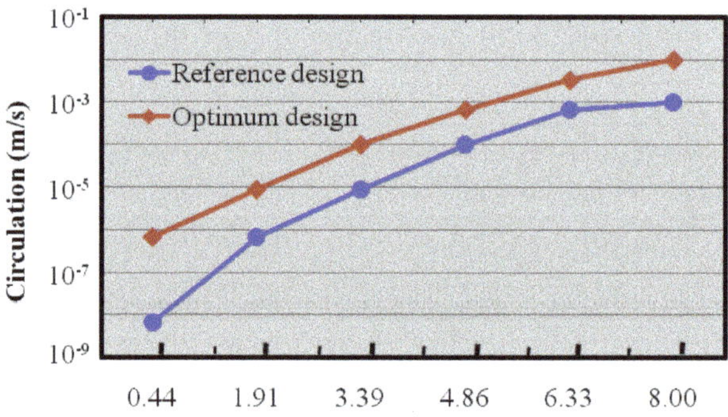

Figure 11. Normalized circulation distributions for reference design and optimum design at Re = 40.

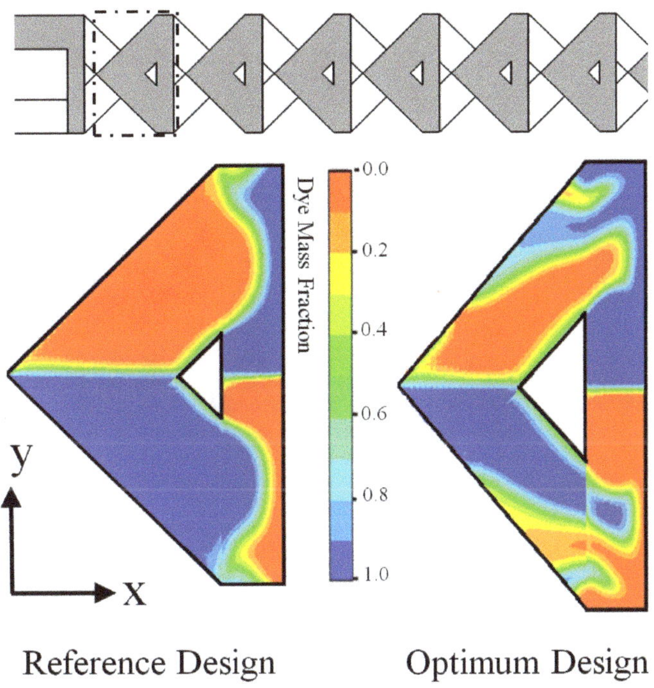

Figure 12. Mass fraction distributions of dye at the middle of the top channel (designated by the dotted lines) of first mixing unit for reference design and the optimum design micromixers at Re = 40.

8. Conclusions

Geometric optimization of a proposed two-layer crossing micromixer was carried out using 3D Navier–Stokes formulas. To estimate the objective function, the well-known RBNN model was used. The optimization study was performed with three design variables; viz., the ratio of the main channel thickness to the pitch length (H/PI), the ratio of the thickness of the diagonal channel to the pitch length (W/PI), and the ratio of the depth of the channel to the pitch length (d/PI). The mixing index for the micromixer (F_{MI}) was considered as an objective function to find the most efficient design.

By this study, one can conclude the following: For Reynolds numbers $\leq$1.0, the objective function was not significantly varied with the geometric parameters. The parametric study representing the objective function values was very much more sensitive to H/PI than to d/PI and W/PI. The maximum mixing index value of 0.72 was found at lowest value of W/PI, where the best matching of the fluids interfacial area and residential time was obtained. Surrogate-based optimization results represented the design variables (H/PI = 1.73, W/PI = 0.42 and d/PI = 0.18) and the objective function (MI = 0.86) for the optimum micromixer. The objective function values of the reference design and predicted optimal design were 0.65 and 0.86, respectively, which confirms a 32% relative increase in the objective function through the surrogate-based optimization procedure. The calculated objective function value using Navier–Stokes equation was compared with the predicted value. The RBNN model represented a 6.2% relative deviation of objective function value from the optimum point. The study represents that the single-objective optimization procedure is favorable for the improvement of micromixer performance. The optimum micromixers could be incorporated into a micro-total analysis system and lab-on-chip (LOC) systems to facilitate the study of reaction kinetics, dilution of fluid samples, and enhancement of reaction selectivity.

Author Contributions: S.H.: Conceptualization, numerical simulation, methodology, data analysis, writing original draft. N.T.T.: Methodology, writing original draft. F.I.: Literature survey, writing original draft. M.K.: Writing–review and editing. P.D.H.B.: Writing–review and editing. M.M.K.B.: Writing–review and editing. M.A.: Writing–review and editing. K.-Y.K.: Writing–review and editing, funding, supervision. All authors have read and agreed to the published version of the manuscript.

Funding: This research received no external funding.

Acknowledgments: This work was supported by the National Research Foundation of Korea (NRF) grant funded by the Korean government (MSIT) (No. 2019R1A2C1007657). The authors gratefully acknowledge this support.

Conflicts of Interest: The authors declare there are no conflict of interest.

References

1. Park, T.; Lee, S.; Seong, G.H.; Choo, J.; Lee, E.K.; Kim, Y.S.; Ji, W.H.; Hwang, S.Y.; Gweon, D.G.; Lee, S. Highly sensitive signal detection of duplex dye-labelled DNA oligonucleotides in a PDMS microfluidic chip: Confocal surface-enhanced Raman spectroscopic study. *Lab Chip* **2005**, *5*, 437–442. [CrossRef] [PubMed]
2. Schulte, T.H.; Bardell, R.L.; Weigl, B.H. Microfluidic technologies in clinical diagnostics. *Clin. Chim. Acta* **2002**, *321*, 1–10. [CrossRef]
3. Rapp, B.E.; Gruhl, F.J.; Länge, K. Biosensors with label-free detection designed for diagnostic applications. *Anal. Bioanal. Chem.* **2010**, *398*, 2403–2412. [CrossRef]
4. Hessel, V.; Löwe, H.; Schönfeld, F. Micromixers—A review on passive and active mixing principles. *Chem. Eng. Sci.* **2005**, *60*, 2479–2501. [CrossRef]
5. Nguyen, N.T. *Micromixers: Fundamentals, Design and Fabrication*; William Andrew Publishers: New York, NY, USA, 2011.
6. Ingham, C.J.; van Hylckama Vlieg, J.E.T. MEMS and the microbe. *Lab Chip* **2008**, *8*, 1604–1616. [CrossRef]
7. Izadi, D.; Nguyen, T.; Lapidus, L.J. Complete Procedure for Fabrication of a Fused Silica Ultrarapid Microfluidic Mixer Used in Biophysical Measurements. *Micromachines* **2017**, *8*, 16. [CrossRef]
8. Aubin, J.; Fletcher, D.F.; Bertrand, J.; Xuereb, C. Characterization of the mixing quality in micromixers. *Chem. Eng. Technol.* **2003**, *26*, 1262–1270. [CrossRef]
9. Hossain, S.; Kim, K.-Y. Mixing analysis in a three-dimensional serpentine split-and-recombine micromixer. *Chem. Eng. Res.Des.* **2015**, *100*, 95–103. [CrossRef]
10. Du, Y.; Zhang, Z.; Yim, C.H.; Lin, M.; Cao, X. A simplified design of the staggered herringbone micromixer. *Biomicrofluidics* **2010**, *4*, 024105. [CrossRef]
11. Johnson, T.J.; Locascio, L.E. Characterization and optimization of slanted well designs for microfluidic mixing under electroosmotic flow. *Lab Chip* **2002**, *2*, 135–140. [CrossRef]
12. Nguyen, N.T.; Wu, Z. Micromixers—A review. *J. Micromech. Microeng.* **2005**, *15*, R1–R16. [CrossRef]
13. Ottino, J.M. *The Kinematics of Mixing: Stretching, Chaos, and Transport*; Cambridge Texts in Applied Mathematics; Cambridge University Press: Cambridge, UK; Oxford, UK, 1989.
14. Liu, R.H.; Stremler, M.A.; Sharp, K.V.; Olsen, M.G.; Santiago, J.G.; Adrian, R.J.; Aref, H.; Beebe, D.J. Passive mixing in a three-dimensional serpentine microchannel. *J. Microelectromechan. Syst.* **2000**, *9*, 190–197. [CrossRef]
15. Stroock, A.D.; Dertinger, S.K.W.; Ajdari, A.; Mezic, I.; Stone, H.A.; Whitesides, G.M. Chaotic mixer for microchannels. *Science* **2002**, *295*, 647–651. [CrossRef] [PubMed]
16. Kim, D.S.; Lee, S.W.; Kwon, T.H.; Lee, S.S. A barrier embedded chaotic micromixer. *J. Micromech. Microeng.* **2004**, *14*, 798–805. [CrossRef]
17. Wang, L.; Yang, J.T.; Lyu, P.C. An overlapping crisscross micromixer. *Chem. Eng. Sci.* **2007**, *62*, 711–720. [CrossRef]
18. Xia, H.M.; Wan, S.Y.M.; Shu, C.; Chew, Y.T. Chaotic micromixers using two-layer crossing channels to exhibit fast mixing at low Reynolds numbers. *Lab Chip* **2005**, *5*, 748–755. [CrossRef]
19. Ansari, M.A.; Kim, K.Y. Shape optimization of a micromixer with staggered herringbone groove. *Chem. Eng. Sci.* **2007**, *62*, 6687–6695. [CrossRef]
20. Lynn, N.S.; Dandy, D.S. Geometrical optimization of helical flow in grooved micromixers. *Lab Chip* **2007**, *7*, 580–587. [CrossRef]
21. Hossain, S.; Ansari, M.A.; Husain, A.; Kim, K.Y. Analysis and optimization of a micromixer with a modified Tesla structure. *Chem. Eng. J.* **2010**, *158*, 305–314. [CrossRef]
22. Cortes-Quiroz, C.A.; Azarbadegan, A.; Zangeneh, M.; Goto, A. Analysis and multi-criteria design optimization of geometric characteristics of grooved micromixer. *Chem. Eng. J.* **2010**, *160*, 852–864. [CrossRef]
23. Hossain, S.; Husain, A.; Kim, K.Y. Shape optimization of a micromixer with staggered-herringbone grooves patterned on opposite walls. *Chem. Eng. J.* **2010**, *162*, 730–737. [CrossRef]
24. Afzal, A.; Kim, K.-Y. Multi-objective optimization of a passive micromixer based on periodic variation of velocity profile. *Chem. Eng. Commun.* **2014**, *202*, 322–331. [CrossRef]

25. Afzal, A.; Kim, K.-Y. Multiobjective optimization of a oicromixer with convergent–divergent sinusoidal walls. *Chem. Eng. Commun.* **2015**, *202*, 1324–1334. [CrossRef]
26. Raza, W.; Ma, S.-B.; Kim, K.-Y. Single and Multi-Objective Optimization of a Three-Dimensional Unbalanced Split-and-Recombine Micromixer. *Micromachines* **2019**, *10*, 711. [CrossRef] [PubMed]
27. Ahmed, F.; Hossain, S.; Ryu, H.; Ansari, M.A.; Khan, M.S.I.; Kim, K.Y. Numerical and Experimental study on Mixing in Chaotic Micromixers with Crossing Structures. *Chem. Eng. Technol.* **2020**, *43–49*, 1866–1875.
28. ANSYS. *Solver Theory Guide, CFX-15.0*; ANSYS Inc.: Canonsburg, PA, USA, 2013.
29. Kim, D.S.; Lee, S.H.; Kwon, T.H.; Ahn, C.H. A serpentine laminating micromixer combining splitting/recombination and advection. *Lab Chip* **2005**, *5*, 739–747. [CrossRef] [PubMed]
30. Hardt, S.; Schonfeld, E. Laminar mixing in different inter-digital micromixers: II. Numerical simulations. *AIChE J.* **2003**, *49*, 578–584. [CrossRef]
31. Van-Doormaal, J.P.; Raithby, G.D. Enhancement of the SIMPLE method for predicting incompressible fluid flows. *Numer. Heat Transf.* **1984**, *7*, 147–163.
32. Kennedy, K. Bridging the Gap between Space-Filling and Optimal Designs. Doctoral Dissertation, Arizona State University, Tempe, AZ, USA, 2013.
33. JMP. *Design of Experiments Guide*; SAS Institute Inc.: Cary, NC, USA, 2005.
34. McKay, M.D.; Beckman, R.J.; Conover, W.J. Comparison of three methods for selecting values of input variables in the analysis of output from a computer code. *Technometrics* **1979**, *21*, 239–245.
35. Deb, K.; Pratap, A.; Agarwal, S.; Meyarivan, T. A fast and elitist multiobjective genetic algorithm: NSGA-II. *IEEE Trans. Evol. Comput.* **2002**, *6*, 182–197. [CrossRef]
36. MATLAB. *The Language of Technical Computing*; Release 14; MathWorks, Inc.: Natick, MA, USA, 2004; Available online: http://www.mathworks.com (accessed on 26 January 2019).
37. Queipo, N.V.; Haftka, R.T.; Shyy, W.; Goel, T.; Vaidyanathan, R.; Tucker, P.K. Surrogate-based analysis and optimization. *Prog. Aerosp. Sci.* **2005**, *41*, 1–28. [CrossRef]
38. Orr, M. Introduction to radial basis function networks. *Univ. Edinbg.* **1996**, 1–7.
39. Stone, M. Cross-validatory choice and assessment of statistical predictions. *J. R. Stat. Soc.* **1974**, *36*, 111–147. [CrossRef]

Article

Numerical Study of T-Shaped Micromixers with Vortex-Inducing Obstacles in the Inlet Channels

Chih-Yang Wu * and Bing-Hao Lai

Department of Mechanical Engineering, National Cheng Kung University, Tainan 701, Taiwan; bc510608@gmail.com

* Correspondence: cywu@mail.ncku.edu.tw; Tel.: +886-6-2757575 (ext. 62151)

Received: 25 November 2020; Accepted: 15 December 2020; Published: 18 December 2020

Abstract: To enhance fluid mixing, a new approach for inlet flow modification by adding vortex-inducing obstacles (VIOs) in the inlet channels of a T-shaped micromixer is proposed and investigated in this work. We use a commercial computational fluid dynamics code to calculate the pressure and the velocity vectors and, to reduce the numerical diffusion in high-Peclet-number flows, we employ the particle-tracking simulation with an approximation diffusion model to calculate the concentration distribution in the micromixers. The effects of geometric parameters, including the distance between the obstacles and the angle of attack of the obstacles, on the mixing performance of micromixers are studied. From the results, we can observe the following trends: (i) the stretched contact surface between different fluids caused by antisymmetric VIOs happens for the cases with the Reynolds number (*Re*) greater than or equal to 27 and the enhancement of mixing increases with the increase of Reynolds number gradually, and (ii) the onset of the engulfment flow happens at $Re \approx 125$ in the T-shaped mixer with symmetric VIOs or at $Re \approx 140$ in the standard planar T-shaped mixer and results in a sudden increase of the degree of mixing. The results indicate that the early initiation of transversal convection by either symmetric or antisymmetric VIOs can enhance fluid mixing at a relatively lower *Re*.

Keywords: microfluidics; T-shaped micromixer; vortex; obstacles; engulfment flow; particle tracking

1. Introduction

Microfluidic mixing has wide applications in biochemical reactions, chemical synthesis and biological analysis [1–3]. Mixing in microfluidic devices is challenging since typical flows in microfluidic devices are laminar in nature and mixing in laminar flows relies mainly on molecular diffusion. Therefore, increasing the interface or shortening the diffusion length between different fluids by handling fluid flows within the mixing channel is of great importance for enhancing mixing in microfluidic devices. The reviews of the related literature reveal that various micromixers, broadly categorized as either active or passive types, have been developed and examined. Passive micromixers utilize system geometry to create favorable hydrodynamics for enhancing mixing and so do not require external energy, except that used to drive the flows. Although the mixing efficiency of active micromixers is better than that of passive micromixers, it is simpler to fabricate passive micromixers and easier to integrate them with microfluidic systems. Thus, passive micromixers have been developed widely [1–3].

Most passive micromixers include a T or Y junction for the confluence of the fluids to be mixed, and a straight mixing channel with square or rectangular cross-section [4–14] or a mixing channel with modified geometry aiming at enhancing the mixing of fluids [15–31]. The T-, Y- or arrow-shaped mixer with a straight mixing channel has been reported to show the so-called engulfment flow and good mixing at a higher Reynolds number (*Re*) defined as $Re = v_m d_h \rho / \mu$, with v_m, d_h, ρ and μ denoting the mean flow speed in the mixing channel, the hydraulic diameter of the mixing channel, the density and the dynamic viscosity of the fluid, respectively. Engler et al. [5] have found that the resulting flow in the mixing channel of a planar T-shaped mixer can be characterized by three flow regimes during steady flow: so-called "stratified" flow, "vortex" flow and "engulfment" flow. They showed that the breakup of symmetry in the flow field at a higher Reynolds number occurred, resulting in the so-called engulfment flow, which was characterized by some fluid from one side reaching beyond the centerline of the micro-T-shaped mixer to engulf the fluid from the other side. It has been found that, in a standard planar T-shaped mixer with two inlets with square cross-section, and a mixing channel of equal combined area where two-opposing planar channel streams join and turn through 90 degrees into the mixing channel, the symmetry breaking of the flow field configuration in the mixing channel starts at a Reynolds number between 138.6 and 140 [7]. The influence of the volume flow rates has been well investigated. Schikarski et al. reported an experimental–computational study of a T-shaped mixer for Reynolds numbers up to 4000 [12]. Besides the Reynolds number, the flow patterns and mixing performance may vary with other flow or geometric parameters. Soleymani et al. proposed a dimensionless number to describe the dependence of the flow regime inside the T-micromixer on the flow rate, aspect ratio and hydrodynamic diameter ratio [8]. Galletti et al. investigated the effect of inlet velocity distributions on the mixing performance [11]. Karvelas et al. investigated the effect of the angle of the Y-shaped micromixer and the effect of different inlet velocity ratios [13]. Recently, Camarri et al. presented an overview study on the effects of the Reynolds number, aspect ratios and mixing angle on the mixing performance of a T-shaped micromixer [14].

Since mixing in a T-, Y- or arrow-shaped mixer with a straight mixing channel is ineffective at lower values of *Re*, various modifications in the geometry of mixing channel have been developed and investigated. These modified mixing channels include introduction of obstacles [15,16] at high Reynolds numbers, serpentine and/or converging–diverging channels [17–20] for *Re* in the range 10–100, and patterned groove microchannels at low Reynolds numbers [21,22]. The other types of micromixers adopted serial lamination [23], split-and-recombine (SAR) channels [24,25], two-layer crossing channels [26], channels with cross-sections other than square or rectangular [30,31] and channel modifications based on combined different mixing mechanisms [27–29,31]. Most existing research has investigated the geometric modification of the mixing channel from the junction to the outlet for enhancing fluid mixing in micromixers, while only a few works take the effects of inlet channel modification into account. Gigras and Pushpavanam have proposed and analyzed the fluid mixing in a micromixer with a curved inlet channel which exploits the early induction of transversal vortices in the inlet channel to enhance mixing [19]. Sultan et al. studied flow behavior and mixing in mixers with T-shaped jet inlets for different geometrical parameters [32,33]. Three-dimensional (3D) T-type micromixers with non-aligned inlets were proposed for convective mixing enhancement [34–36]. Other than the geometric modification of the mixing channel, droplet micromixers [37], micromixers using ultra-hydrophobic surfaces [38] and more designs have been proposed by many researchers (see References [1–3] for reviews of these) to improve the mixing quality in passive micromixers.

A new approach for inlet flow modification by adding obstacles protruding into the flow at an angle of attack in the inlet channels of a T-shaped micromixer is proposed and investigated in this work. The obstacles with a height less than the channel height may induce vortices at the entrance of the mixing channel at a relatively lower Reynolds number and are called the vortex-inducing obstacles (VIOs) in this work. In light of the above-mentioned considerations, it is desirable to explore if the early introduction of vortices before the entrance of the mixing channel can lead to a fundamental improvement in fluid mixing of a T-shaped micromixer. The present work focuses on the effects of geometric modifications of the inlet channel on mixing and does not consider the geometric modification of the mixing channel, such as those proposed in the literature [15–31], or active strategy of mixing, such as the application of magnetic field to the mixing of contaminated water in one stream and water with the magnetic particles in the other stream [39]. The VIOs may be symmetric or antisymmetric, as shown in Figure 1. Numerical simulation has been performed to investigate the mixing behavior and flow characteristics for selected values of geometric parameters of the added VIOs in the inlet channels at different flow rates. To study the effects of the early induction of transversal convection by the VIOs, comparisons of the mixing performance of the proposed mixer with either type of the VIOs and that of a standard T-shaped mixer without VIOs are made. The results indicate that the early initiation of transversal convection by either symmetric or antisymmetric VIOs can enhance fluid mixing at a relatively lower *Re*.

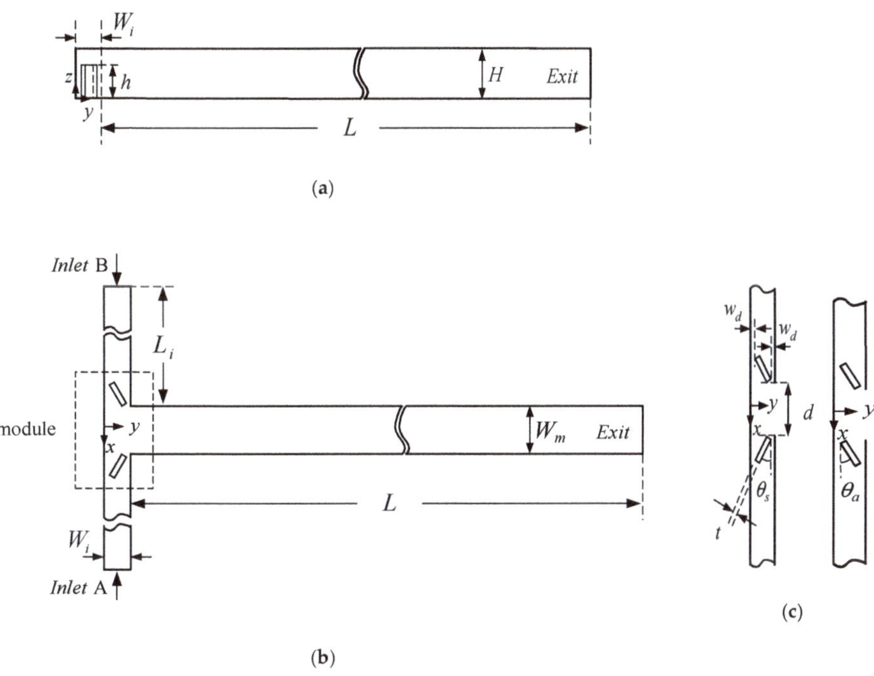

Figure 1. Schematic diagram and geometric parameters of T-shaped micromixers with vortex-inducing obstacles (VIOs) in the inlet channels: (**a**) side view, (**b**) top view, (**c**) inlet channel with symmetric VIOs (**Left**) and that with as antisymmetric VIOs (**Right**).

2. Mixer Geometry, Governing Equations and Computational Procedure

The present micromixers consist of two inlets and a mixing channel, where two opposing streams of different fluids join and turn through 90 degrees into the mixing channel leading towards the outlet. A pair of oblique VIOs are mounted on the bottom of the inlet channels. The pair of VIOs protrudes into the flow at an angle of attack (θ_s or θ_a) symmetrically or anti-symmetrically and has the height of $h < H$, as shown in Figure 1. Here, H is the channel height. The micromixers can be fabricated by using an easy two-step lithography process. There are a very large number of geometrical parameters influencing the performance of the micromixers. The effects of the widths and height of the inlet and mixing channels on fluid mixing have been investigated systematically [8]. The present study focuses on examining the influence of the arrangement of the VIOs, including the distance between the VIOs (d) and the angle of attack (θ_s or θ_a), on fluid mixing; in the meantime, we set the value of H at 120 μm, the width of the mixing channel $W_m = 2H$, the width of the inlet channels $W_i = H$, the thickness of the VIOs $t = 0.25H$, the height of the VIOs $h = 0.75H$, the length of the inlet channel $L_i = 8H$, the length of the mixing channel $L = 15H$, the distance between each obstacle and its neighboring sidewall $w_d = 0.25H$.

The fluid was assumed to be Newtonian, isothermal and incompressible, and the corresponding governing equations of the flow field are the continuity and the Navier–Stokes equations, as follows:

$$\nabla \cdot \overrightarrow{v}^* = 0, \tag{1}$$

$$\left(\overrightarrow{v}^* \cdot \nabla\right)\overrightarrow{v}^* = -\nabla p^* + \frac{1}{Re}\nabla^2 \overrightarrow{v}^*, \tag{2}$$

where, $\overrightarrow{v}^* = \frac{\overrightarrow{v}}{v_m}$ and $p^* = \frac{p-p_0}{\rho v_m^2}$, with $\overrightarrow{v}$, p and p_0 denoting the velocity vector, the pressure and the atmosphere pressure, respectively. In this work, we consider equal flow rate at both inlets, and so the mixing ratio is 1:1. The inlet velocity is set to be a fully developed laminar velocity profile [40]. The no-slip condition is imposed on all solid walls and the pressure at the exit is set to be 1 atm. The concentration field is described by the convective–diffusive equation:

$$\left(\overrightarrow{v}^* \cdot \nabla\right)c^* = \frac{1}{ReSc}\nabla^2 c^* \tag{3}$$

where $c^* = \frac{c-c_B}{c_A-c_B}$, with c denoting the molar concentration per unit volume and the subscripts A and B denoting the inlet A and the inlet B, respectively. $Sc = \mu/\rho D$ is the Schmidt number, with D denoting the diffusion coefficient. The nondimensional molar concentration c^* takes a quantity from 0 to 1. The solid walls are set to be impermeable. The fluids entering the two inlets are a low-concentration solution of Rhodamine B in deionized (DI) water with a diffusion coefficient 3.6×10^{-10} m^2 s^{-1} [41] and pure DI water. Because of the low concentration of the solution, the influence of fluorescent on the fluid behavior can be neglected [5], and so the fluid properties considered are set to be those of the DI water with density, ρ =997 kg m^{-3}, and viscosity, $\mu = 0.00097$ kg s^{-1}m^{-1}. Therefore, the Schmidt number equals 2700.

The pressure and the velocity vector are solved by a grid-based scheme, using the software CFD-ACE+ (CFD Research Corporation, Huntsville, AL, USA). The SIMPLEC (Semi-Implicit Method for Pressure Linked Equations-Consistent) algorithm is used for pressure-velocity coupling and the spatial difference is carried out with the second-order upwind scheme with limiter. When the relative residual of each variable is low to 10^{-5}, the solution is regarded as converged. In this work, we are mainly interested in the effect of transverse convection on fluid mixing, which typically dominates at a high Reynolds number. For such cases with $Sc = 2700$, the value of the Peclet number ($Pe = ReSc$) of the flow in the mixer may be high. The grid-based solutions of the convective–diffusive equation suffer from the false numerical diffusion at a high Peclet number [42]. Therefore, to simulate fluid mixing in the proposed micromixers under various conditions, we adopt the particle tracking method with an approximation diffusion model (ADM), which is virtually free of numerical diffusion in high-Peclet-number flows and takes less computation time than the random-walk particle-tracking method [43,44]. To determine the non-diffused concentration field, we undertake the particle-tracking simulation, which is illustrated by an example shown in Figure 2. The target planes are evenly divided into $(N_{xt} - 1) \times (N_{zt} - 1)$ grid cells and a massless particle is assigned to each grid cell. Then, these particles are backward tracked from the target plane to the two inlets. The advection displacement of a particle can be calculated by using a selected small advection time step and interpolating the velocity data at grid points, which are available from the solution by the software CFD-ACE+. The boundary condition of the concentration field can be accommodated by suitable particle reflections from the boundary [45]. The displacements are repeated until the particle crosses either of the two inlets. The details of the backward particle-tracking simulation were reported by Kuo et al. [44]. Next, the diffused concentration field is obtained by solving the approximation diffusion equation with the non-diffused concentration field as the initial condition. This is a posterior treatment of molecular diffusion proposed by Matsugana et al. [43]. A discretization scheme based on a five-point formula is employed to solve the diffusion equation.

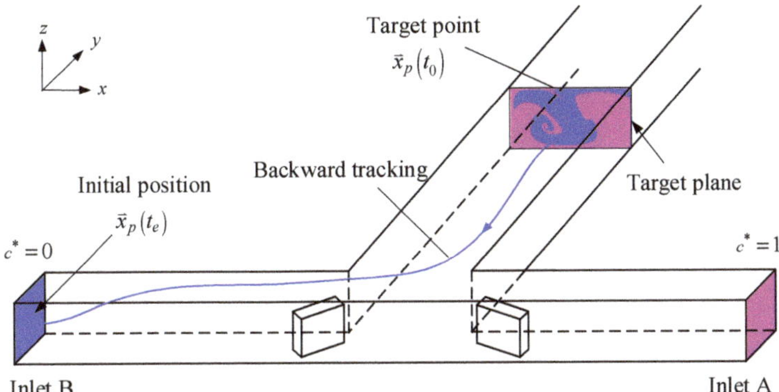

Figure 2. Three-dimensional (3D) micromixer geometry and the sketch of backward fluid particle tracking.

To quantitatively analyze the mixing performance, the degree of mixing at each cross-section of the micromixer is evaluated by:

$$M = 1 - \frac{\sigma}{\sigma_0}, \tag{4}$$

where σ denotes the variance of the concentration at a transverse cross-section, defined as:

$$\sigma^2 = \frac{1}{N_s}\sum_{n=1}^{N_s}(c_n - \bar{c})^2 \tag{5}$$

with N_s denoting the total number of sampling, c_n the concentration at a position on the cross-section considered, $\bar{c}$ the average of c_n and $\sigma_0 = \sqrt{\bar{c}(1-\bar{c})}$ the variance of the concentration in the completely unmixed state [46]. In this work, we estimate the mixing efficiency of a micromixer by calculating the degree of mixing at the exit cross-section. The total number of sampling is the number of grid cells on the target plane, $(N_{yt} - 1) \times (N_{zt} - 1)$, for the particle-tracking simulation with an ADM.

3. Results and Discussion

For the purpose of validation, the results of fluid mixing in a standard planar T-shaped mixer with $H = 300$ μm at $Re' = Hv_m\rho/\mu = 150$ and $Sc = 3200$ [43] are considered first. The distributions of concentration on the cross-section at $y = 1500$ μm, obtained by the particle-tracking method with an ADM, are shown in Figure 3. It is easy to see an excellent agreement of present results obtained by using $N_y \times N_z = 80 \times 40$ grids in the velocity solution and $(N_{yt} - 1) \times (N_{zt} - 1) = 600 \times 300$ particles in the concentration simulation with those reported by Matsunaga et al. [43]. Then, grid size sensitivity tests were performed for mixing flow at $Re = 200$ in a proposed mixer with $H = 120$ μm, $\theta_s = 20°$, $d = 3H$, $h = 0.75H$, $t = 0.25H$ and $w_d = 0.25H$. The velocity components $v^* = v/v_m$ along $z = 150$ μm on the cross-section at $y = 120$ μm obtained by using the grid sizes 3 μm, 4 μm and 5 μm are shown in Figure 4. The discrepancy between those obtained by using the grid sizes of 3 μm and 4 μm is quite small. Particle number sensitivity tests were performed for mixing flow at $Re = 150$ in the same mixer. The distributions of concentration in the present micromixer on the cross-section at $y = 360$ μm, obtained by the particle-tracking method with an ADM using different total numbers of particles, are in excellent agreement, as shown in Figure 5a,b. Figure 5c shows that the distributions of concentration on the cross-section at $y = 360$ μm obtained by using particle numbers 450×225 and 600×300 almost overlap with each other. Since the geometry of the present mixers are more complicated than a standard planar T-shaped mixer, we use the grid size of 3 μm and 600×300 particles launched from a target plane in the following simulation.

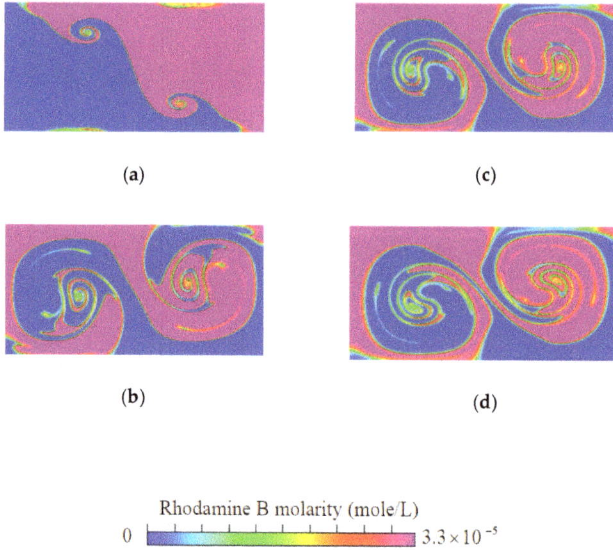

Figure 3. Concentration distributions on the cross-section at (**a**) $y = 300$ µm, (**b**) $y = 1500$ µm, (**c**) $y = 2700$ µm, (**d**) $y = 3900$ µm in a standard planar T-shaped mixer with $H = 300$ µm at $Re' = 150$ and $Sc = 3200$.

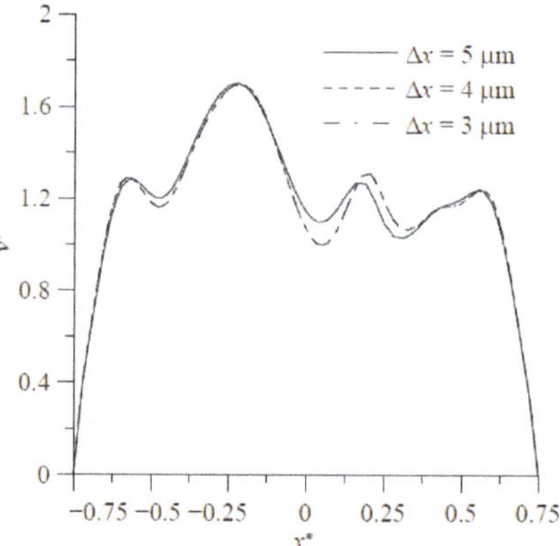

Figure 4. Profiles of velocity component v along the x direction on the middle horizontal on the cross-section at $y = 120$ µm obtained by using various grid sizes.

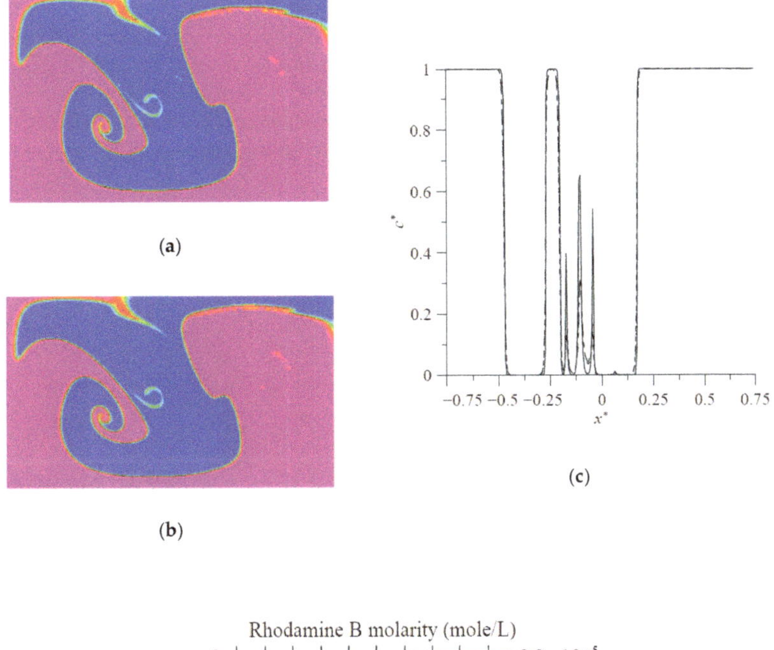

Figure 5. Concentration distributions on the cross-section at $y = 360$ µm obtained by the particle tracking method with an ADM: concentration distributions on the cross-section by using (**a**) $N_{xt} \times N_{zt} = 360 \times 180$ particles and (**b**) $N_{xt} \times N_{zt} = 480 \times 240$ particles, (**c**) concentration profiles along the middle horizontal on the cross-section at $y = 360$ µm by using $N_{xt} \times N_{zt} = 240 \times 120$ particles (—; $N_{xt} \times N_{zt} = 360 \times 180$ particles (– – – –); $N_{xt} \times N_{zt} = 480 \times 240$ particles (— - - —).

The representative path lines and the concentration distributions shown in Figure 6 give an overview of the flow and mixing taking place in the proposed T-shaped micromixers with VIOs in the inlet channels for some selected values of Reynolds numbers between 1 and 150. The selected geometrical parameters associated with the VIOs are $h = 0.75H, t = 0.25H, w_d = 0.25H, \theta_s = 20°$ and $\theta_a = 30°$. It is easy to see that the flow and mixing in the mixer with symmetric VIOs and those in the mixer with antisymmetric VIOs exhibit different patterns.

The left and the right columns of Figure 6a show the flow characteristics and the mixing behavior for fluid flow in the T-shaped micromixers with symmetric VIOs, respectively. There is no formation of any vortex flow due to low inertia force in the fluid streams at $Re = 1$, while the path lines of the cases with $Re = 40{\sim}120$ show symmetric vortex flows with respect to the plane $x = 0$. Thus, the two portions of fluid entering from the two inlet channels remain segregated and mixing occurs only through diffusion. The flow regime of diffusive mixing is characterized by its flow pattern with single reflection symmetry with respect to the plane $x = 0$, which is different from the double reflection symmetry found in the flow of planar T-shaped micromixers with square or rectangular cross-section [5,7–10,14]. The double reflection symmetry even appears in the flow regime of diffusive mixing for flow in planar T-shaped micromixers with non-fully developed flow conditions [11]. The diffusion zone around the vertical plane $x = 0$ becomes thinner as the *Re* becomes larger and the mixing states are not desirable yet because of the slow diffusion mechanism. When the *Re* further increases and reaches a critical value, the onset of the so-called engulfment regime takes place. In this flow regime, the flow pattern becomes asymmetric and each inlet fluid stream reaches the opposite sides of the plane $x = 0$, as indicated by the path lines shown in Figure 6a and the velocity vectors shown in Figure 7. Similar to fluid mixing in planar T-shaped micromixers [5–10,14,32,33], a sudden increase of the degree of mixing can be observed which originates from the intertwining of the path lines. However, the asymmetric flow pattern shown in Figure 7c is different from the rotational-symmetric flow [7], which appears in the steady engulfment regime of flow in a planar T-shaped micromixer. Although the vortical flow induced by the symmetric VIOs is the same in the two inlet channels, its distribution is not symmetric, as shown in Figure 7a, b. Thus, Figure 7c exhibits the asymmetric flow pattern at a location in the mixing channel for the steady engulfment regime.

By contrast with fluid mixing in the T-shaped micromixers with symmetric VIOs, from the path lines and the concentration distributions shown in Figure 6b for fluid flow in the T-shaped micromixers with antisymmetric VIOs, we can observe the following trends: the symmetry of the path lines and the concentration distributions holds in the flow at $Re = 1$, the distortion of the interface and the interaction of the two vortices appears at the cases with $Re \geq 40$ and the enhancement of fluid mixing by the stretched contact surface of the fluids gradually increases with the increase of the Re. This improved mixing is attributed to the asymmetric creation of secondary flow by antisymmetric VIOs in the opposing inlet channels, as shown in Figure 8. The early induced asymmetric lateral convection in the inlet channels breaks up the symmetry of the flow field and augments the contact area between the two streams entering from the two inlet channels, and so enhances fluid mixing at a relatively lower Reynolds number, as shown in Figure 6b. The improved mixing is different from the mixing enhancement caused by the symmetry-breaking bifurcation of fluid flow in the T-shaped micromixer with symmetric VIOs at a higher Reynolds number. A 3D T-mixer with the inlet channels located at different horizontal levels can generate vertical flow into the mixing channel to improve fluid mixing [34–36], and so it also runs without the benefit of the symmetry-breaking bifurcation of fluid flow in the planar T-shaped micromixer.

Before quantitatively comparing the mixing performance of the proposed mixer with symmetric VIOs, the proposed mixer with antisymmetric VIOs and a standard T-shaped mixer, we examine the simulation results for cases with various values of the distance between the VIOs (d) and the angle of attack (θ_s or θ_a) to investigate the influence of these geometrical parameters and to select the appropriate values of these geometric parameters for efficient mixing. Mixing at a microfluidic junction is strongly dependent on Re. Thus, this section reports the results from numerical simulations carried out over a range of Re to examine the performance of the proposed mixer with VIOs. First, to make a comparative study of influence of the distance between the symmetric VIOs, we choose four values of the distance, $d = 2H$, $3H$, $4H$ and $5H$, and the other geometrical parameters associated with the obstacles are kept same. The corresponding values are $\theta_s = 20°$, $h = 0.75H$, $t = 0.25H$ and $w_d = 0.25H$. The conducted numerical simulation for fluid mixing of a T-shaped mixer with symmetric VIOs in the Reynolds number range from 100 to 200 allows for identifying the value of Re corresponding to the transition from the vortex flow regime to the engulfment flow regime. The degree of mixing at the exit (M_{exit}) shows a significant increase at such a value of Re and increases with the increase of Re as the Re increases beyond the critical value of Re corresponding to the transition, as shown in Figure 9a. It is worth noting that for the cases with $d = 2H$, $3H$ and $4H$, the degree of mixing increases suddenly at a Reynolds number less than the critical value of Re reported in the literature for mixing in a standard T-shaped mixer [7]. Moreover, the critical value of Re corresponding to the transition for the case with $d = 3H$ is the smallest and the degree of mixing of the case with $d = 3H$ is greater than or equal to that of the case with $d = 2H$, $4H$ or $5H$, except that the M_{exit} of the case with $d = 2H$ is a little bit greater than that of the case with $d = 3H$ at $Re = 200$. The pressure drop between the mixer inlet and outlet (Δp) monotonically increases with the increase of the Reynolds number in the cases considered. The influence of the distance between the VIOs on the pressure drop is very small and the transition from the vortex flow regime and to the engulfment flow regime has no effect on the pressure drop, as shown in Figure 9b.

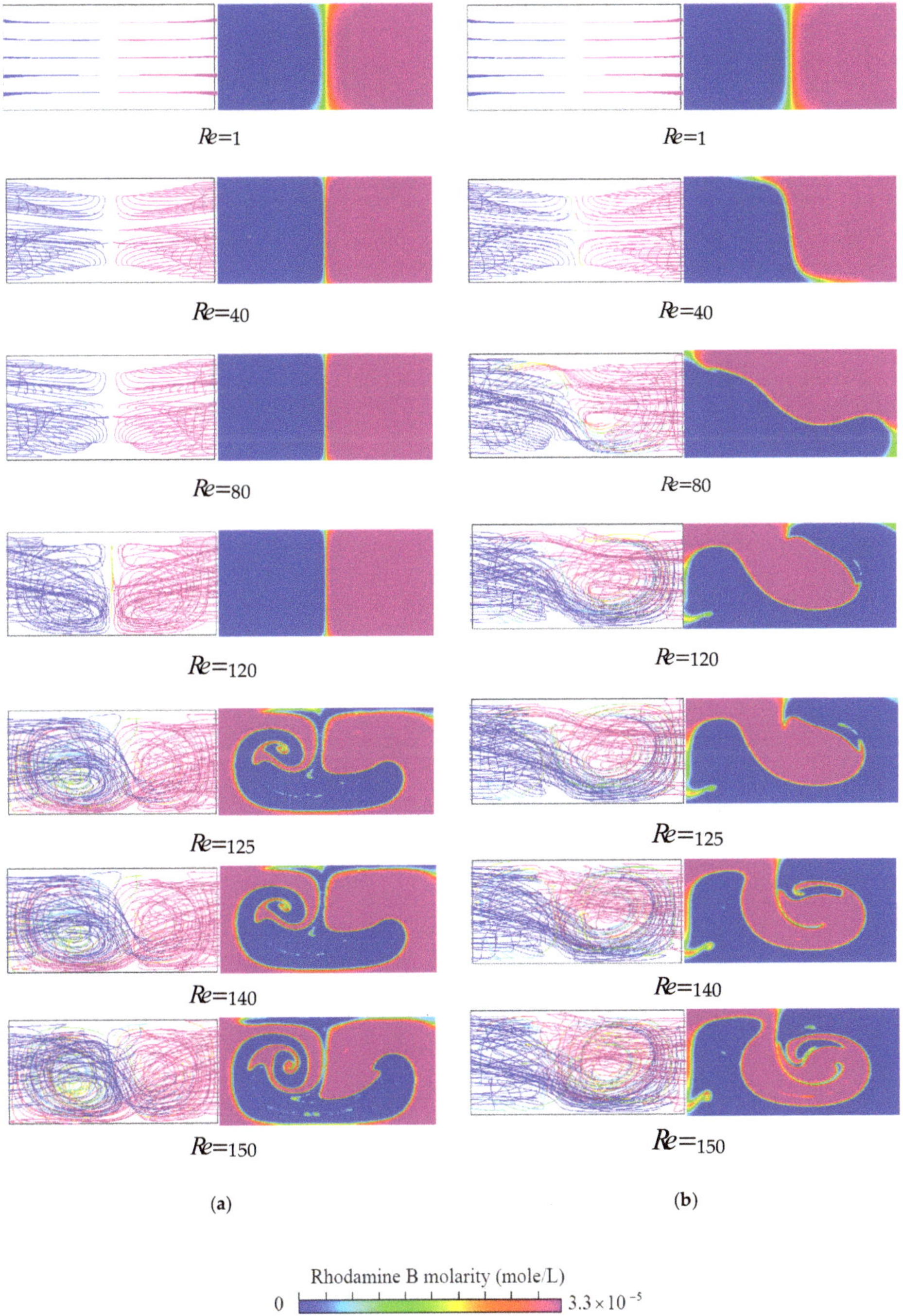

Figure 6. Projected path lines in the T junctions for different mass flows from the left and right inlet channels and the concentration distributions on the cross-section at the exit of (**a**) the mixer with symmetric VIOs and (**b**) the mixer with antisymmetric VIOs. View is into the mixing channel.

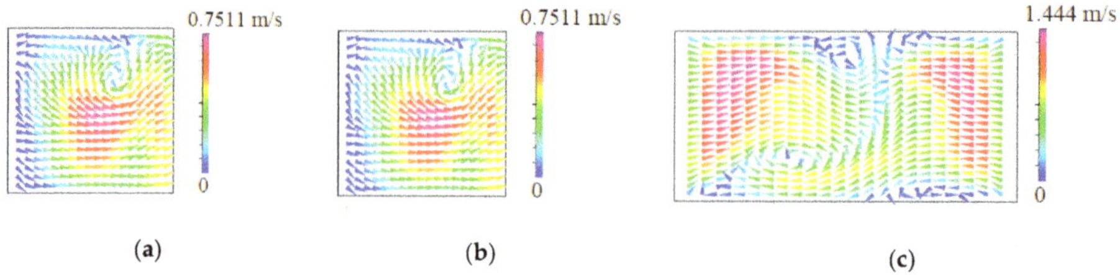

Figure 7. yz-projection of the velocity vectors on the cross-sections at (**a**) $x = H$, (**b**) $x = -H$, (**c**) xz-projection of the velocity vectors on the cross-sections at $y = H$ for the case with $Re = 140$, $\theta_s = 20°$, $h = 0.75H$, $t = 0.25H$ and $w_d = 0.25H$.

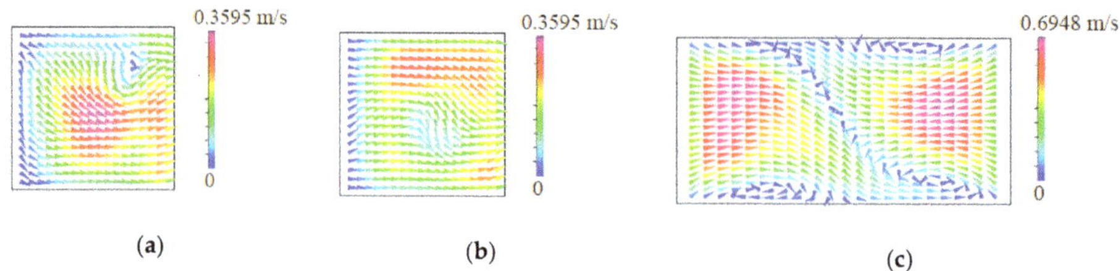

Figure 8. yz-projection of the velocity vectors on the cross-sections at (**a**) $x = H$, (**b**) $x = -H$, (**c**) xz-projection of the velocity vectors on the cross-sections at $y = H$ for the case with $Re = 80$, $\theta_a = 30°$, $h = 0.75H$, $t = 0.25H$ and $w_d = 0.25H$.

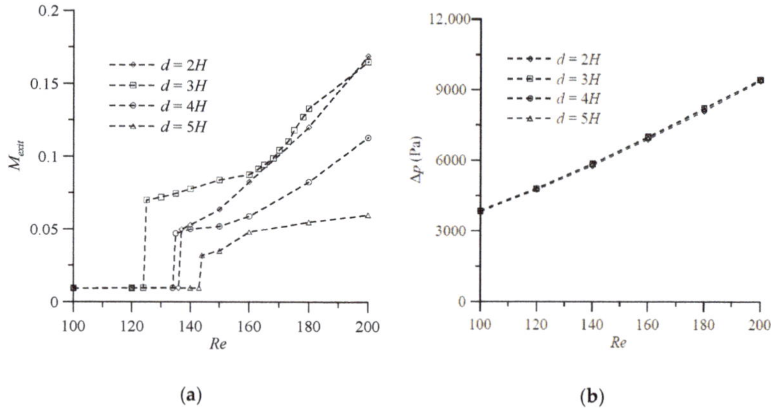

Figure 9. Effect of distance between the symmetric VIOs on (**a**) the degree of mixing and (**b**) the pressure drop in the mixers with $\theta_s = 20°$, $h = 0.75H$, $t = 0.25H$ and $w_d = 0.25H$ for various Reynolds numbers.

Similarly, to make a comparative study of influence of the distance between the antisymmetric VIOs, we choose four values of the distance, $d = 2H$, $3H$, $4H$, and $5H$, and the other geometrical parameters, $\theta_a = 30°$, $h = 0.75H$, $t = 0.25H$ and $w_d = 0.25H$, are kept constant. From the preliminary results shown in Figure 6b for fluid flow in the T-shaped micromixers with antisymmetric VIOs, we know that the distortion of the interface and the interaction of the two vortices may appear at a Re less than 100. Thus, the simulation results for fluid mixing of a T-shaped mixer with antisymmetric VIOs are examined in a wider Reynolds number range from 1 to 200. Figure 10a reveals that the degrees of mixing of the four cases considered increase gradually as the Re increases beyond 30. The enhancement of fluid mixing is caused by the early induction of transversal convection in the inlet channel, which stretches the contact surface of the fluids. The degree of mixing of the case with $d = 2H$ and that of the case with $d = 3H$ are comparable and they are larger than that of the case with $d = 4H$ or $5H$. The effect of the early induction of transversal convection by antisymmetric VIOs in the inlet channel on fluid mixing increases with the increase of Re, as shown in Figure 10a. The distance between the antisymmetric VIOs has no effect on the pressure drop between the mixer inlet and outlet, as shown in Figure 10b. This trend is similar to that observed in T-shaped micromixers with symmetric VIOs in the inlet channels.

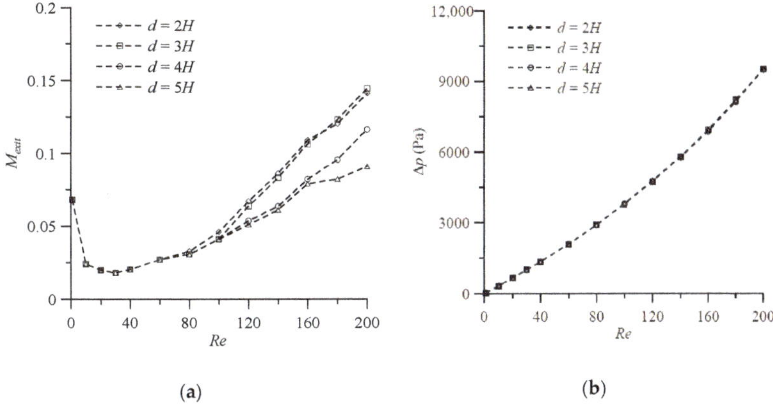

(a) (b)

Figure 10. Effect of the distance between the antisymmetric VIOs on (a) the degree of mixing, and (b) the pressure drop in the mixers with $\theta_a = 30°$, $h = 0.75H$, $t = 0.25H$ and $w_d = 0.25H$ for various Reynolds numbers.

Next, efforts are made to investigate the effect of the angle of attack (θ_s or θ_a) of the VIOs, which is illustrated in Figure 1c. The degrees of mixing at the outlet and the pressure drop between the mixer inlet and outlet for fluid flow in the T-shaped micromixers with symmetric VIOs at $\theta_s = 10°$, $20°$, $30°$ and $40°$ in the Reynolds number range from 100 to 200 are calculated, when the other geometrical parameters, $d = 3H$, $h = 0.75H$, $t = 0.25H$ and $w_d = 0.25H$, are kept constant. The degrees of mixing of the cases with $\theta_s = 20°$ show the smallest value of Re corresponding to the transition from the vortex flow regime to the engulfment flow regime, as shown in Figure 11a. In the engulfment flow regime, the degree of mixing shows a maximum at $\theta_s = 20°$, and then decreases with a further increase or decrease in the θ_s of the VIOs. When W_i, t and w_d are fixed, the length of the VIOs decreases with the increase of θ_s [16]. Thus, the pressure drop decreases with the increase of θ_s, as shown in Figure 11b. In addition, Figure 11b shows that the transition from the vortex flow regime and to the engulfment flow regime has no effect on the pressure drop. To investigate the influence of θ_a, we consider a wider range of Reynolds number ($1 \leq Re \leq 200$) and the value of θ_a varies from 10° to 60°, with an interval of 10°. The simulation results for fluid mixing in T-shaped mixers with antisymmetric VIOs are examined for cases with the six values of θ_a, $d = 3H$, $h = 0.75H$, $t = 0.25H$ and $w_d = 0.25H$. Figure 12a shows that the degrees of mixing of the cases considered increase gradually as the Re increases beyond 30, and the effect of θ_a on fluid mixing becomes noticeable as the Re increases beyond 100. When $Re > 100$, one can see that the degree of mixing increases with the increase of θ_a for the low value of θ_a, shows a maximum at $\theta_a = 30°$ and then decreases with further increase of θ_a. The dependence of Δp on θ_a is similar to that of Δp on θ_s, as shown in Figure 12b. Besides, the comparison of Figures 9b, 10b, 11b and 12b reveals that the effect of θ_s or θ_a on Δp is greater than that of the distance between the antisymmetric VIOs on Δp.

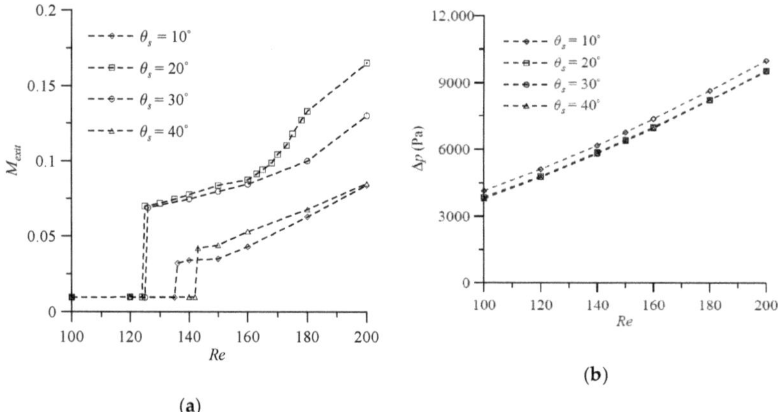

Figure 11. Effect of angle of attack θ_s on (**a**) the degree of mixing, and (**b**) the pressure drop in the mixers with $d = 3H$, $h = 0.75H$, $t = 0.25H$ and $w_d = 0.25H$ for various Reynolds numbers.

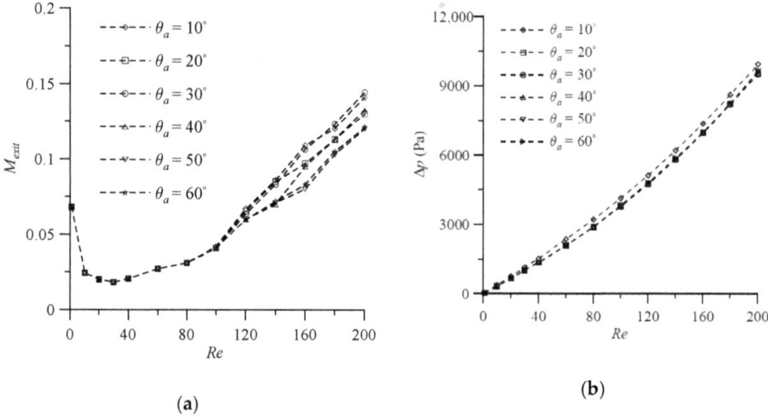

Figure 12. Effect of angle of attack θ_a on (**a**) the degree of mixing, and (**b**) the pressure drop in the mixers with $d = 3H$, $h = 0.75H$, $t = 0.25H$ and $w_d = 0.25H$ for various Reynolds numbers.

As shown in the above study, the dimensions $d = 3H$ and $\theta_s = 20°$ for symmetric VIOs, and $d = 3H$ and $\theta_a = 30°$ for antisymmetric VIOs, are found to be optimum values for mixing performance. Further increase or decrease of these parameters negatively affected the degree of mixing. Figure 13a depicts the variation of the degrees of mixing of the proposed mixers predicted by using these parameter values and that of a standard planar T-shaped mixer over a wide range of *Re*. From Figure 13a, the following features may be noted. First, for the cases with low Reynolds numbers ($Re < 27$), the effects of vortices are not strong enough to enhance mixing and the mixing relies mainly on diffusion in the three T-shaped mixers. For such cases, increasing the value of *Re* reduces the residence time and the mixing efficiency. Second, for the cases with $Re \geq 27$, the effects of the stretched contact surface between different fluids caused by antisymmetric VIOs gradually increases with the increase of *Re*, and so the enhancement of mixing by antisymmetric VIOs also gradually increases with the increase of *Re*. Meanwhile, fluid mixing in the other two T-shaped mixers still relies on diffusion and the mixing efficiency decreases with the increase of *Re* until, at the onset of the engulfment flow, the degree of mixing increases suddenly, which happens at $Re \approx 125$ in the T-shaped mixer with symmetric VIOs and at $Re \approx 140$ in the standard planar T-shaped mixer. Third, with the onset of the engulfment, the mixing efficiency in the T-shaped mixer with symmetric VIOs improved but is not better than that in the T-shaped mixer with antisymmetric VIOs for *Re* in the range 27–175. In addition, the increases of pressure drop caused by both types of VIOs mounted on the bottom of the inlet channels were almost equal and the effect increased with the increase of *Re*, as shown in Figure 13b.

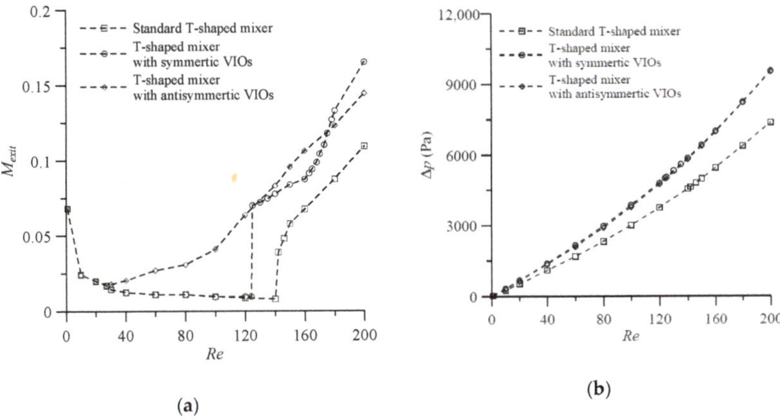

Figure 13. (**a**) Degree of mixing at the exit, and (**b**) pressure drop versus Reynolds number for the proposed mixers with VIOs and a standard planar T-shaped mixer.

4. Conclusions

A new approach for inlet flow modification by adding obstacles protruding into the flow at an angle of attack in the inlet channels of a T-shaped micromixer was proposed and investigated in this work. The dimensions $d = 3H$ and $\theta_s = 20°$ for symmetric VIOs, and $d = 3H$ and $\theta_a = 30°$ for antisymmetric VIOs, were found to be optimum values for mixing performance. Comparisons of the mixing performance of the proposed mixers with the selected values of geometric parameters of VIOs and that of a standard T-shaped mixer without VIOs showed that the antisymmetric VIOs in the inlet channels induced the stretched contact surface between different fluids. The enhancement of mixing by the VIOs gradually increased with the increase of *Re* for the cases with $Re \geq 27$, fluid mixing in the other two T-shaped mixers still relied on diffusion and the mixing efficiency decreased with the increase of *Re* until the onset of the engulfment flow, which happened at $Re \approx 125$ in the T-shaped mixer with symmetric VIOs and at $Re \approx 140$ in the standard planar T-shaped mixer. With the onset of the engulfment, the degree of mixing increased suddenly and so the mixing efficiency improved. For $Re > 175$, the mixing efficiency of the T-shaped mixer with symmetric VIOs was even better than that in the T-shaped mixer with antisymmetric VIOs.

In summary, the early initiation of vortices in the inlet streams of a T-shaped micromixer by either symmetric or antisymmetric VIOs in the inlet channels can lead to improvements in mixing efficiency. The simple but effective geometrical modification can be combined with various geometric modifications of the mixing channel to enhance fluid mixing, and such hybrid micromixers will be considered in future work.

Author Contributions: B.-H.L. performed numerical simulations, analysis of the results, and drafting the manuscript. C.-Y.W. revised the manuscript and gave final authorization of the version to be submitted. All authors have read and agreed to the published version of the manuscript.

Funding: This work was supported by the Ministry of Science and Technology of the Republic of China in Taiwan (grant number: NSC 101-2221-E-006-108-MY3).

Conflicts of Interest: The authors declare there is no conflict of interest.

References

1. Nguyen, N.T.; Wu, Z. Micromixers—A review. *J. Micromech. Microeng.* **2005**, *15*, R1–R16. [CrossRef]
2. Lee, C.-Y.; Wang, W.-T.; Liu, C.-C.; Fu, L.-M. Passive mixers in microfluidic systems: A review. *Chem. Eng. J.* **2016**, *288*, 146–160. [CrossRef]
3. Raza, W.; Hossain, S.; Kim, K.Y. A review of passive micromixers with a comparative analysis. *Micromachines* **2020**, *11*, 455. [CrossRef] [PubMed]
4. Gobby, D.; Angeli, P.; Gavriilidis, A. Mixing characteristics of T-type microfluidic mixers. *J. Micromech. Microeng.* **2001**, *11*, 126–132. [CrossRef]
5. Engler, M.; Kockmann, N.; Kiefer, T.; Woias, P. Numerical and experimental investigations on liquid mixing in static micromixers. *Chem. Eng. J.* **2004**, *101*, 315–322. [CrossRef]
6. Hoffmann, M.; Schlüter, M.; Räbiger, N. Experimental investigation of liquid-liquid mixing in T-shaped micro-mixers using μ-LIF and μ-PIV. *Chem. Eng. Sci.* **2006**, *61*, 2968–2976. [CrossRef]
7. Hussong, J.; Lindken, R.; Pourquie, M.; Westerweel, J. Numerical study on the flow physics of a T-shaped micro mixer. In *IUTAM Symposium on Advances in Micro- and Nanofluidics*; Ellero, M., Hu, X., Fröhlich, J., Adams, N., Eds.; IUTAM Bookseries; Springer: Dordrecht, The Netherlands, 2009; Volume 15, pp. 191–205.
8. Soleymani, A.; Yousefi, H.; Turunen, I. Dimensionless number for identification of flow patterns inside a T-micromixer. *Chem. Eng. Sci.* **2008**, *63*, 5291–5297. [CrossRef]
9. Fani, A.; Camarri, S.; Salvetti, M.V. Investigation of the steady engulfment regime in a three-dimensional T mixer. *Phys. Fluids* **2013**, *25*, 064102. [CrossRef]
10. Andreussi, T.; Galletti, C.; Mauri, R.; Camarri, S.; Salvetti, M.V. Flow regimes in T-shaped micro-mixers. *Comput. Chem. Eng.* **2015**, *76*, 150–159. [CrossRef]
11. Galletti, C.; Roudgar, M.; Brunazzi, E.; Mauri, R. Effect of inlet conditions on the engulfment pattern in a T-shaped micro-mixer. *Chem. Eng. J.* **2012**, *185–186*, 300–313. [CrossRef]
12. Schikarski, T.; Trzenschiok, H.; Peukert, W.; Avila, M. Inflow boundary conditions determine T-mixer efficiency. *React. Chem. Eng.* **2019**, *4*, 559–568. [CrossRef]
13. Karvelas, E.; Liosis, C.; Benos, L.; Karakasidis, T.; Sarris, I. Micromixing efficiency of particles in heavy metal removal processes under various inlet conditions. *Water* **2019**, *11*, 1135. [CrossRef]
14. Camarri, S.; Mariotti, A.; Galletti, C.; Brunazzi, E.; Mauri, R.; Salvetti, M.V. An overview of flow features and mixing in micro t and arrow mixers. *Ind. Eng. Chem. Res.* **2020**, *59*, 3669–3686. [CrossRef]
15. Lin, Y.C.; Chung, Y.C.; Wu, C.Y. Mixing enhancement of the passive microfluidic mixer with J-shaped baffles in the tee channel. *Biomed. Microdevices* **2007**, *9*, 215–221. [CrossRef]
16. Hsiao, K.-Y.; Wu, C.-Y.; Huang, Y.-T. Fluid mixing in a microchannel with longitudinal vortex generators. *Chem. Eng. J.* **2014**, *235*, 27–36. [CrossRef]
17. Vijayendran, R.A.; Motsegood, K.M.; Beebe, D.J.; Leckband, D.E. Evaluation of a three-dimensional micromixer in a surface-based biosensor. *Langmuir* **2003**, *19*, 1824–1828. [CrossRef]
18. Jiang, F.; Drese, K.S.; Hardt, S.; Küpper, M.; Schönfeld, F. Helical flows and chaotic mixing in curved micro channels. *AIChE J.* **2004**, *50*, 2297–2305. [CrossRef]
19. Gigras, A.; Pushpavanam, S. Early induction of secondary vortices for micromixing enhancement. *Microfluid. Nanofluidics* **2008**, *5*, 89–99. [CrossRef]
20. Wu, C.-Y.; Tsai, R.-T. Fluid mixing via multidirectional vortices in converging-diverging meandering microchannels with semi-elliptical side walls. *Chem. Eng. J.* **2013**, *217*, 320–328. [CrossRef]
21. Johnson, T.J.; Ross, D.; Locascio, L.E. Rapid microfluidic mixing. *Anal. Chem.* **2002**, *74*, 45–51. [CrossRef]
22. Stroock, A.D.; Dertinger, S.K.W.; Ajdari, A.; Mezic, I.; Stone, H.A.; Whitesides, G.M. Chaotic mixer for microchannels. *Science* **2002**, *295*, 647–651. [CrossRef]
23. Branebjerg, J.; Gravesen, P.; Krog, J.P.; Nielsen, C.R. Fast mixing by lamination. In Proceedings of the Ninth International Workshop on Micro Electromechanical Systems, San Diego, CA, USA, 11–15 February 1996; pp. 441–446.
24. Schönfeld, F.; Hessel, V.; Hofmann, C. An optimised split-and-recombine micro-mixer with uniform 'chaotic' mixing. *Lab Chip* **2004**, *4*, 65–69. [CrossRef] [PubMed]
25. Hossain, S.; Kim, K.Y. Mixing analysis of passive micromixer with unbalanced three-split rhombic sub-channels. *Micromachines* **2014**, *5*, 913–928. [CrossRef]

26. Xia, H.M.; Wan, S.Y.M.; Shu, C.; Chew, Y.T. Chaotic micromixers using two-layer crossing channels to exhibit fast mixing at low Reynolds numbers. *Lab Chip* **2005**, *5*, 748–755. [CrossRef] [PubMed]
27. Lee, J.; Kwon, S. Mixing efficiency of a multilamination micromixer with consecutive recirculation zones. *Chem. Eng. Sci.* **2009**, *64*, 1223–1231. [CrossRef]
28. Chen, J.J.; Lai, Y.R.; Tsai, R.T.; Lin, J.D.; Wu, C.-Y. Crosswise ridge micromixers with split and recombination helical flows. *Chem. Eng. Sci.* **2011**, *66*, 2164–2176. [CrossRef]
29. Raza, W.; Kim, K.Y. Asymmetrical split-and-recombine micromixer with baffles. *Micromachines* **2019**, *10*, 844. [CrossRef]
30. Kockmann, N.; Engler, M.; Föll, C.; Woias, P. Liquid mixing in static micro-mixers with various cross sections. In Proceedings of the First International Conference on Microchannels and Minichannels, Rochester, NY, USA, 24–25 April 2003.
31. Clark, J.; Kaufman, M.; Fodor, P.S. Mixing Enhancement in Serpentine Micromixers with a Non-Rectangular Cross-Section. *Micromachines* **2018**, *9*, 107. [CrossRef]
32. Sultan, M.A.; Fonte, C.P.; Dias, M.M.; Lopes, J.C.B.; Santos, R.J. Experimental study of flow regime and mixing in T-jets mixers. *Chem. Eng. Sci.* **2012**, *73*, 388–399. [CrossRef]
33. Sultan, M.A.; Pardilho, S.L.; Brito, M.S.C.A.; Fonte, C.P.; Dias, M.M.; Lopes, J.C.B.; Santos, R.J. 3D mixing dynamics in t-jet mixers. *Chem. Eng. Technol.* **2019**, *42*, 119–128. [CrossRef]
34. Cortes-Quiroz, C.A.; Azarbadegan, A.; Zangeneh, M. Characterization and optimization of a three dimensional T-type micromixer for convective mixing enhancement with reduced pressure loss. In Proceedings of the ASME 2010 8th International Conference on Nanochannels, Microchannels and Minichannels, Montreal, QB, Canada, 1–5 August 2010; pp. 1357–1364, Paper No. FEDSM-ICNMM2010-31257.
35. Ansari, M.A.; Kim, K.-Y.; Anwar, K.; Kim, S.M. Vortex micro T-mixer with non-aligned inputs. *Chem. Eng. J.* **2012**, *181–182*, 846–850. [CrossRef]
36. Matsunaga, T.; Nishino, K. Swirl-inducing inlet for passive micromixers. *RSC Adv.* **2014**, *4*, 824–829. [CrossRef]
37. Chen, C.; Zhao, Y.; Wang, J.; Zhu, P.; Tian, Y.; Xu, M.; Wang, L.; Huang, X. Passive mixing inside microdroplets. *Micromachines* **2018**, *9*, 160. [CrossRef]
38. Ou, J.; Moss, G.; Rothstein, J. Enhanced mixing in laminar flows using ultrahydrophobic surfaces. *Phys. Rev. E.* **2007**, *76*, 016304. [CrossRef] [PubMed]
39. Liosis, C.; Karvelas, E.G.; Karakasidis, T.; Sarris, I.E. Numerical study of magnetic particles mixing in waste water under an external magnetic field. *J. Water Supply Res. T.* **2020**, *69*, 266–275. [CrossRef]
40. Shah, R.K.; London, A.L. *Laminar Flow Forced Convection in Ducts: A Source Book for Compact Heat Exchanger Analytical Data*; Academic Press: New York, NY, USA, 1978; pp. 196–222.
41. Rani, S.A.; Pitts, B.; Steward, P.S. Rapid diffusion of fluorescent tracers into staphylococcus epidermidis biofilms visualized by time lapse microscopy. *Antimicrob. Agents Chemother.* **2005**, *49*, 728–732. [CrossRef]
42. Hoffman, J.D. *Numerical Methods for Engineers and Scientists*; McGraw-Hill: New York, NY, USA, 1993; pp. 504–514.
43. Matsunaga, T.; Lee, H.-J.; Nishino, K. An approach for accurate simulation of liquid mixing in a T-shaped micromixer. *Lab Chip* **2013**, *13*, 1515–1521. [CrossRef]
44. Kuo, M.-Y.; Wu, C.-Y.; Hsu, K.-C.; Chang, C.-Y.; Jiang, W. Numerical investigation of high-Peclet-number mixing in periodically-curved microchannel with strong curvature. *Heat Transf. Eng.* **2019**, *40*, 1736–1749. [CrossRef]
45. Szymczak, P.; Ladd, A.J.C. Boundary conditions for stochastic solutions of the convection-diffusion equation. *Phys. Rev. E* **2003**, *68*, 036704. [CrossRef]

46. Bothe, D. Evaluating the quality of a mixture: Degree of homogeneity and scale of segregation. In *Micro and Macro Mixing: Analysis, Simulation and Numerical Calculation*; Bockhorn, H., Mewes, D., Peukert, W., Warnecke, H.J., Eds.; Springer: Berlin, Germany, 2010; pp. 17–35.

Publisher's Note: MDPI stays neutral with regard to jurisdictional claims in published maps and institutional affiliations.

Article

Enhancement of Fluid Mixing with U-Shaped Channels on a Rotating Disc

Chi-Wei Hsu, Po-Tin Shih and Jerry M. Chen *

Department of Mechanical Engineering, National Chung Hsing University, Taichung 402, Taiwan; ty2jac@yahoo.com.tw (C.-W.H.); quajzmshdg@gmail.com (P.-T.S.)
* Correspondence: jerryc@dragon.nchu.edu.tw

Received: 3 November 2020; Accepted: 13 December 2020; Published: 15 December 2020

Abstract: In this study, centrifugal microfluidics with a simple geometry of U-shaped structure was designed, fabricated and analyzed to attain rapid and efficient fluid mixing. Visualization experiments together with numerical simulations were carried out to investigate the mixing behavior for the microfluidics with single, double and triple U-shaped structures, where each of the U-structures consisted of four consecutive 90° bends. It is found that the U-shaped structure markedly enhances mixing by transverse secondary flow that is originated from the Coriolis-induced vortices and further intensified by the Dean force generated as the stream turns along the 90° bends. The secondary flow becomes stronger with increasing rotational speed and with more U-shaped structures, hence higher mixing performance. The mixing efficiency measured for the three types of mixers shows a sharp increase with increasing rotational speed in the lower range. As the rotational speed further increases, nearly complete mixing can be achieved at 600 rpm for the triple-U mixer and at 720 rpm for the double-U mixer, while a maximum efficiency level of 83–86% is reached for the single-U mixer. The simulation results that reveal detailed characteristics of the flow and concentration fields are in good agreement with the experiments.

Keywords: centrifugal microfluidics; micromixer; U-shaped channel; Coriolis force; flow visualization; mixing efficiency

---

1. Introduction

Microfluidics has been widely employed in microtechnology for applications in chemical and biochemical analyses [1–3]. Among these applications, micromixing of two or more of fluids often plays a key role in the processes for the analyses [4], such as in chemical selectivity [5], reactors [6,7], extraction [8,9], DNA amplification [10] and DNA microarray [11]. The mixing mechanism in a microscale channel where flow is typically laminar, however, differs from that in a macroscale channel. Without assistance of turbulence, mixing at the microscale occurs mainly via molecular diffusion at the fluid interface. Consequently, it may take a much longer time for complete mixing in microsystems than in macrosystems.

Both active and passive methods have been proposed to enhance mixing at microscale. An extensive review on the recent development of various types and designs of active and passive mixing methods can be found in the articles of Nguyen and Wu [12], Chang and Yang [13] and Ward and Fang [14], and very recently in Raza et al. [15]. The passive method appears to be more appealing than the active one in the applications to chemical and biochemical analyses. This is mainly because in most of active micomixers, it could be more complex and difficult to implement and control the external forcing sources. On the other hand, in order to enhance mixing via secondary flow and possibly chaotic advection, most of passive micromixers are constructed in a rather complex three-dimensional (3D) geometry such as staggered herringbone, serpentine structure and double layers [16–21]. Fluid stretching, splitting, folding and recombining are observed in these complex micromixers. However, complex channel structures may hinder the fabrication of the micromixer and its integration with other components. In contrast to 3D structure, Kochmann et al. [22] reported numerical simulations and experiments of high throughput convective micromixing for 2D channels with configurations including T-shaped, U-shaped and tangential elements. They found that the micromixer with U-shaped elements having four successive 90° bends achieved the most effective micromixing mainly due to the vortices generated in the bend flow and the mixing efficiency grew with an increase of Reynolds number up to as high as $Re \geq 270$. In any case, the complex channel structure tends to raise flow resistance, which leads to increasing the residence time of mixing fluids at the cost of pressure loss. A recent comparative analysis of five types of passive micromixers by Raza et al. [15] shows that a 2D micromixer with split-and-recombination (SAR) Tesla structure outperforms other 2D serpentine and SAR mixers in the intermediate and high Reynolds number ranges ($1 < Re \leq 40$ and $40 < Re \leq 120$, respectively). Nevertheless, a 2D micromixer with Tesla structure, which utilizes the Coanda effect to generate the transverse dispersion resulting in strong cross-channel convection of the mixing, still suffers high pressure drop [15,23,24].

Different from the aforementioned passive mixing techniques with complex channel structure, mixing taking place in centrifuge-driven microchannels of rather simpler geometry promises to be an attractive method with less research [25–27]. Centrifugal microfluidics are fabricated onto a compact disc (CD)-based substrate, which only requires a simple, low power motor to spin the disc [28–30]. Pumping in such a way is relatively insensitive to the physicochemical properties of the working fluids. Fluid mixing may be performed on the CD-based microfluidics with the advantages of safety as well as low-cost, easy operation, parallel detection, and fast response. Accordingly, applications of centrifugal microfluidic devices to biomedical analysis and point-of-care diagnostics have received increasing attention. Comprehensive reviews on centrifugal microfluidic platforms for biomedical applications has been recently made by Gilmore et al. [31] and Tang et al. [32]. In centrifugal microfluidics, effective sample mixing methods that facilitate chemical or biochemical reaction and reduce the time of the assay is still highly demanded for improving the bioassay devices among with the unit operations of valving, switching, metering, and sequential loading. Notably, the fluids driven by the centrifugal force increases their flow velocity significantly with increasing rotational speed [33]. An increase of rotational speed tends to reduce the residence time of mixing fluids and may result in poor performance without a dedicated design of microfluidics. As the centrifugal microfluidics rotates fast enough, a number of studies have shown that the Coriolis force-induced transverse secondary flow significantly enhances the mixing [30,34,35]. La et al. [35], in a study of a centrifugal serpentine micromixer with relatively long circumferential channels, displayed the significant enhancement of mixing caused by the alternating transverse secondary flows at a high rotational speed of 2000 rpm. Kuo and Jiang [36] compared the CD-based centrifugal micromixers with different curved microchannels and found the square-wave microchannel to have the best mixing efficiency. They further explored the optimal design of the square-wave micromixer in a rotational speed range of 1000–5000 rpm for mixing efficiency as high as 93%. Kuo and Li [37] later studied centrifugal micromixers with three-dimensional square-wave structure for plasma mixing and found that a mixing efficiency of more than 91% could be achieved at 1000 rpm despite a notable decease in the efficiency with a further increase of rotational speed from 1200 to 1600 rpm. Such a higher mixing efficiency is due mainly to the transverse secondary flows and the stirring at the corners of the square-wave channel, which is similar to the mixing enhancement mechanism employed by La et al. [35]. Shamloo et al. [38] recently investigated the Dean flow effect in addition to the already existing Coriolis force in a numerical study of centrifugal micromixer with repeated S-shaped channel for a broad range of rotational speeds of 72–3340 rpm (7.5–350 rad/s). The mixing efficiency computed in their simulations shows a sharp fall with increasing rotational speed in the lower range until it reaches a minimum near 480 rpm. Then the efficiency grows continuously with increasing rotational speed as the transverse secondary flows dominate the diffusion mixing. A continuous growth of mixing efficiency with increasing rotational speed in the range 480–1910 rpm (50–200 rad/s) was also reported by Shamloo et al. [39] in their numerical simulations of centrifugal serpentine micromixers with square-wave structure. The aforementioned studies indicate that dependence of mixing efficiency on rotation speed can vary for centrifugal micromixers with different channel structures and arrangements. Dependence of mixing efficiency on rotational speed is critically important to the design of centrifugal micromixer and needs further investigation. Such a dependence may relate closely to combined effects on the secondary flows caused by the Coriolis and Dean forces, particularly for the microchannel with a simple geometry of 2D structure and yet to be clarified.

The present study aimed to design and fabricate centrifugal microchannels with a simple geometry of U-shaped structure to attain rapid and efficient mixing. The mixers with single, double and triple U-shaped channels fabricated on a disc substrate were investigated using the flow visualization technique. The fluids employed for mixing experiments were ferric chloride and ammonium thiocyanate solutions. The chemical reaction of the two fluids produces a blood-red color that can be employed for flow visualization and quantification of the index of mixing efficiency. Effects of rotational speed on the mixing efficiency and the corresponding flow patterns were carefully examined in the experiments together with numerical simulations. The experimental and numerical studies also examined differences in the flow and concentration fields resulting from counterclockwise and clockwise rotations.

2. Experiments

Figure 1 shows the schematic configuration of the centrifugal U-shaped micromixer and the experimental setup for flow visualization of the mixing fluids. The micromixer is composed of two semicircle reservoirs for loading fluids A and B, a T-junction for bringing the fluids into contact, and the U-shaped channel with consecutive four 90° bends. The semicircle reservoirs have a radius of 10 mm with outlets (x_c = 0) located at 30 mm from the center of the disc. A capillary valve with an expansion angle of 90° was fabricated ahead of the T-junction (x_c = 1.3 mm) to stop the flow in the hydrophilic channel before rotation of the disc [28,40]. The burst rotational speed for the present centrifugal micromixers was 360 rpm. It should be noted that the U-shaped channels (beginning from x_u = 3.0 mm) are short and straight with an arc curve at the outer corner while keeping a sharp 90° bend at the inner corner. Such a design enables to rapidly alternate the flow direction for generating strong centrifugal force locally through the consecutive 90° bends. The U-shaped microfluidic structure was manufactured using a micro-CNC machine on a polymethylmethacrylate (PMMA) substrate of 10 cm in diameter and 2 mm in thickness. In addition to the single U-shaped structure shown in Figure 1b, micromixers with double and triple U-shaped structures were fabricated for comparison. All the microfluidic channels have the same cross-section of 300 μm in width (b) and 300 μm in depth (h). The total radial channel length is 20 mm for all the three types of U-shaped micromixers. The microchannels with double and triple U-structures also begin their first 90° bend from x_u = 3.0 mm and an additional U-shaped element adds a length of 1.2 mm in the x-direction. Since the static pressure of the liquid flow in the rotating microchannels is essentially below the atmospheric pressure [33], the structured microfluidics was simply covered with transparent Scotch tape (3040-4PK, 3M, St. Paul, MN, USA) to provide good sealing as well as excellent optical access for flow visualization.

In order to acquire clear images of the mixing phenomenon in the rotating microchannel for the present experiments, the two fluid streams for mixing were ferric chloride ($FeCl_3 \cdot 6H_2O$) and ammonium thiocyanate (NH_4SCN) dissolved in deionized (DI) water at the same concentration of 0.2 mol/kg. Ferric chloride solution is pale yellow and ammonium thiocyanate solution is colorless. When these two solutions come into contact, the ferric ions bind with the thiocyanate ions instead of the chloride ions to produce the blood-red color [41]. The intensity of the red color described by the RGB values for each of pixels could represent the amount of fluids that have mixed and reacted. The normalized color intensity was quantified as the index of mixing efficiency between the two fluids.

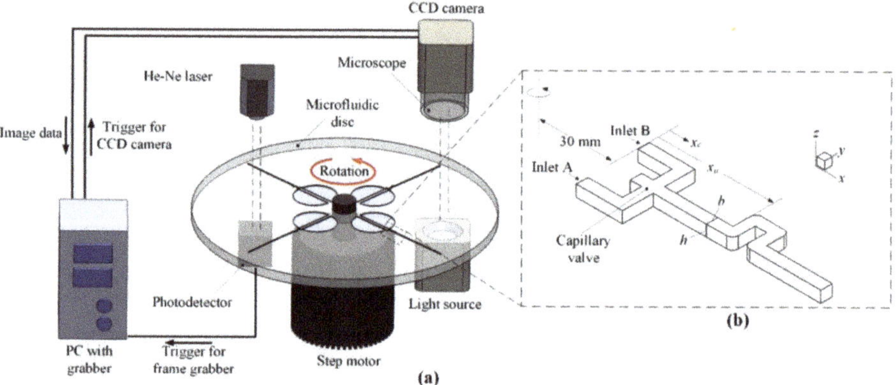

Figure 1. Schematic of the centrifugal U-shaped micromixer: (**a**) experimental arrangement for flow visualization; (**b**) the microchannel geometry and coordinates for single U-shaped structure.

A schematic diagram of the experimental setup for flow visualization of the mixing phenomenon on a rotating disc is illustrated in Figure 1. The apparatus used for the flow visualization experiments included the rotation platform, the microimage-capturing unit and the light source. The microfluidic disc was spun by a step motor having a speed accuracy of ±0.2%. The microimage-capturing unit was composed of a CCD camera (CV-M71CL, 768 × 576 pixels, 60 frames/s, JAI, Yokohama, Kanagawa, Japan), a frame grabber (Metero-II/CL, Matrox, Dorval, QC, Canada), and a microscope (2× objective) above the disc. To synchronize the image capturing unit with the rotational motion, the CCD camera was triggered through a photo diode (wavelength 320–730 nm). A He-Ne laser (wavelength 633 nm) of 5 mW was used as the triggering light source. A halogen light (MHF-G150LR, Moritex, Saitama, Japan) was placed beneath the disc to produce sufficient light for illuminating the flow field undergoing a rapid rotational motion. In synchronization with the rotational motion, the CCD camera could be triggered to allow one shot of the targeted object on the rotating disc per revolution.

3. Numerical Simulation

Numerical simulations were employed to reveal details of the flow patterns and mixing behavior for the fluid streams in the rotating microchannels as shown in Figure 1. All the channels have a total length of 20 mm in the x-direction and the same width b = 300 μm and depth h = 300 μm. Fluids A and B inlet the channels at 30 mm from the center of rotation and then contact at the T-junction (x_c = 1.3 mm) without a capillary valve. The fluids were assumed to have the same constant density as water ρ = 1000 kg/m^3 and the same viscosity μ = 0.001 kg/m^2·s. The flow field of fluids A and B in the microchannel was computed on the rotating frame. The origin of the computational coordinates is located at the center of the rotating disc. The flow field is governed by the steady, three-dimensional continuity and Navier–Stokes equations:

$$\nabla \cdot \vec{V} = 0, \tag{1}$$

$$\rho\vec{V} \cdot \nabla\vec{V} = -\nabla p + \mu\nabla^2\vec{V} - 2\rho\vec{\Omega} \times \vec{V} - \rho\vec{\Omega} \times \left(\vec{\Omega} \times \vec{r}\right), \tag{2}$$

where $\vec{V}$ is the velocity vector on the reference frame rotating at a constant angular velocity $\vec{\Omega} = \Omega\vec{e}_z$, p is the pressure, and $\vec{r}$ is the position vector. The source terms on the right-hand side of Equation (2) include sequentially the pressure force, the viscous force, the Coriolis force and the centrifugal force.

Gravitational effects are ignored. The species concentration for calculating the mixing performance is governed the convection-diffusion equation:

$$\vec{V} \cdot \nabla C = D\nabla^2 \vec{V}, \tag{3}$$

where D is the diffusion coefficient, and C is the species concentration with $C = 1$ representing only fluid A and $C = 0$ for only fluid B. As the two fluids are completely mixed, $C = 0.5$ The diffusion coefficient for the present simulations, unless otherwise specified, was taken to be at a constant value of 3.0×10^{-10} m^2/s, which was based on the study of Kochman et al. [22].

The finite-volume-based CFD software ANSYS Fluent 15.0 was used to solve Equations (1)–(3) for the two-phase (fluids A and B) flow and concentration fields. The inlet and outlet flow boundary conditions were set at the atmospheric pressure $p = 0$. On the walls of the channels, a no-slip condition was imposed for the Navier-Stokes equations and no-flux condition for the species-concentration equation. Total grids of about 5.6×10^5–6.8×10^5 were employed for the computations.

The mixing efficiency was calculated based on the concentration distribution on a designated cross-sectional plane perpendicular to the fluid stream [18]:

$$\eta_{\text{sim}} = \left(1 - \frac{1}{C_\infty}\sqrt{\frac{1}{n}\sum_{i=1}^{n}(C_i - C_\infty)^2}\right) \times 100\%, \tag{4}$$

where C_∞ is the completely mixed concentration, C_i is the concentration at a grid i, and n is the number of grids computed on the cross-sectional plane. The mixing efficiency varies from zero (no mixing at all) to unity (complete mixing) is generally a function of the streamwise position. For the present numerical simulations, the mixing efficiency was evaluated on Section 5 as indicated in Figure 1, which was located at the immediate exit of the most downstream U-shaped element ($x_c = 4.2$, 5.4 and 6.6 mm for single, double and triple U-shaped mixers, respectively).

4. Results and Discussion

4.1. Simulation Results

Figure 2 shows the top view of the single-U mixer with directions of the body forces and the simulation result of concentration for the mixer rotating at 600 rpm counterclockwise (ccw). The fluids in the rotating U-shaped channel can experience three types of body forces, namely the centrifugal force f_ω generated through the system rotation, the Coriolis force f_C, and the Dean force f_D, which is the centrifugal force locally generated through the consecutive 90° bends [36,38]. The magnitude of these three forces may be estimated as follows [42]:

$$f_\omega = \rho r \Omega^2, \tag{5}$$

$$f_C = 2\rho\Omega U, \tag{6}$$

$$f_D = \frac{\rho U^2}{R}, \tag{7}$$

where r is the distance of the fluid to the center of rotation, U is the stream velocity in the channel, and R is the radius of the stream bended in the channel. The directions of Coriolis and Dean forces vary rapidly along the U-shaped channel as indicted in Figure 2a, where the stream is driven by system's centrifugal force under a counterclockwise rotation. Notably the Coriolis and Dean forces alternate not only in direction but also in position between Sections 2 and 4. The concentration distribution (with values between 0 and 1) at the midplanes is shown in Figure 2b. At the inlets, red color represents fluid A of species concentration 1.0, and blue color fluid B of concentration 0. As the two fluids fully mix, the color turns green representing a concentration of 0.5. It can be seen that the Coriolis-induced transverse secondary flow begins to influence the interface in the straight channel region, moving fluid A toward upper and lower walls and fluid B toward sidewall A to form a "C-shaped" interface in Section 1 at $x_c = 2.7$ mm. Subsequently, the 90° bends appear to stir the fluids by flipping fluids A and B in Section 2. This can be seen first in Section 2 changing from AB to BA and flipping again in Section 3. Then in Section 4 where fluid A splits to sandwich fluid B forming ABA. In addition to flipping, interface folding also becomes more significant as the flow moves to Section 4. After flowing through these bends, the two fluids are well mixed in Section 5 ($x_c = 4.2$ mm). Similar patterns of flipping, splitting, folding and stirring were also observed in the study by La et al. [35] but with a much longer circumferential channel length of 5.2 mm (17 times the channel width) at a much higher rotational speed $\Omega = 2000$ rpm. It should be noted that there is no Dean force generated in a long circumferential channel ($R \to \infty$) [38], in which the secondary flow is essentially induced by the Coriolis force.

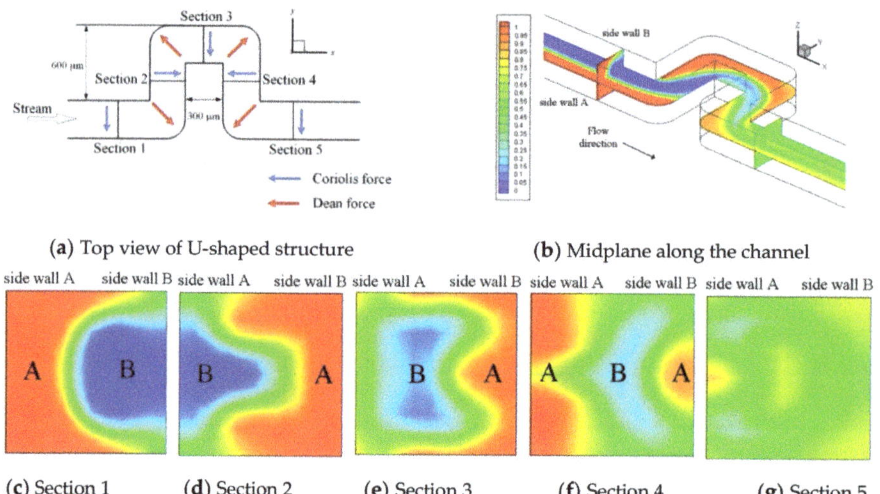

Figure 2. Top view of the U-shaped channel and concentration distribution of fluids A and B for the single-U mixer rotating at $\Omega = 600$ rpm (counter-clockwise; ccw): (**a**) top view with detailed dimensions and cross-sections selected for examination as well as the directions of Coriolis and Dean forces acting on the fluids; (**b**) distribution on mid-plane along the channel; (**c**–**g**) on cross-sections 1–5 normal to the fluid stream as indicated in (**b**).

Figures 3 and 4 show streamlines and velocity vectors, respectively, together with directions of the Coriolis and Dean forces on cross-sections 2–5 for the single U-mixer rotating at $\Omega = 360$ and 600 rpm ccw. There are at least a pair of symmetric, counter-rotating vortices induced by the Coriolis and Dean forces on all the cross-sections for both two rotational speeds. At $\Omega = 360$, these secondary vortices reverse their directions of rotation after the flow makes each of the 90° turns and then a saddle point can be found in Section 5. The reversed rotation of the secondary vortices after the 90° turns indicates effects of the Dean force, which as shown in Figure 4a alternates its direction pointing from side walls B toward A (B→A) to A→B on Sections 3 and 4. In the meantime, the Coriolis force remains unchanged pointing from side walls B toward A (B→A) through the four 90° turns. At a higher rotational speed $\Omega = 600$ rpm, one more pair of symmetric, counter-rotating vortices appear in Sections 3 and 5, indicating that the flow is significantly affected not only by the Coriolis force but also by the Dean forces in the transverse direction during the 90° turns. The significant effect of the Dean force can be seen in Figure 4b displaying larger transverse velocities on Sections 3 and 4 as compared those at the lower rotational speed of $\Omega = 360$ rpm. The strong transverse secondary flow initiated by the Coriolis force and then intensified by the Dean force through the 90° bends markedly enhances the fluid mixing as shown in Figure 2.

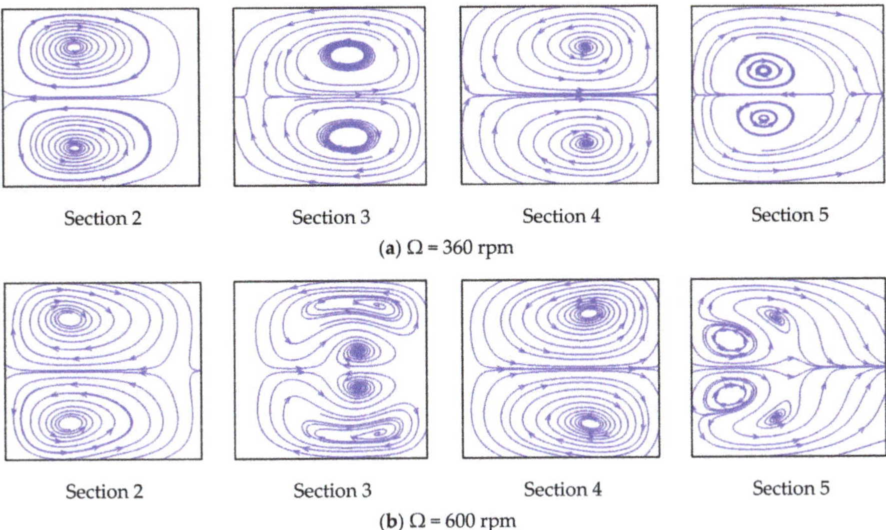

Figure 3. Two-dimensional streamlines on cross-sections 2–5 as indicated in Figure 2a for the single U-shaped mixer: (**a**) rotating at Ω = 360 rpm (ccw); (**b**) rotating at Ω = 600 rpm (ccw).

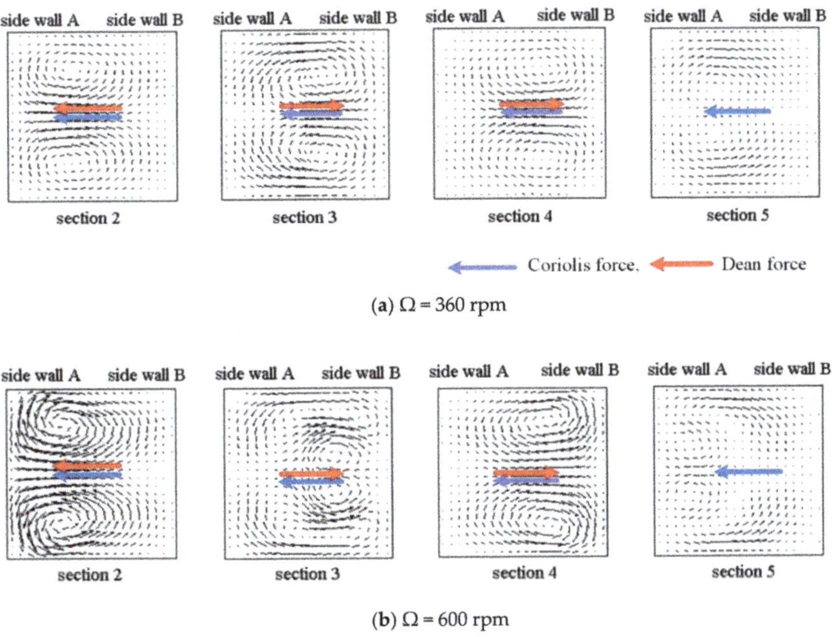

Figure 4. Velocity vector field with the directions of Coriolis and Dean forces on cross sections 2–5 as indicated in Figure 2a for the single U-shaped mixer: (**a**) rotating at Ω = 360 rpm (ccw); (**b**) rotating at Ω = 600 rpm (ccw).

The coupled Coriolis–Dean effects may be examined from the ratio γ_c of the two forces given in Equations (6) and (7) as [42]:

$$\gamma_c = \frac{f_D}{f_C} = \frac{U}{2R\Omega}, \tag{8}$$

where the stream velocity U is a function of rotational speed Ω. Then U can be taken to be the mean velocity U_m and expressed as:

$$U_m \propto \Omega^k, \tag{9}$$

with k representing the power-relation exponent. For a straight and round channel ($R \to \infty$) on a rotating disc, the power-relation exponent $k = 2$ approximately [26]. For the single-U mixer employed in the present study, $k \approx 1.84$ in the lower rotational speed range ($\Omega \leq 480$ rpm) and the exponent becomes smaller $k \approx 1.41$ in the higher range ($\Omega \geq 600$ rpm). Taking the case of $\Omega = 600$ rpm as an example, with $U_m = 0.38$ m/s and an average radius $R \approx (2^{1/2})b = 424$ µm, the ratio of the Dean force to the Coriolis force is approximated to be $\gamma_c \approx 7.1$. This means that the Dean force dominates the Coriolis force in generating the secondary flow as the stream turns along the consecutive 90° bends, and the Dean–Coriolis force ratio increases with increasing rotational speed.

In addition, it is worth looking into the effect of the system's centrifugal force on secondary flows, which play a major role in enhancing mixing of the present study, as compared to those of the Coriolis force and the Dean force. For a radial straight microchannel of rectangular cross-section, the ratio of the system's centrifugal force to the Coriolis force may be scaled as [30,34]:

$$\frac{f_\omega}{f_C} = \frac{r\Omega}{2U} = \frac{4\mu}{\rho b^2 \Omega}, \tag{10}$$

where the mean stream velocity U_m instead of the maximum stream velocity U_{max} is used to scale the Coriolis force and a characteristic scaling $f_C \propto \Omega^2$ is invoked. As already mentioned, the mean stream velocity U_m depends on Ω, and the exponent k in Equation (9) tends to decrease in the higher rotational speed range for the single-U mixer. It should be noted that in Equation (10), the Coriolis force is perpendicular to the radial direction of the system's centrifugal force. The transverse secondary flow accompanied by the streamwise vorticity in the radial straight channel is mainly due to the Coriolis force, while the system's centrifugal force contributes to drive the stream flow along the channel [43,44]. When examining the effect of the system's centrifugal force on secondary flows in the circumferential channel section (orthogonal to the radial direction), great care is required because f_ω is parallel to f_C. This effect may be understood based on vorticity consideration by rewriting Equation (2) for a constant angular velocity Ω as [43,44]:

$$\rho\vec{V}\cdot\nabla\vec{V} = -\nabla p^* + \mu\nabla^2\vec{V} - 2\rho\vec{\Omega}\times\vec{V}, \tag{11}$$

where the centrifugal force is incorporated into the modified pressure p^* given by

$$p^* = p - \frac{1}{2}\rho\Omega^2(x^2 + y^2). \tag{12}$$

The vorticity equation for incompressible flow can then be obtained by taking the curl of Equation (11) as [45,46]:

$$(\vec{V}\cdot\nabla)\vec{\omega} = (\vec{\omega}\cdot\nabla)\vec{V} + \nu\nabla^2\vec{\omega} - 2\nabla\times(\vec{\Omega}\times\vec{V}), \tag{13}$$

where $\vec{\omega} = \nabla\times\vec{V}$ is the vorticity and ν is the kinematic viscosity. The three source terms on the right-hand side of Equation (13) represent the vortex stretching, the viscous diffusion of vorticity and the Coriolis-force production of vorticity, respectively. Note that the modified pressure does not appear in the vorticity equation since the curl of the gradient of any scalar is zero. This indicates that the system's centrifugal force does not 'directly' contribute to generation of vorticity in the interior of the channel [46,47]. However, this does not mean that the system's centrifugal force plays no role in generation of streamwise vorticity in a channel section that is orthogonal to the radial direction. The system's centrifugal force, nevertheless, has important influences in helping vorticity generation in microchannel flow through change of the boundary geometry (e.g., a sudden expansion or a curved bend) and variation in pressure gradient along the channel's solid surface [47]. For the flow in the U-shaped channel structure with a very short circumferential section, the streamwise vorticities that lead to producing and intensifying the transverse secondary flow are generated mainly from stretching and turning of vortex lines through the consecutive 90° bends of the channel as well as from the Coriolis force. The streamwise vorticities generated when the flow turning along the consecutive bends are, in the present study, accounted for as the Dean–Coriolis force effect, which is indeed originated from the flow driven by the system's centrifugal force.

Effects of the coupled Dean–Coriolis forces can also be revealed through comparison of centrifugal mixing characteristics between the U-shaped mixer and a T-type mixer. Figure 5a illustrates schematic of the centrifugal T-type micromixer employed for the present simulations. The T-type micromixer with fluid inlets located at 30 mm from the center of the disc is simply a straight channel having the same cross-section ($b = h = 300$ μm), radial length (exit at $x_c = 20$ mm) and junction position ($x_j = 1.3$ mm) as the U-mixer. Figure 5b displays the cross-sectional concentration distributions and velocity vectors computed at $x_c = 4.2$, which is the same x-position as Section 5 of the U-mixer, for two rotational speeds $\Omega = 360$ ad 600 rpm (ccw). A C-shaped interface of the fluids can be clearly seen in the cross-section for both rotational speeds. In the T-mixer, the transverse secondary flow is induced essentially by the Coriolis force alone [30] since there is no Dean force in the straight channel ($R \rightarrow \infty$). It can be seen from the transverse velocity vectors that the secondary flow drives the fluids from side walls B toward A (B→A) in the middle of the cross-section neighboring $z = 0$, where the maximum radial velocity occurs, and A→B near the top and bottom walls. When comparing these two cross-sectional distributions, the C-shaped interface for the higher rotational speed of 600 rpm appears to be a bit wider and lighter near the corners on side wall B than that of 360 rpm. This indicates a slightly better mixing for the higher rotational speed due to a stronger Coriolis-induced secondary flow that compensates for the shorter residence time of the mixing fluid stream.

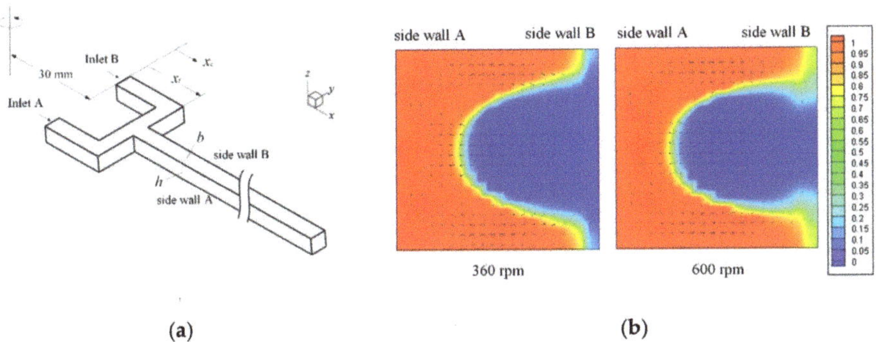

Figure 5. Centrifugal T-type micromixer and cross-sectional concentration distribution: (**a**) schematic of the T-type micromixer showing the channel geometry and coordinates; (**b**) concentration distributions and velocity vectors at $x_c = 4.2$ mm for $\Omega = 360$ and 600 rpm (ccw).

Figure 6 compares mixing efficiency of the single U-shaped mixer with the T-type mixer at different rotational speeds ranging from 120 to 1200 rpm (ccw). The mixing efficiency was calculated using Equation (4) based on the cross-sectional concentration distribution in Section 5 ($x_c = 4.2$ mm) for the U-mixer and at the same x-position ($x_c = 4.2$ mm) for the T-mixer. It can be seen that mixing efficiency for the U-mixer increases rapidly with rotational speed, from 20% at $\Omega = 120$ rpm to 82%

at Ω = 840 rpm. Beyond Ω = 840 rpm, the mixing efficiency appears to level off around 83–84%. The flat level efficiency in the higher speed range is due mainly to the high flow velocity that increases rapidly with rotational speed. The mean velocities at the U-mixer exit were found to be 0.67 m/s for Ω = 900 rpm and 1.0 m/s for Ω = 1200 rpm, corresponding to Reynolds numbers Re = 200 and 300. The rapidly increasing velocity reduces the fluid residence time and appears to offset the growing influence of transverse secondary flow. It is worth mentioning that the simulations with different diffusion coefficients (1.0×10^{-12} and 3.0×10^{-9} m^2/s) also exhibit a similar trend of mixing efficiency variation with rotational speed as in Figure 6 for the same U-shaped mixer, but give a slightly higher efficiency (by an amount of 5–6%) for the larger diffusion coefficient and slightly lower efficiency (by an amount of 4–5%) for the smaller diffusion coefficient in the level-off range Ω = 900–1200 rpm. In other words, changes in mixing efficiency contributed from the diffusion coefficient variation by an order of magnitude are limited. This indicates that advection plays the key role in enhancing fluid mixing for the present centrifugal U-shaped microchannel.

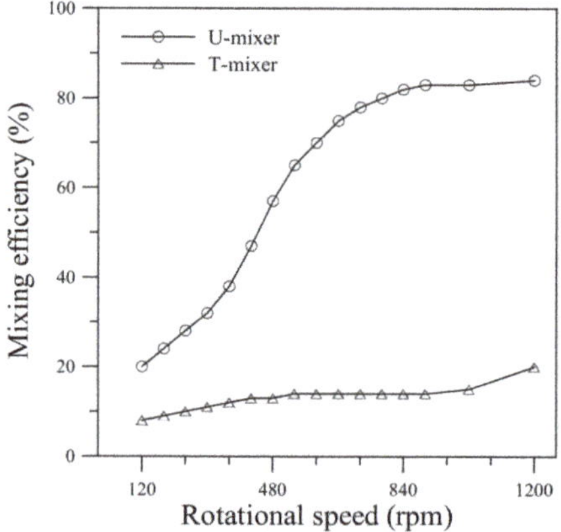

Figure 6. Comparison of mixing efficiency for the single U-shaped mixer and T-type mixer rotating at different speeds from 120 to 1200 rpm (ccw).

In Figure 6, the T-mixer also shows an overall growth in mixing efficiency with increasing rotational speed due mainly to the Coriolis-induced secondary flow. However, the fluids flowing through the straight channel of the T-mixer without U-turns have a much lower mixing efficiency (8–20%) than that observed in the U-mixer. The comparison in Figure 6 points out that Coriolis-induced vortices alone may not help mixing the fluids so much in a rotating microchannel. It is necessary to have the Dean flow generated during rapid turns of fluid stream along the consecutive bends to strengthen the transverse secondary flow for further enhancement of the mixing. The simulation results also indicate that the centrifugal U-mixer that enables effective assistance of fluid mixing within a short channel length is very suitable for use in CD-based microfluidic systems. Moreover, an additional U-shaped structure can be expected to further enhance the fluid mixing by maintaining a large value of γ_c for a longer channel length.

4.2. Experimental Results and Comparison

Figure 7 shows flow visualization of top view images for the single-U mixer rotating at different speeds from 360 to 900 rpm (ccw). The flow enters the U-shaped channel from the left. At a lower speed of 360 rpm, the colored fluid appears on the upper side of the entrance region due to the effect of Coriolis force. The colored area indicates the mixed interface where the two testing fluids contact and react. The Coriolis force, in this counterclockwise rotating case, points downward on the top view images. As a result, the colored area in the entrance region appears in the upper half of the image, reflecting the C-shaped interface observed on Section 1 of Figure 2c. The mixed area becomes wider but lighter, signifying stretching of the interface, as flow turns along the first and second 90° bends. Then along the third and fourth bends, the mixed interface clearly shows a twisting shape accompanied by the stretching phenomenon. At the exit of the U-shaped structure, the mixed area increases largely than at the entrance. As the speed increases to 480 rpm, the mixed area is enlarged in the entrance region where stretching and twisting of the interface take place slightly earlier than at the lower speed. Arriving at the fourth 90° bend, the colored lines' entanglement implies folding and twisting of the interface. Thereafter the testing fluids mix well at the exit of the U-shaped structure. As the speed further increases to 600 and 900 rpm, folding and twisting of the interface occurs even earlier. As a result, the colored area covers nearly the whole downstream of the U-structure. It can be seen that the mixing performance is significantly enhanced at the exit for Ω = 900 rpm. Note that, at the higher speeds, the colored area appears to spread from the upper wall toward the down side but with lighter color. This is a result of large ratio of Dean–Coriolis forces that tends to widen the interface but the faster flow velocity, particularly in the neighborhood of the channel centerline, allows less time for the fluid interface to diffuse.

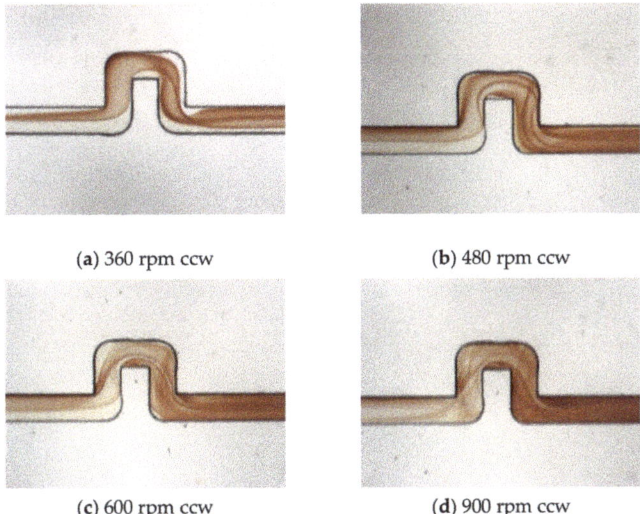

(a) 360 rpm ccw **(b)** 480 rpm ccw **(c)** 600 rpm ccw **(d)** 900 rpm ccw

Figure 7. (**a**–**d**) Top view visualization of the mixed fluids for the single U-shaped mixer rotating from Ω = 360 to 900 rpm (ccw), where the flow enters from the left.

Under the effects of the Coriolis force, the present U-shaped mixer is not fully symmetric between the clockwise and counterclockwise rotations. Figure 8 shows flow visualization of top view images for the same U-mixer of Figure 7 but rotating clockwise (cw) at speeds from 360 to 900 rpm. The Coriolis force now points upward on the top view images for the flow in the x-direction. Therefore the colored interface develops on the lower side of the image in the entrance region and becomes wider but lighter as the rotational speed increases. At lower speeds Ω = 360 and 480 rpm, it is observed that the interface turns around the second and third 90° bends without touching the upper wall of the radial channel between Sections 2 and 4 because of the upward Coriolis force, unlike those shown in Figure 7a,b for the counterclockwise rotation. At higher speeds Ω = 600 and 900 rpm, the interface broadens and covers most of the channel after Section 2 due mainly to the stronger Dean force produced as the stream with a higher velocity turns along the 90° bends. It is also observed that at these higher velocities, stretching, folding and twisting of the fluid interface appear to display a lot more entangled lines and surfaces. As a result, the mixing at the exit becomes much better than at the lower speeds.

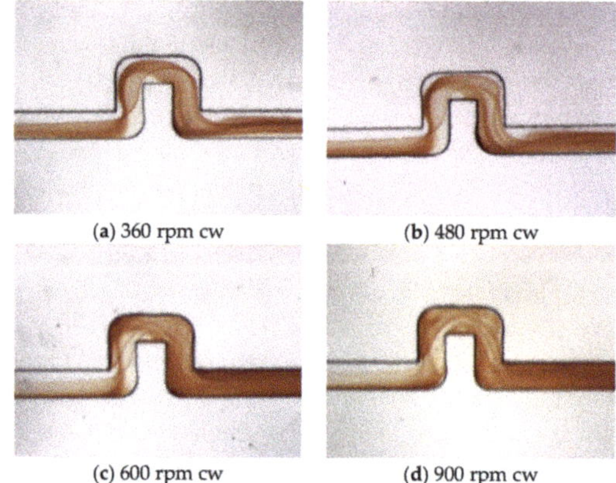

Figure 8. (**a**–**d**) Top view visualization of the mixed fluids for the single U-shaped mixer rotating from Ω = 360 to 900 rpm (clockwise; cw), where the flow enters from the left.

Figure 9 shows the top view images of numerical simulations displaying the fluid mixing phenomenon for Ω = 360 and 600 rpm undergoing both counterclockwise and clockwise rotations. Each of these simulation images are composed of the species concentration distributions on the x-y plane in the upper half of the channel (from $z = 0$ to 150 μm). The concentration distributions are shown here in gray scale with black representing the complete mixing (C = 0.5) and white for unmixed fluids (C = 0 or 1.0). The composite distribution images were made for comparison with the visualization images of Figures 7 and 8, which accumulated the entire color changes through the depth of the channel. For both lower and higher rotational speeds, the simulation images clearly demonstrate the interface shape that is affected by the Coriolis force and the Dean force as the stream turns along the 90° bends. The Coriolis effects caused by counterclockwise rotation can be distinct from those by clockwise rotation. For the higher rotational speed Ω = 600 rpm, the twisted and overlapped lines are largely increased to exhibit a darker surface, particularly near the exit of the U-shaped structure, indicating a higher mixing performance. The simulation images of Figure 9 are consistent with and closely resemble those observed in the experiments having the same speeds and directions of rotation.

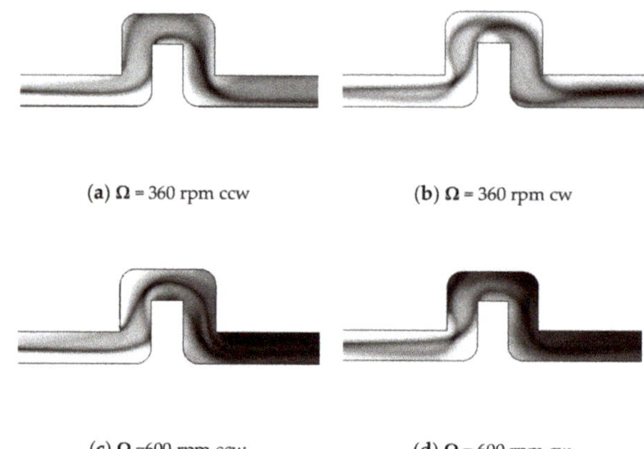

Figure 9. (**a**–**d**) Top view images of fluid mixing obtained from numerical simulations for the single U-shaped mixer at Ω = 360 and 600 rpm undergoing both counterclockwise and clockwise rotations, where the flow enters from the left.

Figure 10 shows flow visualization of the double-U mixer at rotational speeds of 360 and 600 rpm. Both clockwise and counterclockwise rotations are presented for comparison. At the lower speed Ω = 360 rpm, with one more U-structure, it can be seen that the colored lines and surfaces representing the mixed fluid interface become more stretching and twisting in the second U-structure. For the higher speed Ω = 600 rpm, the colored lines and surfaces are largely entangled. For both lower and higher rotational speeds, the mixed area at the exit of the second U-shaped structure becomes darker than that observed in the single U-shaped structure. The second U-structure apparently further enhances the mixing through the intensified transverse secondary flow caused by the Coriolis force and the Dean force in particular during the four more 90° turns. In the clockwise rotation cases, the Coriolis force that points upward causes the mixed interface to move toward the lower half of the entrance channel. The stretching, twisting and folding phenomenon of the interface appears to be similar to that in the clockwise single-U mixer but larger in strength. The mixed area at the exit of the second U-structure for the higher rotational speed Ω = 600 rpm is observed to be darker than that for the single U-shaped channel.

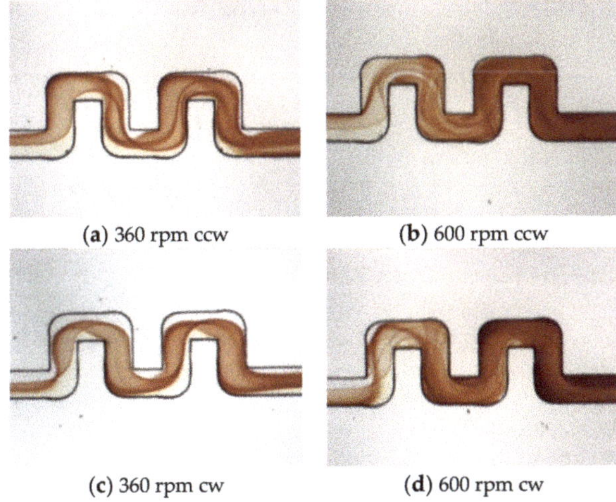

Figure 10. (**a–d**) Top view visualization of the mixed fluids for the double-U mixer undergoing counterclockwise (ccw) and clockwise (cw) rotations at 360 and 600 rpm, where the flow enters from the left.

Figure 11 presents the simulation results for the variations of mixing efficiency with the downstream channel distance beginning from the T-junction for the double-U mixer at Ω = 360, 600 and 900 rpm (ccw). At the lower speed Ω = 360 rpm, the mixing efficiency increases significantly at Sections 2–4 of the first U-structure and is further enhanced in at Sections 2–4 of the second U-structure. There is a small decline between Sections 4 and 5 of the second U-structure, which may be due to the fact that the rapidly changing flow directions are not perpendicular to the cross-sections during the turns, causing the concentration collected on the cross-section to have a small variation with the flow angle. At the higher speeds Ω = 600 and 900 rpm, the mixing efficiency dramatically grows at Sections 2–5 of the first U-structure. Then the efficiency still maintains a significant growth but with a less steep slope in the second U-structure. The dramatic growth followed by a significant rise in mixing efficiency is due largely to the strong Dean flow caused by the high Dean-Coriolis-force ratio ($\gamma_c \approx 6.2$ and 7.4 for Ω = 600 and 900, respectively).

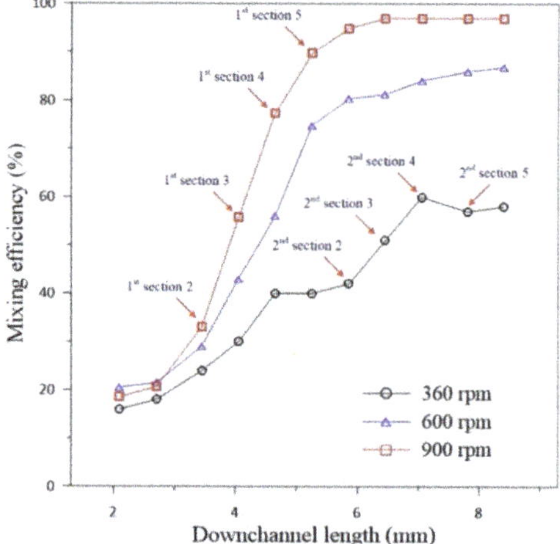

Figure 11. Variations of simulation mixing efficiency in downstream channel distance for the double-U mixer undergoing counterclockwise rotations at 360, 600 and 900 rpm with indication of downstream positions for the cross-sections of the first and second U-structures.

Figure 12 shows top-view mixing images of flow visualization and numerical simulations of the triple-U mixer at a rotational speed of 360 rpm for counterclockwise and clockwise rotations. For the flow visualization images, only flow in the second and third U-structures are displayed because the visualization was designated to target the exit of the third U-structure for quantifying mixing efficiency. It can be observed that colored surfaces of the mixed-fluid interface in the second and third U-structures are even more stretching and twisting than those in double-U mixer at the same speed, indicating more stirring. It can also be seen from the visualization images that the counterclockwise rotation mixing appears to better the clockwise case. This could be due to the combined effects of Dean and Coriolis forces, which are toward more opposite directions between Sections 2 and 4 for the counterclockwise case, resulting in larger stirring of the mixing fluids than the clockwise case. The simulation images, displaying the flow along the whole three U-structures, clearly demonstrate the difference in interface shape between the counterclockwise and clockwise rotation cases as those shown in the flow visualization. The twisted and overlapped interface representing the mixed fluids become more entangled and darker in the second and third U-structures, indicating that mixing enhancement can be achieved with more U-structures.

Figure 13 compares the mixing efficiencies measured from the visualization experiments with those obtained from the numerical simulations for the single-, double- and triple-U mixers undergoing counterclockwise rotation at different speeds. A similar comparison for the three mixers undergoing clockwise rotation is presented in Figure 14. Quantification of the mixing performance in the present experiments is based on the concentrations measured in the imaged area located at the exit of the U-shaped structure. The imaged area was approximately 450 μm × 400 μm with its left side located at x_c = 4.2, 5.4 and 6.6 mm for the single-, double- and triple-U mixer, respectively. The mixing efficiency computed from the pixel intensity of averaged RGB values in the imaged area is given by [16]:

$$\eta_{exp} = \frac{I - I_{min}}{I_{max} - I_{min}} \times 100\%. \quad (14)$$

The maximum red intensity I_{max} is observed in a fully mixed image and the minimum intensity I_{min} is observed in an image of DI water. The technique of quantifying mixing efficiency from concentration measurements was employed previously by Chen et al. [48]. For all the three U-mixers, the mixing efficiency measured from the visualization experiments increases sharply with increasing rotational speed in the lower range, Ω = 360 to 600 rpm for the single- and double-U mixers and an even lower range of Ω = 360 to 480 rpm for the triple-U mixer. As the speed further increases to 720 and 900 rpm, the efficiency gradually levels off for the single-U mixer. Nearly complete mixing is attained at Ω = 720 rpm for the double-U mixer and at an even lower speed of Ω = 600 rpm for the triple-U mixer. This increasing trend is observed in both counterclockwise and clockwise cases and is well predicted by the simulations with a constant diffusion coefficient of 3×10^{-10} m^2/s.

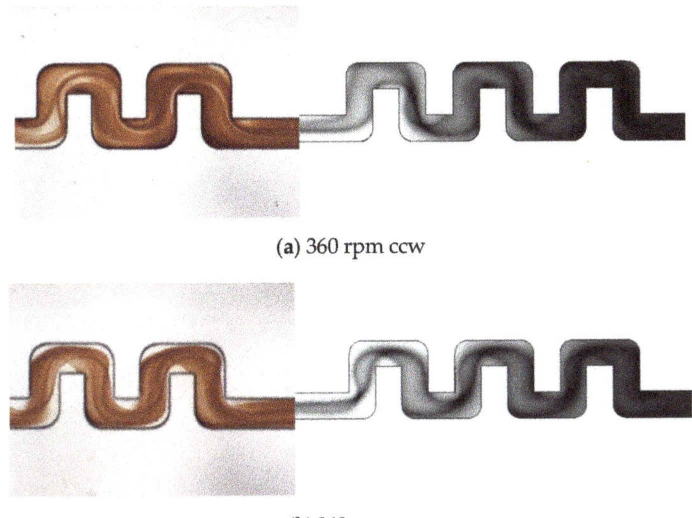

(**a**) 360 rpm ccw

(**b**) 360 rpm cw

Figure 12. Top view images of fluid mixing from visualization experiments (**left**, where only the second and third U-structures are displayed) and numerical simulations (**right**) for the triple-U mixer at Ω = 360 rpm: (**a**) counterclockwise rotation; (**b**) clockwise rotation. The flow enters from the left.

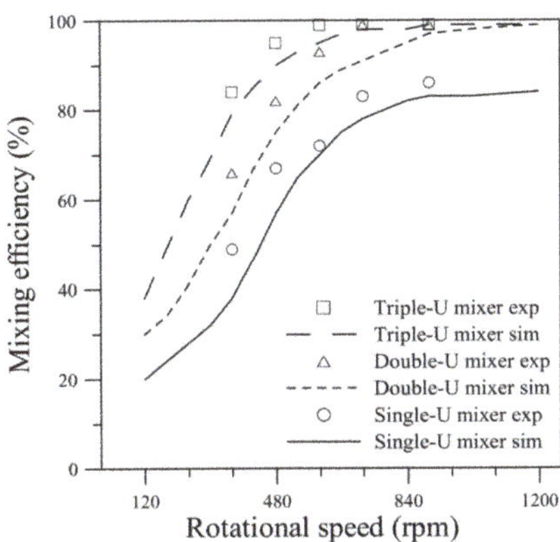

Figure 13. Comparison of mixing efficiency between experiments and simulations for the single-, double- and triple-U mixers undergoing counterclockwise rotation in the range 120–1200 rpm (360–900 rpm for experiments).

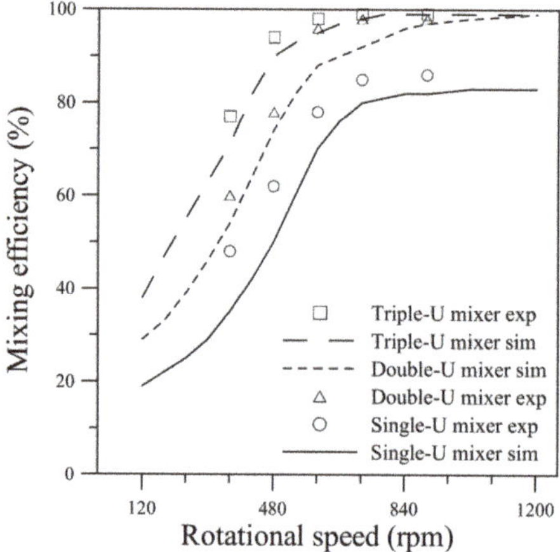

Figure 14. Comparison of mixing efficiency between experiments and simulations for the single-, double- and triple-U mixers undergoing clockwise rotation in the range 120–1200 rpm (360–900 rpm for experiments).

For the single-U mixer undergoing counterclockwise rotation, the measured mixing efficiency increases from 48% at Ω = 360 rpm to 72% at Ω = 600 rpm and reaches 86% at Ω = 900 rpm. The clockwise rotation case is slightly lower than the counterclockwise rotation by about 5% at Ω = 480 rpm. But at Ω = 600 rpm, the clockwise rotation jumps to lead the counterclockwise rotation by approximately 6% and then approaches a maximum of 86% as the speed further increases. For the double-U mixer, the measured mixing efficiency is apparently higher than that for the single-U mixer by an amount of 10–20%. The efficiency increases from 60–65% at Ω = 360 rpm to reach nearly 99% at Ω = 720 and 900 rpm for both clockwise and counterclockwise cases. This is consistent with the observation of Figure 10. An additional U-shaped structure further enhances the mixing through the intensified transverse secondary flow caused by the Coriolis and Dean forces with four more 90° turns of the stream. In the meantime, the mean velocity at the exit of the double-U mixer is found to be slower than that of the single-U mixer by 13–17%, which leads to an increase of fluid residence time giving help to the mixing. The mean velocity was estimated from the visualization images beginning from the burst of flow until the fluid reservoirs become empty. It was found that the mean velocity reduced from 0.16 m/s for the single-U mixer to 0.14 m/s for the double-U mixer at the burst rotational speed Ω = 360 rpm, and from 0.74 to 0.64 m/s at the highest speed Ω = 900 rpm. For the triple-U mixer under counterclockwise rotation, the measured mixing efficiency was found to be higher than 80% even at a lower speed Ω = 360 rpm, subsequently grows to 95% at Ω = 480 rpm and then reaches a nearly complete mixing value of 99% at Ω = 600 rpm and beyond. In the clockwise rotation case, efficiency for the triple-U mixer also significantly better than those for single- and double-U mixers, attaining about 95% at Ω = 480 and 600 rpm, and rising to near complete mixing at Ω = 720 and 900 rpm. For a lower speed of 360 rpm, the efficiency appears to be 7% lower than that of counterclockwise rotation at the same speed, as observed in Figure 12. The simulations that cover a rotational range of 120–1200 rpm show in good agreement with measurements. Overall, the simulation efficiencies appear slightly lower than the measured values with an average difference of about 4%. At Ω = 900, the simulation mixing efficiency approaches 97% for the double-U mixer and 99% for the triple-U mixer. With one or two more U-structures, the simulation mean velocity was also found to be slower becoming 0.59 and 0.53 m/s at Ω = 900 rpm for the double- and triple-U mixers, respectively. Moreover, the simulations show no discernible difference in mixing efficiency between clockwise and counterclockwise cases, for which the difference in mean velocity is very small (~1%) at the same rotational speed.

5. Conclusions

Centrifugal microfluidics with a simple geometry of U-shaped channel was investigated for achievement of rapid and efficient mixing of the fluids. Three types of microfluidics, namely microchannels with single-, double- and triple-U structures, were examined in the present study. Experimentally, the testing fluids of ferric chloride and ammonium thiocyanate solutions were employed for flow visualization of the mixing. The blood-red color produced as the two fluids in contact and reaction provides clear images for qualitative interface visualization as well as quantitative evaluation of mixing efficiency. Numerical simulations were also carried out to reveal detailed characteristics of the flow and concentration fields. It is found that the fluid mixing is remarkably enhanced by the transverse secondary flow. The secondary flow originated from the Coriolis-induced vortices due to the channel flow on a rotating disc is further intensified by the Dean force generated as the stream turns along each of the 90° bends of the U-shaped structures. A scaling parameter γ_c, similar to the one used by Zhang et al. [42], is introduced to examine effects of the Dean force as compared to the Coriolis force. Stretching, twisting and folding of mixing interface caused by the transverse secondary flow were observed in flow visualization and closely resembled in numerical simulation as well. It is also found that the secondary flow becomes stronger with increasing rotational speed and with more U-shaped structures, resulting in a larger area of mixed fluids and hence higher mixing performance. The mixing efficiency measured for the three types of U-shaped mixers shows a sharp increase with increasing rotational speed in the lower range $\Omega \leq 600$ rpm. As the speed further increases, nearly complete mixing can be achieved at $\Omega = 600$ for the triple-U mixer and at $\Omega = 720$ rpm for the double-U mixer, while a maximum level between 83 and 86% is reached for the single-U mixer at $\Omega = 720$ and 900 rpm. The variation of simulated mixing efficiency with rotational speed in a wider range (120–1200 rpm) agrees well with the measurements. Moreover, both the simulation and measurement results show no discernible difference in mixing efficiency between clockwise and counter-clockwise cases. The centrifugal U-shaped micromixers presented in this study enabling to assist fluid mixing effectively within a short channel length are especially suitable for use in CD-based microfluidic systems.

Author Contributions: Conceptualization, J.M.C. and C.-W.H.; experiment, C.-W.H.; simulation, C.-W.H. and P.-T.S.; validation, C.-W.H. and J.M.C.; formal analysis, C.-W.H. and P.-T.S.; investigation, C.-W.H. and J.M.C.; resources, J.M.C.; data curation, C.-W.H. and P.-T.S.; writing—original draft preparation, C.-W.H.; writing—review and editing, J.M.C.; visualization, C.-W.H.; supervision, J.M.C.; project administration, J.M.C.; funding acquisition, J.M.C. All authors have read and agreed to the published version of the manuscript.

Funding: The authors are grateful for the financial support from the Minister of Science and Technology of Taiwan under Contract Number MOST 106-2221-E-005-049.

Conflicts of Interest: The authors declare no conflict of interest.

References

1. Whitesides, G. The lab finally comes to the chip! *Lab Chip* **2014**, *14*, 3125–3126. [CrossRef] [PubMed]
2. Rapp, B.; Gruhl, F.J.; Länge, K. Biosensors with label-free detection designed for diagnostic applications. *Anal. Bioanal. Chem.* **2000**, *398*, 2403–2412. [CrossRef] [PubMed]
3. Rajendran, S.T.; Scarano, E.; Bergkamp, M.H.; Capria, A.M.; Cheng, C.H.; Sanger, K.; Ferrari, G.; Nielsen, L.H.; Hwu, E.T.; Zór, K.; et al. Modular, lightweight, wireless potentiostat-on-a-disc for electrochemical detection in centrifugal microfluidics. *Anal. Chem.* **2019**, *91*, 11620–11628. [CrossRef]
4. Jeong, G.S.; Chung, S.; Kim, C.B.; Lee, S.H. Applications of micromixing technology. *Analyst* **2010**, *135*, 460–473. [CrossRef] [PubMed]
5. Yoshida, J.; Nagaki, A.; Iwasaki, T. Enhancement of chemical selectivity by microreactors. *Chem. Eng. Technol.* **2005**, *28*, 259–266. [CrossRef]
6. Doku, G.; Verboom, W.; Reinhoudt, D.; van den Berg, A. On-microchip multiphase chemistry—A review of microreactor design principles and reagent contacting modes. *Tetrahedron* **2005**, *61*, 2733–2742. [CrossRef]

7. Su, Y.; Chen, G.; Yuan, Q. Ideal micromixing performance in packed microchannels. *Chem. Eng. Sci.* **2011**, *13*, 2912–2919. [CrossRef]
8. Benz, K.; Jackel, K.P.; Regenauer, K.J.; Schiewe, J.; Drese, K.; Ehrfeld, W.; Hessel, V.; Lowe, H. Utilization of Micromixers for Extraction Processes. *Chem. Eng. Technol.* **2001**, *24*, 11–17. [CrossRef]
9. Okubo, Y.; Toma, M.; Ueda, H.; Maki, T.; KazuhiroMae, K. Microchannel devices for the coalescence of dispersed droplets produced for use in rapid extraction processes. *Chem. Eng. J.* **2004**, *101*, 39–48. [CrossRef]
10. Zhang, Y.; Ozdemir, P. Microfluidic DNA amplification—A review. *Anal. Chim. Acta* **2009**, *638*, 115–125. [CrossRef]
11. Nguyen, N.T.; Wu, Z. Micromixers—A review. *J. Micromech. Microeng.* **2005**, *15*, R1–R16. [CrossRef]
12. Wang, L.; Li, P.C.H. Microfluidic DNA microarray analysis: A review. *Anal. Chim. Acta* **2011**, *687*, 12–27. [CrossRef] [PubMed]
13. Chang, C.C.; Yang, R.J. Electrokinetic mixing in microfluidic systems. *Microfluid. Nanofluidics* **2007**, *3*, 501–525. [CrossRef]
14. Ward, K.; Fan, Z.H. Mixing in microfluidic devices and enhancement methods. *J. Micromech. Microeng.* **2015**, *25*, 094001. [CrossRef] [PubMed]
15. Raza, W.; Hossain, S.; Kim, K.Y. A review of passive micromixers with a comparative analysis. *Micromachines* **2020**, *11*, 455. [CrossRef] [PubMed]
16. Liu, R.H.; Stremler, M.A.; Sharp, K.V.; Olsen, M.G.; Santiago, J.G.; Adrian, R.J.; Aref, H.; Beeb, D.J. Passive mixing in a three-dimensional serpentine microchannel. *J. Microelectromech. Syst.* **2000**, *9*, 190–197. [CrossRef]
17. Stroock, A.D.; Dertinger, S.K.W.; Ajdari, A.; Mezic, I.; Stone, H.A.; Whitesides, G.M. Chaotic mixer for microchannels. *Science* **2005**, *295*, 647–651. [CrossRef]
18. Xia, H.M.; Wan, S.Y.M.; Shu, C.; Chew, Y.T. Chaotic micromixers using two-layer crossing channels to exhibit fast mixing at low Reynolds numbers. *Lab Chip* **2005**, *5*, 748–755. [CrossRef]
19. Park, J.M.; Kim, D.S.; Kang, T.G.; Kwon, T.H. Improved serpentine laminating micromixer with enhanced local advection. *Microfluid. Nanofluidics* **2008**, *4*, 513–523. [CrossRef]
20. Ansari, M.A.; Kim, K.Y. Parametric study on mixing of two fluids in a three-dimensional serpentine microchannel. *Chem. Eng. J.* **2009**, *146*, 439–448. [CrossRef]
21. Hossain, S.; Lee, I.; Kim, S.M.; Kim, K.Y. A micromixer with two-layer serpentine crossing channels having excellent mixing performance at low Reynolds numbers. *Chem. Eng. J.* **2017**, *327*, 268–277. [CrossRef]
22. Kockmann, N.; Kiefer, T.; Engler, M. Silicon microstructures for high throughput mixing devices. *Microfluid. Nanofluidics* **2006**, *2*, 327–335. [CrossRef]
23. Hong, C.C.; Choi, J.W.; Ahn, C.H. A novel in-plane passive microfluidic mixer with modified tesla structures. *Lab Chip* **2004**, *4*, 109–113. [CrossRef] [PubMed]
24. Hossain, S.; Ansari, M.A.; Husain, A.; Kim, K.Y. Analysis and optimization of a micromixer with a modified Tesla structure. *Chem. Eng. J.* **2010**, *158*, 305–314. [CrossRef]
25. Madou, M.; Zoval, J.; Jia, G.; Kido, H.; Kim, J.; Kim, N. Lab on a CD. *Annu. Rev. Biomed. Eng.* **2006**, *8*, 601–628. [CrossRef]
26. Ducrée, J.; Haeberle, S.; Lutz, S.; Pausch, S.; von Stetten, F.; Zengerle, R. The centrifugal microfluidic Bio-Disk platform. *J. Micromech. Microeng.* **2007**, *17*, S103–S115. [CrossRef]
27. Gorkin, R.; Park, J.; Siegrist, J.; Amasia, M.; Lee, B.S.; Park, J.M.; Kim, J.; Kim, H.; Madou, M.; Cho, Y.K. Centrifugal microfluidics for biomedical applications. *Lab Chip* **2010**, *10*, 1758–1773. [CrossRef]
28. Duffy, D.C.; Gills, H.L.; Lin, J.; Sheppard, N.F.; Kellogg, G.J. Microfabricated centrifugal microfluidic systems: Characterization and multiple enzymatic assays. *Analy. Chem.* **1999**, *71*, 4669–4678. [CrossRef]
29. Madou, M.J.; Lee, L.J.; Daunert, S.; Lai, S.; Shih, C.-H. Design and fabrication of CD-like microfluidic platforms for diagnostics: Microfluidic functions. *Biomed. Microdevices* **2001**, *3*, 245–254. [CrossRef]
30. Ducree, J.; Haeberle, S.; Brenner, T.; Glatzel, T.; Zengerle, R. Patterning of flow and mixing in rotating radial microchannels. *Microfluid. Nanofluidics* **2006**, *2*, 97–105. [CrossRef]
31. Gilmore, J.; Islam, M.; Martinez-Duarte, R. Challenges in the use of compact disc-based centrifugal microfluidics for healthcare diagnostics at the extreme point of care. *Micromachines* **2016**, *7*, 52. [CrossRef] [PubMed]
32. Tang, M.; Wang, G.; Siu-Kai Kong, S.K.; Ho, H.P. A review of biomedical centrifugal microfluidic platforms. *Micromachines* **2016**, *7*, 26. [CrossRef] [PubMed]
33. Kim, D.S.; Kwon, T.H. Modeling, analysis and design of centrifugal force-driven transient filling flow into a circular microchannel. *Microfluid. Nanofluidics* **2006**, *2*, 125–140. [CrossRef]
34. Chiu, H.C.; Chen, J.M. Numerical study fluid mixing in a microchannel with a circular chamber on a rotating disk. *Adv. Mat. Res.* **2012**, *542*, 1113–1119. [CrossRef]

35. La, M.; Park, S.J.; Kim, H.W.; Park, J.J.; Ahn, K.T.; Ryew, S.M.; Kim, D.S. A centrifugal force-based serpentine micromixer (CSM) on a plastic lab-on-a-disk for biochemical assays. *Microfluid. Nanofluidics* **2013**, *15*, 87–98. [CrossRef]
36. Kuo, J.N.; Jiang, L.R. Design optimization of micromixer with square-wave microchannel on compact disk microfluidic platform. *Microsyst. Technol.* **2014**, *20*, 91–99. [CrossRef]
37. Kuo, J.N.; Li, Y.S. Centrifugal-based micromixer with three-dimensional square-wave microchannel for blood plasma mixing. *Microsyst. Technol.* **2017**, *23*, 2343–2354. [CrossRef]
38. Shamloo, A.; Vatankhah, P.; Akbari, A. Analyzing mixing quality in a curved centrifugal micromixer through numerical simulation. *Chem. Eng. Process.* **2017**, *116*, 9–16. [CrossRef]
39. Shamloo, A.; Madadelahi, M.; Akbari, A. Numerical simulation of centrifugal serpentine micromixers and analyzing mixing quality parameters. *Chem. Eng. Process.* **2016**, *104*, 243–252. [CrossRef]
40. Chen, J.M.; Huang, P.C.; Lin, M.G. Analysis and experiment of capillary valves for microfluidics on a rotating disk. *Microfluid. Nanofluidics* **2008**, *4*, 427–437. [CrossRef]
41. Shakhashiri, B.Z. *Chemical Demonstrations: A Handbook for Teachers of Chemistry*; University of Wisconsin Press: Madison, WI, USA, 1983; Volume 1.
42. Zhang, J.; Guo, Q.; Liu, M.; Yang, J. A lab-on-CD prototype for high-speed blood separation. *J. Micromech. Microeng.* **2008**, *18*, 125025. [CrossRef]
43. Lee, G.H.; Baek, J.H. A numerical study of the similarity of fully developed laminar flows in orthogonally rotating rectangular ducts and stationary curved rectangular ducts of arbitrary aspect ratio. *Comput. Mech.* **2002**, *29*, 183–190. [CrossRef]
44. Dai, Y.J.; Huang, W.X.; Xu, C.X.; Cui, G.X. Direct numerical simulation of turbulent flow in a rotating square duct. *Phys. Fluids* **2015**, *27*, 065104. [CrossRef]
45. Currie, I.G. *Fundamental Mechanics of Fluids*; McGraw-Hill: New York, NY, USA, 1974.
46. Batchelor, G.K. *An Introduction to Fluid Dynamics*; Cambridge University Press: Cambridge, UK, 1967.
47. Panton, R.L. *Incompressible Flow*; John Wiley & Sons: New York, NY, USA, 1984.
48. Chen, J.M.; Horng, T.L.; Tan, W.Y. Analysis and measurements of mixing in pressure-driven microchannel flow. *Microfluid. Nanofluidics* **2006**, *2*, 455–469. [CrossRef]

Article

Liquid Mixing Based on Electrokinetic Vortices Generated in a T-Type Microchannel

Chengfa Wang 

Department of Marine Engineering, Dalian Maritime University, No.1 Linghai Road, Dalian 116026, China; wangcf08@dlmu.edu.cn

Abstract: This article proposes a micromixer based on the vortices generated in a T-type microchannel with nonuniform but same polarity zeta potentials under a direct current (DC) electric field. The downstream section (modified section) of the outlet channel was designed with a smaller zeta potential than others (unmodified section). When a DC electric field is applied in the microchannel, the electrokinetic vortices will form under certain conditions and hence mix the solution. The numerical results show that the mixing performance is better when the channel width and the zeta potential ratio of the modified section to the unmodified section are smaller. Besides, the electrokinetic vortices formed in the microchannel are stronger under a larger length ratio of the modified section to the unmodified section of the outlet channel, and correspondingly, the mixing performance is better. The micromixer presented in the paper is quite simple in structure and has good potential applications in microfluidic devices.

Keywords: micromixer; electrokinetic vortices; T-type microchannel; zeta potential ratio; length ratio

Citation: Wang, C. Liquid Mixing Based on Electrokinetic Vortices Generated in a T-Type Microchannel. *Micromachines* **2021**, *12*, 130. https://doi.org/10.3390/mi12020130

Academic Editor: Kwang-Yong Kim
Received: 31 December 2020
Accepted: 22 January 2021
Published: 26 January 2021

1. Introduction

Microfluidic devices have been widely used in biological or chemical analysis because of the significant advantages of easy operation, low cost, and less sample consumption [1]. Since the liquid is usually laminar in microchannels due to the small Reynolds number, the effects of diffusion and convection are quite weak and the liquid mixing is a challenge in microchannels. Therefore, micromixers are the key components of microfluidic devices and have been studied extensively.

The reported micromixers are usually divided into two categories: Active micromixers and passive micromixers. Active micromixers utilize external energy sources such as electric field [2–4] magnetic field [5–7], and acoustic field [8–10] applied in microchannels to improve the solution mixing performance. The channel structures of these micromixers are relatively simple. By contrast, external energy sources are not required for passive micromixers, but this type of micromixer generally needs to have a complicated channel structure [11–19] or set obstacles [20–23] in the microchannel to disturb the laminar flow for solution mixing under a relatively high pressure generated by a syringe pump. The mixing efficiency of passive micromixers is high, but the sealing of a microfluidic chip under high pressure is a challenge. The details can be found in the review articles [24,25].

It is generally known that when a charged solid surface is in contact with an aqueous solution, the surface electrostatic charges will attract the counter-ions in the solution and then the electric double layer (EDL) forms on the surface [26]. The zeta potential is an important parameter to evaluate the surface charge density. Once a DC electric field is applied along the surface, the electroosmotic flow (EOF) will form in the EDL and drive the liquid near the surface to flow. If the channel walls have different polarity zeta potentials, electrokinetic vortices can be easily generated due to the different directions of electroosmotic flows formed on the walls. Vortices can stir the liquid, thereby enhancing the effects of diffusion and convection, so the electrokinetic vortex is an effective method to solve the mixing problem in microfluidic devices and can be used to design different types of electrokinetically-driven active micromixer. It is widely known that a metal object immersed in an aqueous solution will be induced by an external electric field. As a result, one side of the metal object obtains positive charges and another side obtains equal amounts of negative charges. The details can be found in the reference [27]. Therefore, the electrokinetic vortex can be easily formed by the induced charge electroosmotic flow (ICEO) generated on a metal object surface [27–30]. Based on electrokinetic vortices generated by ICEO, Wu and Li [31,32] designed a micromixer and conducted a systematically numerical and experimental investigation. The advantage of this kind of micromixer is that the inhomogeneous zeta potential surfaces can be obtained directly without surface modification. However, under a high electric field strength, the "Bipolar Electrochemistry" phenomenon [33] generated on conducting surfaces is unfavorable to the vortex generation, thereby affecting the mixing performance.

A metal surface can obtain different polarity zeta potentials without surface treatment. However, a non-conductive surface usually has a uniform zeta potential, so surface modification technology should be used to make the wall obtain inhomogeneous zeta potentials, then electrokinetic vortices can be formed in the microchannel. Hau et al. [34] and Stroock et al. [35] experimentally investigated the electroosmotic flow formed on a surface with different polarity zeta potentials and demonstrated that electrokinetic vortices can form near this kind of surface. Erickson and Li [36–38] carried out numerical studies on the electrokinetic vortices formed in a straight microchannel by designing various patterns on the microchannel walls with different polarity zeta potentials and found the degree of pattern heterogeneity greatly affect the electrokinetic vortex generation. However, it is not easy to manufacture the surface with different polarity zeta potentials since a surface in the same aqueous solution usually has the same polarity zeta potentials [39].

As described, electrokinetic vortices can easily form on a surface with different polarity zeta potentials. If the surface has different value but same polarity zeta potentials, can electrokinetic vortices still form? Recently published studies show that electrokinetic vortices can also form on such surfaces under certain conditions [40,41]. Song et al. [41] utilized polybrene (PB) coating technology to design a two-section straight microchannel with two same polarity zeta potentials (one section was modified with a smaller absolute value of negative zeta potential, and another section was not modified). and studied the evolution of the flow field in the microchannel. The results show that electrokinetic vortices can form in such a microchannel when the length ratio of the modified section to the unmodified section is large enough.

Active micromixers based on electrokinetic vortices have the advantages of easy operation and sample structure. Biddiss et al. [42] carried out an investigation on the liquid mixing performance near a non-conductive surface with different value and different polarity zeta potentials. However, so far very few published works have presented the liquid mixing based on electrokinetic vortices generated on a surface with different value but same polarity zeta potentials. In this paper, a T-type micromixer was designed utilizing electrokinetic vortices formed in the microchannel with different value but same polarity zeta potentials. The dependencies of surface zeta potentials, channel width, and other factors on the mixing performance of the micromixer were numerically investigated. The advantage of the presented micromixer is that the microchannel structure and the work system are quite simple. The micromixer only needs a small battery to work, which is conducive to the integration and miniaturization of microfluidic devices. This work provides a certain theoretical basis for practical application in the future.

2. Structure and Working Principle of the Micromixer

Figure 1 displays the structure and working principle of the proposed micromixer. The structure of the micromixer is a T-type microchannel with one outlet and two inlets. The lengths of the inlet and outlet channels are L_{inlet} and L_{outlet}, and the width and height of the channel are equal, namely, $W = H$. In this study, the dimensionless lengths of the inlet and outlet channels were set to be 10 and 30, respectively, and the characteristic length is L_{ref} = 10 μm. The outlet channel is divided into an unmodified section and a modified section. The lengths of the unmodified section and the modified section are L_u and L_m, and the parameter of $\beta = L_m/L_u$ is defined as the length ratio of the modified section to the unmodified section. Two different DC voltages of V_{inlet} and V_{outlet} are applied at the two inlets and the outlet of the microchannel. The walls of the inlet channel and the unmodified section of the outlet channel have the same negative zeta potential (ζ_u), and the negative zeta potential (ζ_m) of the modified section is designed with a smaller absolute value. The parameter of $\alpha = \zeta_m/\zeta_u$ is defined as the zeta potential ratio of the modified section to the unmodified section.

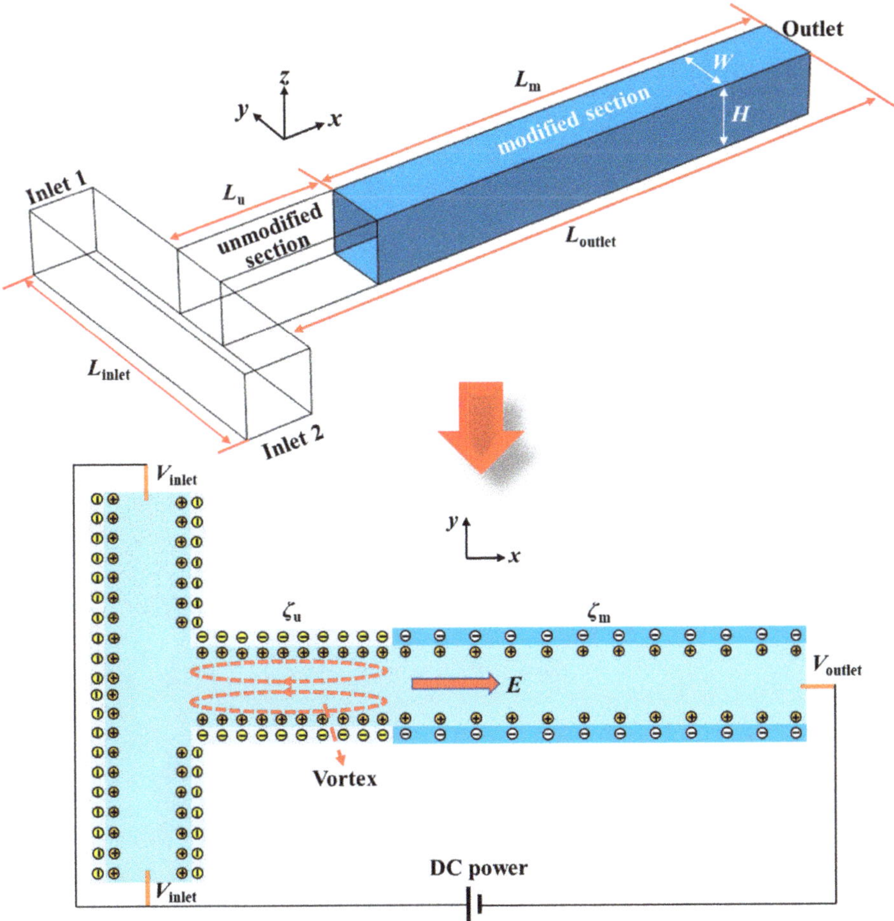

Figure 1. Schematic diagram of the micromixer structure and working principle. The zeta potentials of the unmodified section wall (ζ_u) and the modified section (ζ_m) wall are all negative, but the absolute value of ζ_m is smaller than that of ζ_u.

Once the charged channel walls are in contact with an aqueous solution, electric double layers (EDL) will form along the walls. When the voltages of V_{inlet} and V_{outlet} are applied, the electroosmotic flow (EOF) will be generated in the microchannels. Under the same electric field, the velocity of EOF is determined by the surface zeta potential. The larger the absolute value of surface zeta potential is, the higher the velocity of EOF is. Since the absolute value of ζ_m is smaller than that of ζ_u, the velocity of EOF formed in the modified section is smaller than that in the unmodified section. Therefore, the liquid in the modified section will impede the flow of liquid in the unmodified section. When the EOF velocity difference of the unmodified section to the modified section is large enough, the "impede effect" of the modified section liquid will become strong. As a result, the liquid in the unmodified section will flow back, hence the electrokinetic vortex generates in the microchannel, as shown in Figure 1. The formed electrokinetic vortices stir the laminar flow in the microchannel and then realize liquid mixing. The parameters of the zeta potential ratio (α) and the length ratio (β) of the modified section to the unmodified section have an important influence on the "impede effect" of modified section liquid, thereby affecting the strength of electrokinetic vortices and the corresponding mixing performance.

3. Theoretical Model

After a DC electric field is applied in the microchannels, electrokinetic vortices will form, thereby mixing the solution. Therefore, this model contains an electric field, a flow field, and a concentration field.

3.1. Electric Field

The DC electric potential (V) field in the microchannels is described by Laplace's equation:

$$\nabla^2 V = 0 \tag{1}$$

The relationship between the electric potential and the electric field strength ($\boldsymbol{E}$) is as follows:

$$\boldsymbol{E} = -\nabla V \tag{2}$$

Since all the channel walls are electrically insulated, the boundary condition is

$$\boldsymbol{n} \cdot \boldsymbol{E} = 0 \tag{3}$$

where $\boldsymbol{n}$ is the unit normal vector on the boundary walls.

The voltages of V_{inlet} and V_{outlet} are applied at two inlet boundaries and one outlet boundary, respectively, and the boundary conditions are

$$V = V_{\text{inlet}} \tag{4}$$

$$V = V_{\text{outlet}} \tag{5}$$

In this work, the outlet boundary was grounded, namely, $V_{\text{outlet}} = 0$. The electric field strength ($\boldsymbol{E}$) in the outlet channel is set as a constant value of 50 V/cm by adjusting the inlet voltage of V_{inlet}.

3.2. Flow Field

The Navier–Stokes equation and the continuity equation govern the distribution of the flow field in the microchannels. Their expressions in steady-state are as follows:

$$\rho \boldsymbol{U} \cdot \nabla \boldsymbol{U} = -\nabla p + \mu \nabla^2 \boldsymbol{U} \tag{6}$$

$$\nabla \cdot \boldsymbol{U} = 0 \tag{7}$$

where ρ = 1000 kg/m^3 is the solution density, $\boldsymbol{U}$ is the velocity vector, μ = 0.001 Pa·s is the dynamic viscosity of the solution, and p is the pressure.

In this study, the flow of the solution in the microchannels is dependent on the EOFs formed on the charged walls by the DC electric field, and there is no pressure-driven flow. Thus, the boundary conditions of the inlets and outlet are

$$p = 0 \tag{8}$$

In this model, the net charge density is zero except in EDLs, but compared with the scale of the microchannel, the EDL thickness is negligible. Thus, this model does not consider the body force acting in the EDL by the DC electric field and the distribution of the flow field in the EDL. Instead, a slip boundary condition (Helmholtz–Smoluchowski velocity [26,43]) was adopted to reflect the EOFs generated on the channel walls. Thus, the boundary conditions of the flow field on the channel walls are as follows:

$$\boldsymbol{U} = -\frac{\varepsilon_0 \varepsilon \zeta_{\mathrm{m}}}{\mu} \boldsymbol{E} \tag{9}$$

on the channel walls of the modified section;

$$\boldsymbol{U} = -\frac{\varepsilon_0 \varepsilon \zeta_{\mathrm{u}}}{\mu} \boldsymbol{E} \tag{10}$$

on the other channel walls, where $\varepsilon_0 = 8.85 \times 10^{-12}$ F/m is the vacuum permittivity, $\varepsilon = 80$ is the relative dielectric constant of solution.

Considering that polydimethylsiloxane (PDMS) is a widely-used material in the fabrication of microchannels and its zeta potential can be controlled in the range of −1.7 mV and −57 mV by surface treatment [44], this model took $\zeta_{\mathrm{U}} = -50$ mV as a characteristic value of the zeta potential on the unmodified section walls.

The Reynolds number in this study is less than 0.1 due to the small microchannel size and the electric field intensity. With the increase of the electric field, the velocity of EOFs and the corresponding Reynolds number will increase. The electric field and the micromixer size can be adjusted according to the actual needs. However, it should be noted that the Joule heating effect from a strong electric field will play a negative effect on the biological cells in the solutions. Thus, a small electric field is used in this work.

3.3. Concentration Field

Assuming that the concentration is not affected by any reactions and there is no migration of ionic species, the convection-diffusion equation, which describes the concentration of the solute in water in a steady-state, is as below [45]:

$$\nabla \cdot (-D\nabla c) + \boldsymbol{U} \cdot \nabla c = 0 \tag{11}$$

where c is the solute concentration, $D = 10^{-11}$ m^2/s is the diffusion coefficient of the solute.

Since all channel walls are insulated for the solute, no flux condition (Equation (12)) was applied to the walls.

$$-\boldsymbol{n} \cdot (-D\nabla c + \boldsymbol{U}c) = 0 \tag{12}$$

At Inlet 1 (see Figure 1) the solute concentration is c_0 = 1 mol/m^3; at Inlet 2 the concentration is zero, that is

$$c = c_0 \tag{13}$$

at Inlet 1;

$$c = 0 \tag{14}$$

at Inlet 2.

The solute is transported out of the outlet boundary by liquid motion, so convection dominates the solute transport and the diffusive transport can be ignored. Thus, the boundary condition at Outlet is

$$\boldsymbol{n} \cdot D\nabla c = 0 \tag{15}$$

To evaluate the mixing performance of the micromixer, a mixing index is defined as follows:

$$\gamma = \frac{c_{\min}}{c_{\max}} \tag{16}$$

where c_{max} and c_{min} are the maximum value and the minimum value of the solute concentration at the Outlet (see Figure 1). The value of γ is in the range of 0 to 1. It is obvious that the larger the value of γ, the better the mixing performance.

COMSOL MULTIPHYSICS 5.2a ® was chosen to calculate the above 3D numerical model in this work. The electric field, flow field, and concentration field in the model are mutually coupled and solved simultaneously by the software. The simulation process is as follows. According to the above-mentioned governing equations, the appropriate physical fields are firstly chosen from the "Physics" module of the software. Secondly, the computational domain is drawn via the "Geometry" module. Thirdly, the boundary conditions are set in the software according to the theoretical model. Finally, the mesh is built and the model is calculated.

Mesh quality greatly affects the accuracy of the numerical result. To avoid this problem, mesh independence was examined by calculating the mixing index of γ under different mesh numbers. The result shows that the rate of change in γ is less than 0.6% when the mesh number is increased from 361,345 to 433,680. Therefore, the mesh number was set to not less than 361,345 in the simulation.

4. Results and Discussion

4.1. Distribution of Concentration Field in the Microchannels

Figure 2 shows the distribution of the concentration field in the microchannels in steady-state under different conditions. Assuming that all the walls are unmodified, namely, all the walls have the same zeta potential, the concentration field distribution in the microchannels is shown in Figure 2A. It is obvious that the mixing effect of the solute is poor in this case. By contrast, when the downstream of the outlet is designed with a smaller absolute value of zeta potential (see Figure 1), the distribution of the concentration field is shown in Figure 2B (The zeta potential ratio and the length ratio of the modified section to the unmodified section were set to be $\alpha = 0.1$ and $\beta = 6$, respectively). It can be easily seen that the mixing performance, in this case, is much better than that in Figure 2A. The main reason is the generation of the electrokinetic vortices due to the "impede effect" of the low flow velocity liquid in the modified section to the high flow velocity liquid in the unmodified section, as displayed in the inset of Figure 2B. The electrokinetic vortices enhance the convection and diffusion effect and the corresponding mixing performance.

The modified section and unmodified section of the outlet channel can be seen as a mixing unit. There is one mixing unit in Figure 2B. If two mixing units were set in the outlet channel, what would happen? As shown in Figure 2C, two same size mixing units were set. The length of the outlet channel, the zeta potential ratio, and the length ratio of the two mixing units in Figure 2C are the same as that in Figure 2B. It is obvious that extra vortices form in the unmodified section of the mixing unit 2, as displayed in the insets of Figure 2C. Therefore, the mixing performance in Figure 2C is better than that in Figure 2B. However, since the extra vortices in Figure 2C are weak, the improvement of the mixing effect is limited.

Besides, to evaluate the presented micromixer, a similar micromixer reported by other researchers is compared. Biddiss et al. [42] designed a similar micromixer to the micromixer presented in this paper and investigated the liquid mixing performance in the microchannel. They set a complex patch patterning on the walls of a T-type microchannel and the patch patterning have different value and different polarity zeta potentials. Their study shows that for a 95% mixture, the required mixing length of the outlet channel is more than 2 mm under the conditions of the diffusion coefficient 4.37×10^{-10} m^2/s and the electric field 280 V/cm (The Reynolds number is about 0.35.). To evaluate the introduced micromixer, the required mixing length for a 95% mixture is also calculated under the same diffusion coefficient and Reynolds number. The numerical result shows that the required mixing length of the introduced micromixer for a 95% mixture is less than 1 mm. The micromixer reported by Biddiss et al. needs to set the channel walls with a complex patch patterning via surface modification technology. By contrast, for the micromixer presented in the paper, only the downstream part of the outlet channel needs to be treated. Thus, the fabrication of the introduced micromixer is relatively easier.

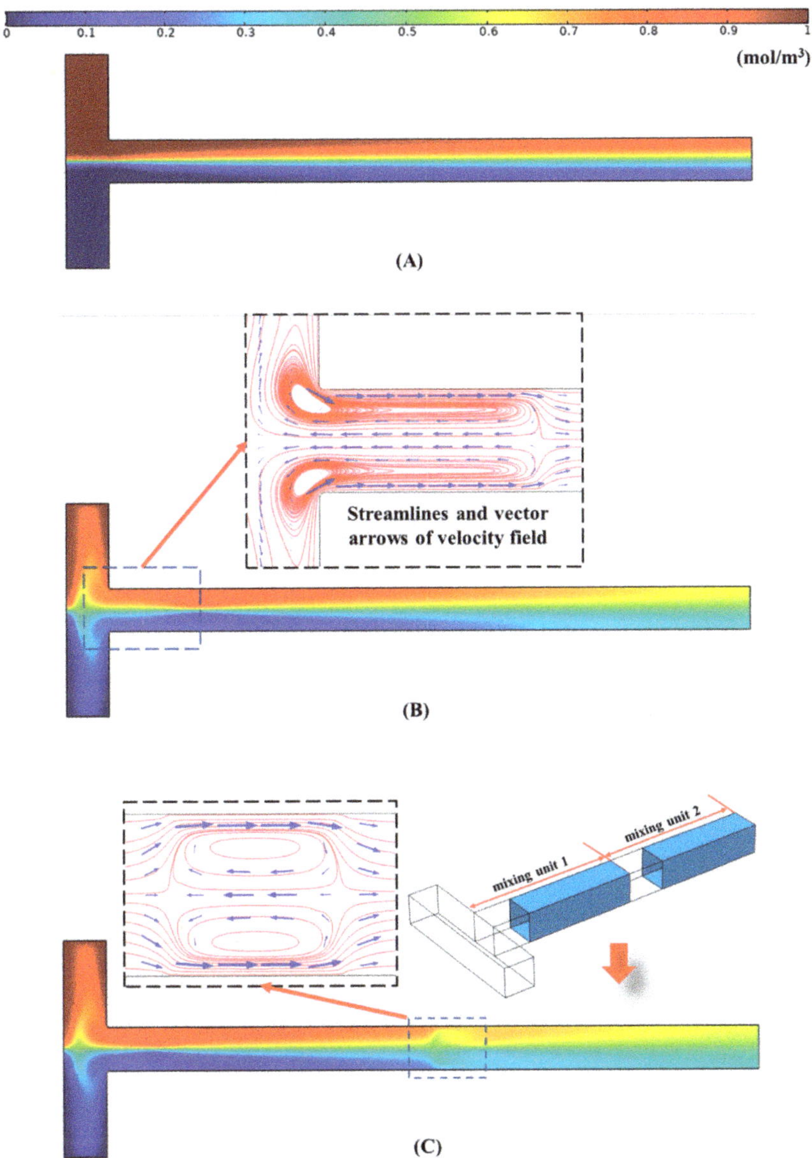

Figure 2. Distribution of the concentration field on the center cut plane of the microchannel under different cases: (**A**) The channel wall has uniform zeta potential (−50 mV); (**B**) the channel wall has non-uniform zeta potential (one mixing unit: $\zeta_u = -50$ mV, $\alpha = 0.1$, $\beta = 6$); (**C**) the channel wall has non-uniform zeta potential (two mixing units: $\zeta_u = -50$ mV, $\alpha = 0.1$, $\beta = 6$).

4.2. Effect of Zeta Potential Ratio (α) on Mixing Index

The dependence of the zeta potential ratio (α) on the mixing index (γ) is shown in Figure 3 under the conditions of the length ratio $\beta = 7$ and the dimensionless channel width

$W^* = 2$ ($W^* = W/L_{ref}$, $L_{ref} = 10$ μm). It is clear in Figure 3 that the mixing index (γ) of the solute declines with the increase of the zeta potential ratio (α) of the modified section to the unmodified section. It could be explained as follows:

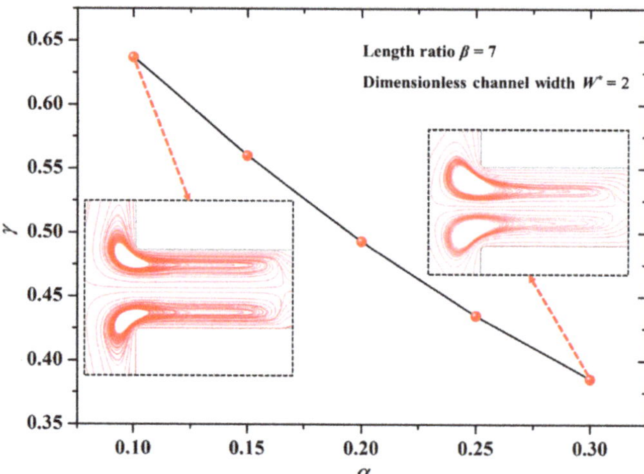

Figure 3. Dependence of the zeta potential ratio on the mixing index under the conditions of the length ratio $\beta = 7$ and the dimensionless channel width $W^* = 2$ ($W^* = W/L_{ref}$, $L_{ref} = 10$ μm).

The larger value of α means the absolute value difference of the zeta potentials between the unmodified section and the modified section is smaller, accordingly, the velocity difference of EOFs formed in the two sections is smaller. As a result, the "impede effect" of the low flow velocity liquid in the modified section to the high flow velocity liquid in the unmodified section is weaker, hence, the electrokinetic vortices formed in the channel are weak, as shown in the insets of Figure 3. Therefore, it is not difficult to understand the changing trend of γ with α in Figure 3.

4.3. *Effect of Length Ratio (β) on Mixing Index*

Figure 4 shows the changing trend of the mixing index (γ) with the length ratio (β) of the modified section to the unmodified section. The other parameters were set to be the zeta potential ratio $\alpha = 0.1$ and the dimensionless channel width $W^* = 2$. One can easily see in Figure 4 that the larger the value β, the higher the mixing index (γ). Namely, the mixing performance is better under the larger length ratio (β) of the modified section to the unmodified section.

As mentioned above, the liquid in the modified section with a smaller EOF velocity will impede the flow of liquid in the unmodified section with a higher EOF velocity. The larger value of β signifies the length difference between the modified section and the unmodified section is larger. As a result, the "impede effect" of modified section liquid to the liquid in the unmodified section will become stronger, hence, the rotation velocity of vortices formed in the channel will be higher, as displayed in the insets of Figure 4. Thus, the mixing performance becomes better with the increase of the length ratio (β).

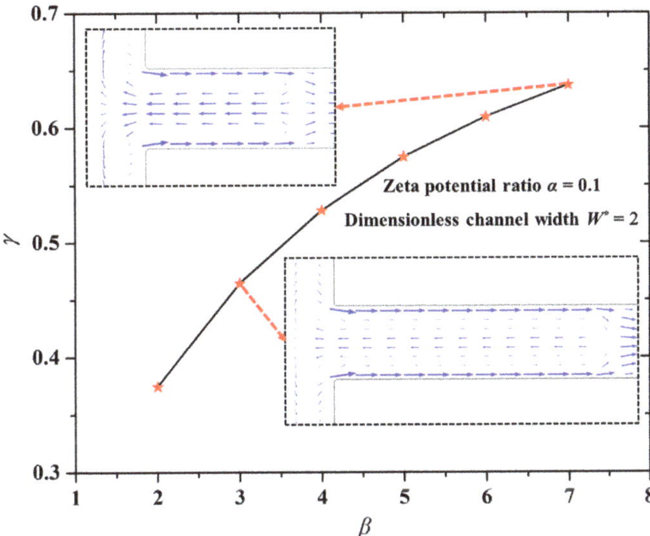

Figure 4. Dependence of the length ratio on the mixing index. The other parameters were set to be the zeta potential ratio $\alpha = 0.1$ and the dimensionless channel width $W^* = 2$.

4.4. Effect of Channel Width (W) on Mixing Index

The channel width greatly affects the mixing performance of the solute, as shown in Figure 5 under the conditions of the zeta potential ratio $\alpha = 0.1$ and the length ratio $\beta = 4$. From Figure 5 it is found that the mixing index (γ) is smaller under the larger dimensionless value of the channel width (W^*). In other words, the mixing performance is better in the microchannel with a smaller channel width. The explanation is as follows.

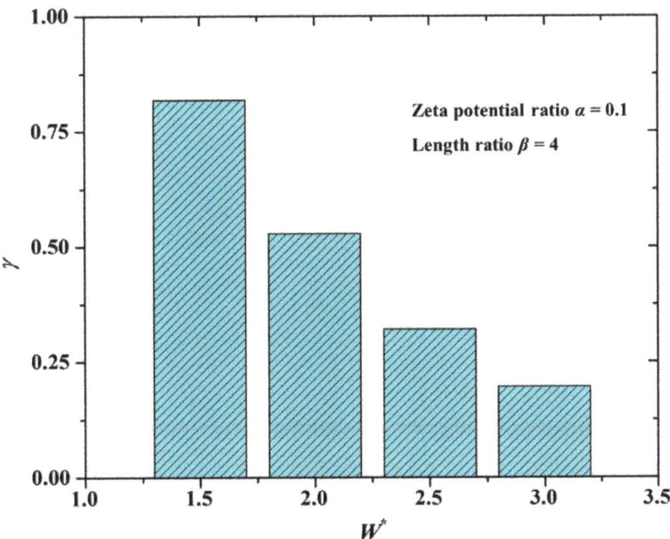

Figure 5. Dependence of the channel width on the mixing index under the conditions of the zeta potential ratio $\alpha = 0.1$ and the length ratio $\beta = 4$.

The "impede effect" of the modified section liquid to the unmodified section liquid determines the formation of electrokinetic vortices and hence the mixing performance. Because the velocity of EOF formed on the walls of the modified section is low, with the decrease of the channel width, the channel walls will squeeze the liquid, thereby enhancing the "impede effect" of the modified section liquid. Assuming that the channel width is quite large, the above-mentioned "squeeze effect" from the modified section walls will become quite weak. As a result, the rotation velocity of the vortices is very low, even the vortices disappear. Therefore, the mixing index (γ) decreases with the increase of the channel width, as shown in Figure 5.

5. Conclusions

Liquid mixing based on the electrokinetic vortices generated in a T-type microchannel with nonuniform but same polarity zeta potentials is numerically investigated in this work. The outlet channel of the T-type microchannel is composed of a modified section with a smaller zeta potential and an unmodified section with a larger zeta potential. The results show that the "impede effect" of the modified section liquid to the unmodified section liquid determines the formation of electrokinetic vortices and hence the mixing performance. The larger the EOF velocity difference of the unmodified section to the modified section is, the stronger the "impede effect" is. The "impede effect" can be enhanced by decreasing the channel width and the zeta potential ratio of the modified section to the unmodified section (namely, increasing the zeta potential difference of the unmodified section to the modified section) and increasing the length ratio of the modified section to the unmodified section, thereby improving the mixing performance.

The work provides a basic understanding of liquid mixing in a T-type microchannel with nonuniform but same polarity zeta potentials. The presented micromixer is quite simple in structure and can be fabricated easily. Since the micromixer only needs a small battery to work, the work system of this micromixer is also simple. Therefore, the above advantages of this micromixer are conducive to the integration and miniaturization of microfluidic devices.

Funding: This study was supported by the Fundamental Research Funds for the Central Universities (Grant No. 3132019192, 3132019336), and the Project funded by the China Postdoctoral Science Foundation (Grant No. 2019M661083) and the National Natural Science Foundation of China (Grant No. 52001050).

Conflicts of Interest: The author declares no conflict of interest.

References

1. Yang, Y.; Noviana, E.; Nguyen, M.P.; Geiss, B.J.; Dandy, D.S.; Henry, C.S. Paper-Based Microfluidic Devices: Emerging Themes and Applications. *Anal. Chem.* **2017**, *89*, 71–91. [CrossRef] [PubMed]
2. Rashidi, S.; Bafekr, H.; Valipour, M.S.; Esfahani, J.A. A review on the application, simulation, and experiment of the electrokinetic mixers. *Chem. Eng. Process. Process Intensif.* **2018**, *126*, 108–122. [CrossRef]
3. Deng, Y.; Zhou, T.; Liu, Z.; Wu, Y.; Qian, S.; Korvink, J.G. Topology optimization of electrode patterns for electroosmotic micromixer. *Int. J. Heat Mass Transf.* **2018**, *126*, 1299–1315. [CrossRef]
4. Chen, L.; Deng, Y.; Zhou, T.; Pan, H.; Liu, Z. A novel electroosmotic micromixer with asymmetric lateral structures and DC electrode arrays. *Micromachines* **2017**, *8*, 105. [CrossRef]
5. Chen, Y.; Kim, C.N. Numerical analysis of the mixing of two electrolyte solutions in an electromagnetic rectangular micromixer. *J. Ind. Eng. Chem.* **2018**, *60*, 377–389. [CrossRef]
6. Wang, Y.; Zhe, J.; Chung, B.T.F.; Dutta, P. A rapid magnetic particle driven micromixer. *Microfluid. Nanofluidics* **2008**, *4*, 375–389. [CrossRef]
7. Dallakehnejad, M.; Mirbozorgi, S.A.; Niazmand, H. A numerical investigation of magnetic mixing in electroosmotic flows. *J. Electrostat.* **2019**, *100*, 103354. [CrossRef]
8. Rasouli, M.R.; Tabrizian, M. An ultra-rapid acoustic micromixer for synthesis of organic nanoparticles. *Lab Chip* **2019**, *19*, 3316–3325. [CrossRef]

9. Shang, X.; Huang, X.; Yang, C. Vortex generation and control in a microfluidic chamber with actuations. *Phys. Fluids* **2016**, *28*, 122001. [CrossRef]
10. Ahmed, D.; Mao, X.; Juluri, B.K.; Huang, T.J. A fast microfluidic mixer based on acoustically driven sidewall-trapped microbubbles. *Microfluid. Nanofluidics* **2009**, *5*, 727–731. [CrossRef]
11. Stroock, A.D.; Dertinger, S.K.W.; Ajdari, A.; Mezić, I.; Stone, H.A.; Whitesides, G.M. Chaotic mixer for microchannels. *Science* **2012**, *295*, 647–651. [CrossRef] [PubMed]
12. Mondal, B.; Mehta, S.K.; Patowari, P.K.; Pati, S. Numerical study of mixing in wavy micromixers: Comparison between raccoon and serpentine mixer. *Chem. Eng. Process.-Process Intensif.* **2019**, *136*, 44–61. [CrossRef]
13. Raza, W.; Kim, K.-Y. Unbalanced Split and Recombine Micromixer with Three-Dimensional Steps. *Ind. Eng. Chem. Res.* **2019**. [CrossRef]
14. Bazaz, S.R.; Mehrizi, A.A.; Ghorbani, S.; Vasilescu, S.; Asadnia, M.; Warkiani, M.E. A hybrid micromixer with planar mixing units. *RSC Adv.* **2018**, *8*, 33103–33120. [CrossRef]
15. Wang, H.; Shi, L.; Zhou, T.; Xu, C.; Deng, Y. A novel passive micromixer with modified asymmetric lateral wall structures. *Asia-Pacific J. Chem. Eng.* **2018**, *13*, e2202. [CrossRef]
16. Zhou, T.; Xu, Y.; Liu, Z.; Joo, S.W. An enhanced one-layer passive microfluidic mixer with an optimized lateral structure with the dean effect. *J. Fluids Eng.* **2015**, *137*, 091102. [CrossRef]
17. Chen, X.; Li, T. A novel passive micromixer designed by applying an optimization algorithm to the zigzag microchannel. *Chem. Eng. J.* **2017**, *313*, 1406–1414. [CrossRef]
18. Afzal, A.; Kim, K. Convergent–divergent micromixer coupled with pulsatile flow. *Sensors Actuators B Chem.* **2015**, *211*, 198–205. [CrossRef]
19. Khaydarov, V.; Borovinskaya, E.S.; Reschetilowski, W. Numerical and experimental investigations of a micromixer with chicane mixing geometry. *Appl. Sci.* **2018**, *8*, 2458. [CrossRef]
20. Raza, W.; Kim, K. Asymmetrical Split-and-Recombine Micromixer with Baffles. *Micromachines* **2019**, *10*, 844. [CrossRef]
21. Wang, L.; Ma, S.; Wang, X.; Bi, H.; Han, X. Mixing enhancement of a passive microfluidic mixer containing triangle baffles. *Asia-Pacific J. Chem. Eng.* **2014**, *9*, 877–885. [CrossRef]
22. Chen, X.; Zhao, Z. Numerical investigation on layout optimization of obstacles in a three-dimensional passive micromixer. *Anal. Chim. Acta* **2017**, *964*, 142–149. [CrossRef] [PubMed]
23. Tsai, R.-T.; Wu, C.-Y. An efficient micromixer based on multidirectional vortices due to baffles and channel curvature. *Biomicrofluidics* **2011**, *5*, 014103. [CrossRef]
24. Bayareh, M.; Ashani, M.N.; Usefian, A. Active and passive micromixers: A comprehensive review. *Chem. Eng. Process.-Process Intensif.* **2020**, *147*, 107771. [CrossRef]
25. Cai, G.; Li, X.; Zhang, H.; Lin, J. A review on micromixers. *Micromachines* **2017**, *8*, 274. [CrossRef] [PubMed]
26. Hunter, R.J. *Zeta Potential in Colloid Science: Principles and Applications*; Academic Press: London, UK, 1981; ISBN 1483214087.
27. Squires, T.M.; Bazant, M.Z. Induced-charge electro-osmosis. *J. Fluid Mech.* **2004**, *509*, 217–252. [CrossRef]
28. Daghighi, Y.; Sinn, I.; Kopelman, R.; Li, D. Experimental validation of induced-charge electrokinetic motion of electrically conducting particles. *Electrochim. Acta* **2013**, *87*, 270–276. [CrossRef]
29. Wang, C.; Song, Y.; Pan, X.; Li, D. A novel microfluidic valve controlled by induced charge electro-osmotic flow. *J. Micromech. Microeng.* **2016**, *26*, 075002. [CrossRef]
30. Song, Y.; Wang, C.; Li, M.; Pan, X.; Li, D. Focusing particles by induced charge electrokinetic flow in a microchannel. *Electrophoresis* **2016**, *37*, 666–675. [CrossRef]
31. Wu, Z.; Li, D. Micromixing using induced-charge electrokinetic flow. *Electrochim. Acta* **2008**, *53*, 5827–5835. [CrossRef]
32. Wu, Z.; Li, D. Mixing and flow regulating by induced-charge electrokinetic flow in a microchannel with a pair of conducting triangle hurdles. *Microfluid. Nanofluidics* **2008**, *5*, 65–76. [CrossRef]
33. Fosdick, S.E.; Knust, K.N.; Scida, K.; Crooks, R.M. Bipolar electrochemistry. *Angew. Chemie Int. Ed.* **2013**, *52*, 10438–10456. [CrossRef] [PubMed]
34. Hau, W.L.W.; Lee, L.M.; Lee, Y.-K.; Wong, M.; Zohar, Y. Experimental investigation of electrokinetically generated in-plane vorticity in a microchannel. In Proceedings of the TRANSDUCERS'03. 12th International Conference on Solid-State Sensors, Actuators and Microsystems, IEEE. Boston, MA, USA, 9–12 June 2003; Volume 1, pp. 651–654.
35. Stroock, A.D.; Weck, M.; Chiu, D.T.; Huck, W.T.S.; Kenis, P.J.A.; Ismagilov, R.F.; Whitesides, G.M. Patterning electro-osmotic flow with patterned surface charge. *Phys. Rev. Lett.* **2000**, *84*, 3314–3317. [CrossRef] [PubMed]
36. Erickson, D.; Li, D. Three-dimensional structure of electroosmotic flow over heterogeneous surfaces. *J. Phys. Chem. B* **2003**, *107*, 12212–12220. [CrossRef]
37. Erickson, D.; Li, D. Influence of surface heterogeneity on electrokinetically driven microfluidic mixing. *Langmuir* **2002**, *18*, 1883–1892. [CrossRef]
38. Erickson, D.; Li, D. Microchannel flow with patchwise and periodic surface heterogeneity. *Langmuir* **2002**, *18*, 8949–8959. [CrossRef]
39. Kirby, B.J.; Hasselbrink, E.F., Jr. Zeta potential of microfluidic substrates: 1.Theory, experimental techniques, and effects on separations. *Electrophoresis* **2004**, *25*, 187–202. [CrossRef]

40. Wang, C.; Song, Y.; Pan, X. Electrokinetic-vortex formation near a two-part cylinder with same-sign zeta potentials in a straight microchannel. *Electrophoresis* **2020**, *41*, 793–801. [CrossRef]
41. Song, Y.; Wang, C.; Li, J.; Li, D. Vortex generation in electroosmotic flow in a straight polydimethylsiloxane microchannel with different polybrene modified-to-unmodified section length ratios. *Microfluid. Nanofluidics* **2019**, *23*, 24. [CrossRef]
42. Biddiss, E.; Erickson, D.; Li, D. Heterogeneous surface charge enhanced micromixing for electrokinetic flows. *Anal. Chem.* **2004**, *76*, 3208–3213. [CrossRef]
43. Wang, C.; Li, M.; Song, Y.; Pan, X.; Li, D. Electrokinetic motion of a spherical micro particle at an oil−water interface in microchannel. *Electrophoresis* **2018**, *39*, 807–815. [CrossRef] [PubMed]
44. Song, Y.; Li, J.; Li, D. Zeta potentials of polydimethylsiloxane surfaces modified by polybrene of different. *Electrophoresis* **2016**, *37*, 567–572. [CrossRef] [PubMed]
45. Zhou, T.; Wang, H.; Shi, L.; Liu, Z.; Joo, S.W. An enhanced electroosmotic micromixer with an efficient asymmetric lateral structure. *Micromachines* **2016**, *7*, 218. [CrossRef] [PubMed]

Article

Understanding Interdependencies between Mechanical Velocity and Electrical Voltage in Electromagnetic Micromixers

Noori Kim [1,*], Wei Xuan Chan [2], Sum Huan Ng [3], Yong-Jin Yoon [4] and Jont B. Allen [5]

1 Department of Electrical and Electronic Engineering, Newcastle University in Singapore, 172A Ang Mo Kio Avenue 8, ♯05-01 SIT@NYP Building, Singapore 567739, Singapore
2 Department of Biomedical Engineering, National University of Singapore, 21 Lower Kent Ridge Rd, Singapore 119077, Singapore; biecwx@nus.edu.sg
3 Singapore Institute of Manufacturing Technology, 2 Fusionopolis Way, Singapore 138634, Singapore; shng@simtech.a-star.edu.sg
4 Department of Mechanical Engineering, Korea Advanced Institute of Science and Technology, Daejeon 34141, Korea; yongjiny@kaist.ac.kr
5 Department of Electrical and Computer Engineering, University of Illinois at Urbana-Champaign, Urbana, IL 61801, USA; jontalle@illinois.edu
* Correspondence: noori.kim@ncl.ac.uk

Received: 18 May 2020; Accepted: 25 June 2020; Published: 29 June 2020

Abstract: Micromixers are critical components in the lab-on-a-chip or micro total analysis systems technology found in micro-electro-mechanical systems. In general, the mixing performance of the micromixers is determined by characterising the mixing time of a system, for example the time or number of circulations and vibrations guided by tracers (i.e., fluorescent dyes). Our previous study showed that the mixing performance could be detected solely from the electrical measurement. In this paper, we employ electromagnetic micromixers to investigate the correlation between electrical and mechanical behaviours in the mixer system. This work contemplates the "anti-reciprocity" concept by providing a theoretical insight into the measurement of the mixer system; the work explains the data interdependence between the electrical point impedance (voltage per unit current) and the mechanical velocity. This study puts the electromagnetic micromixer theory on a firm theoretical and empirical basis.

Keywords: micromixers; acoustic micromixers; active micromixers; electromagnetic micromixers; voice-coil mixers; mixers; anti-reciprocity; electrical impedance; mechanical velocity; gyrator; electro-mechanical systems

---

1. Introduction

Fluid mixing techniques are ubiquitous in microfluidic applications. The spectrum of use has been broadened in many areas, such as biological, medical, and chemical research and industries [1,2]. Micromixers are typical devices for mixing a micro amount of fluids with various mixing principals. There are two types of micromixers, active and passive. The passive micromixers do not need power sources; they use pressure to guide fluid blending. However, active micromixers require actuators, which use an external source of energy in the form of acoustic, electrokinetic, electrowetting, magnetic, electromagnetic, etc. [3].

Electromagnetic mixers utilise the Lorentz force; alternating voltages are applied to positive and negative electrical terminals of the devices to induce the alternating current. The fluctuating electrical motions cause complicated fluid motions, which generate multi-micro streams and enhance the mixing efficiency [4]. An electromagnet was used to generate transient interactive flows to enhance micromixing by Chih-Yung Wen and Lung-Ming [5]; the authors observed interesting miscible maze-like patterns. They presented a ferrohydrodynamic micromixer that utilises low-voltage and low-frequency properties. Numerical studies on electromagnetic mixers were also performed to characterise magneto-hydrodynamic micro-mixers [4,6,7]. Yiping Chen and Kim [7] introduced an electromagnetic fluidic device to enable both mixing and pumping functions by the Lorentz force, which was induced by the current and applied magnetic field. Pengwang et al. [8] reviewed and compared many types of actuators to be used in the micro-electromechanical system (MEMS) area, including electromagnetic actuators. Although the main objective of the article focuses on MEMS-based scanning micromirrors, general actuator principles apply the same in the MEMS area. Compared to other types of actuators, the electromagnetic actuator requires lower driving voltage, but can achieve a more significant driving force. However, they may need external magnets, which can increase the system size and create electromagnetic interference.

The acoustic mixing techniques provide highly efficient and controlled mixing results with little restrictions on the types of fluids used. The acoustic micromixers blend fluids using devices that induce micro-streams. They use various mediums, such as trapped micro air bubbles [9], a thin solid plate [10], or a piezoelectric membrane [11–15] to transfer the acoustic energy. Chan et al. [16] demonstrated controlling acoustic energy propagation in a microfluidic chip via frequency selectivity. They employed a voice coil as an electromagnetic force-driven acoustic actuator. The electromagnetic and acoustic domains were coupled by the "moving-coil" component.

Kim et al. [17] presented a microfluidic mixer, which consisted of a chamber and an acoustic actuator. The voice coil actuator was electromagnetically and mechanically coupled to the cylindrical chamber. It converted a periodic input electrical signal at an optimum operating frequency to vibratory mechanical stress entering the chamber. They also demonstrated that an optimum operating frequency of the input electrical signal could be determined by sweeping frequencies, measuring the corresponding impedance in the frequency range, and selecting the sweet-spot operating frequency based on the impedance-frequency plot.

Commonly, the mixing efficiency of micromixers is analysed by mixing parameters such as time, length, and the mixing index [18,19], and these parameters depend on the ability of the transducer system (which includes the energy source, transducers, and membrane/transmission materials). To optimise transducer efficiency, there are generally a few approaches. As a theoretical approach, system modelling is performed for numerical simulations using various software such as ANSYS/CFX, FLUENT, and COMSOL Multiphysics or written codes, for instance the lattice Boltzmann technique [18]. The computing simulation is critical to estimating and optimizing all mixing parameters, especially for the experiments' initial conditions. As a direct practical approach, a laser Doppler vibrometer or a micro-force sensor is employed. However, these approaches require additional costs such as computational resources or system installation. With an indirect approach, the efficiency of the transducer is combined with the other parameters influencing the mixing efficiency, and measurement using a tracer (such as a colour dye) combined with an image processing technique is used to estimate the transducer efficiency [17]. Although mixing parameters based on the experiment would predict a sufficiently accurate result, dead zones may exist; the image-based tracer analysis can be a suitable method for ascertaining the degree of mixing. External image monitoring devices can be considered as add-ons to support mixing efficiency analysis; however, it adds costs to the system installation [16,18]. Moreover, more trial and error are required to optimise both the transducer and the mixing system together.

One unique approach introduced by Kim et al. [17] was to use input impedance measurement from electrical terminals of the systems to measure the velocity response of the membrane. They demonstrated the mixing degree detection from the electrical impedance measurement. From a practical point of view, the experimental installation was simple and affordable compared to the conventional methods (of tracking the mixing degree) discussed above. In this paper, we elucidate the physical foundation underlying the micromixer's electromagnetic coupling by adopting the principle of a component called a gyrator. Starting by introducing two electrical coupling devices, a transformer and a gyrator, we derive a correlation between the electrical driving point impedance and mechanical velocity in a two port electromagnetic system, the acoustic voice-coil micromixer. We also discuss experimentally measured results to support the theoretical concept.

2. Theories and Methods

A systematic way to understand different fields in engineering or physics begins by generalising the areas. Fields (or domains or modalities, i.e., electric, acoustic, mechanic, etc.) are analogous to each other. Two conjugate variables exist in pairs in each field, a generalised force and a flow. The two variables are used to characterise a modality by their product and their ratio. They could be either a **vector** (in bold) or a scalar and also can vary spatially. A product of these two variables defines the power, while a ratio establishes the impedance, which is usually determined in the frequency domain. Some examples of the conjugate variables in each modality are described in Table 1.

Table 1. Examples of modalities and their conjugate variables. Upper case symbols are used for the frequency domain variables except in the electromagnetic (EM) case, as its traditional notation uses capital letters for the time-domain analysis (i.e., $\mathbf{E}(t) = -\nabla\phi(t)$, where $\phi(t)$ is the electric scaler potential and the voltage in the time domain.

Modality	Conjugate variables (vector in bold)	
	Generalised force (unit)	Flow (unit)
Electric	Voltage (Φ) (V)	Current (I) (A)
Mechanic	Force (**F**) (N)	Particle velocity (**U**) (m/s)
Acoustic	Pressure (P) (N/m^2)	Volume velocity (**V**) (1/ms)
Electromagnetic	Electric field (**E**) (V/m)	Magnetic field (**H**) (A/m)

One can define a system using a single modality (one port system) or a combination of them. Typical examples of the combined modality systems are electroacoustics and electromechanics, speakers, and motors, respectively. An electromagnetic system is under a subcategory of the electro-mechanical systems. It couples electrical and mechanical domains through the electromagnetic field (Table 1).

To combine the modalities, the port network concept and coupling components are required. Regardless of the modalities, there is a law that applies to the every system, the law of the conservation of energy [20,21]:

$$e(t) \equiv \int_{-\infty}^{t} power(t)dt \geq 0, \tag{1}$$

where the total delivered energy $e(t)$, which is an integration of the power over time, should be greater than (or equal to) zero, and $power(t)$ is work done per unit time defined as a potential multiplied by a net flow. In other words, Equation 1 means we cannot have more energy than we supply.

A transformer and a gyrator are two standard modality-coupling units. Both of them are defined as electrical components having transmission (ABCD) matrices; cascading ABCD matrices of circuit components are a popular tool to analyse circuits [22–25]. The transformer is a typical element that links one modality to another in a one-to-one manner; a flow in one modality is linear to the flow in the other field. Ideal and non-ideal cases of the transformer's theory and the ABCD matrices are established well in classic works in the literature [23]. In the non-ideal case, mutual coupling factors between two adjacent circuits are considered [20,23,26].

McMillan [27] introduced a two port network that violates reciprocity; two years later, Tellegen [28] defined an ideal gyrator as the 5th electrical component to support the anti-reciprocal characteristics of the system. A unique thing about this gyrator is that it can model an electromagnetic (EM) network without using the mobility analogy, which is effective mathematically, but an anti-intuitive method to describe an EM system. The variables of each modality can be represented without modification as they have a gyrator as an EM coupler between the electrical and mechanical terminals.

The mobility (dual) analogy must be appreciated to model an EM system with a transformer. Figure 1 explains the dual method; the two circuit representations are equivalent to each other. If the gyrator is not used in the circuit, a series combination of the mass, damping, and stiffness in the mechanical side variables becomes a parallel relationship with the anti-reciprocation of each variable [29]. Furthermore, the flow and potential in one domain (i.e., $\mathbf{U}, \mathbf{F}$) become a potential and flow on the other side, respectively (i.e.,$\mathbf{\Phi}, \mathbf{I}$). Despite the usefulness of using a gyrator in modelling EM systems, this component has not become mainstream in EM circuit modelling and analysis.

Other than [25], only a few approaches have been made in EM system modelling with gyrators, mainly limited in power electronic fields. Yan and Lehman [30] demoted the benefit of using a gyrator in the EM modelling approach. In their work, a simplified modelling method was introduced for DC/DC converters using an extended capacitor-gyrator (C-G) modelling technique. They showed their modelling feasibility in complex core and winding structures. Young et al. [31] used the C-G to model a continuously variable series reactor (CVSR), which requires prudent planning to design. The authors also highlighted the convenience of using gyrators in their system modelling. Zhang et al. [32] demonstrated an improved C-G modelling method in power systems by taking the eddy current effect into account. They claimed that classical C-G modelling was not suitable; the eddy current impact must be considered in EM modelling. This point also has been demonstrated by Kim and Allen [25].

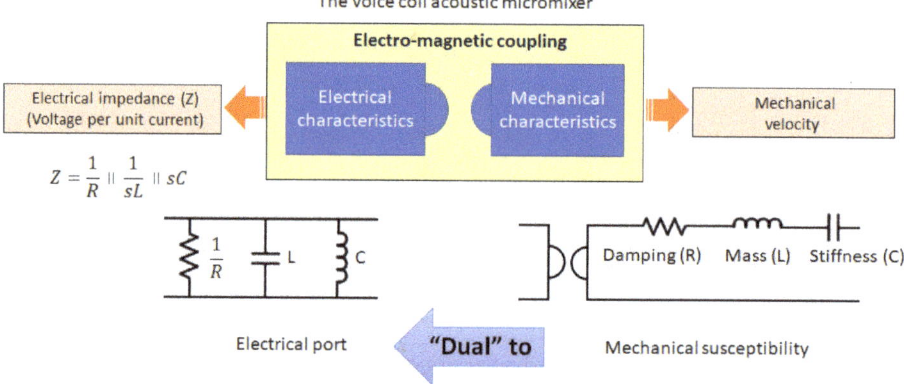

Figure 1. Mobility networking through electromagnetic coupling in the voice-coil acoustic micromixer. Without employing the gyrator for EM circuits, the mechanical components become dual when seen on the electrical side of the network.

A two-port network, such as an electro-mechanical system, has Φ, I, F, and U as the system's variables. A gyrator exists to couple the electric and mechanical sides. To specify this property, the impedance matrix of an ideal gyrator is employed:

$$Z_{gyrator} = \begin{bmatrix} 0 & -G \\ G & 0 \end{bmatrix}, \tag{2}$$

where $G = B_0 l$ is the gyration coefficient, B_0 is the DC magnetic field, and l is the length of the wire. Thus:

$$\begin{bmatrix} \Phi(\omega) \\ F(\omega) \end{bmatrix} = \begin{bmatrix} 0 & -B_0 l \\ B_0 l & 0 \end{bmatrix} \begin{bmatrix} I(\omega) \\ U(\omega) \end{bmatrix}, \tag{3}$$

namely,

$$\Phi(\omega) = -B_0 l U(\omega) \text{ and } F(\omega) = B_0 l I(\omega). \tag{4}$$

Taking the ratio of the two equations in Equation (4), an impedance is derived:

$$\begin{aligned} \frac{\Phi(\omega)}{B_0 l I(\omega)} &= \frac{-B_0 l U(\omega)}{F(\omega)} \\ Z(\omega) &= \frac{\Phi(\omega)}{I(\omega)} \end{aligned} \tag{5}$$

When defining an impedance, the flow direction is defined as into the terminals; thus, U is defined as going into the network. Therefore, the minus sign of U in Equation (4) follows from Lenz's law. Note that Equation (4) explains an ideal gyrator, considering only a DC magnetic field. Appreciating that the impedance is a concept in the frequency domain, an angular frequency symbol, ω, is omitted from this part and onward.

Considering only a single ideal gyrator element, we can obtain the velocity response via measuring the electrical impedance while performing a constant current sweep across frequencies,

$$Z_{idealG}(\omega) = \frac{-B_0 l U(\omega)}{I} \tag{6}$$

2.1. The Non-Ideal Gyrator

The non-ideal case of Equation (4) is considered from the basics of electromagnetism. Ulaby [33], Kim and Allen [34] described the induced emf (voltage $\phi(t)$) as the sum of a transformer component ($\phi_{tr}(t)$) and a motional component ($\phi_{mot}(t)$), namely,

$$\phi(t) = \phi_{tr}(t) + \phi_{mot}(t). \tag{7}$$

The transformer voltage is:

$$\begin{aligned} \phi_{tr}(t) &= -(-\int \frac{\partial \mathbf{B}(t)}{\partial t} \cdot \mathbf{dA}) \\ &= \frac{\partial \psi(t)}{\partial t} \end{aligned} \tag{8}$$

where $\psi(t)$ is the magnetic flux. The voltage has an opposite direction from the emf, emf $\equiv \oint \mathbf{E} \cdot \mathbf{dl} = \int_a^b \mathbf{E}' \cdot \mathbf{dl} = -\phi(t)$. In the static case, the time-varying term is zero.

ϕ_{mot} represents the motion of the electrical voltage [33]. The voltage is associated with the motion from the other port (i.e., mechanical). In other words, the signal is observed from the mechanical side (motional voltage due to u). Note that this concept can be applied only in two port (or higher order) systems.

Derivation of ϕ_{mot} starts from the Lorentz magnetic force ($\mathbf{f}_m$), acting on a moving charge q inside a magnetic field $\mathbf{B}$ with a velocity $\mathbf{U}$,

$$\mathbf{f}_m = q(\mathbf{U} \times \mathbf{B}). \tag{9}$$

Then, the motion of the magnetic force from the electrical field $\mathbf{E}_{mot}$ is $\mathbf{f}_m = q\mathbf{E}_{mot}$.The unit of q is in Coulombs (C) and $\mathbf{E}_{mot}$ in (V/m) = (N/C) as 1 (V) $\equiv$ 1 (J/C) and 1 (N) = 1 J/m. Therefore, $q\mathbf{E}$ stands for force with a unit of N. A positive charge ($q > 0$, proton) is 1.602×10^{-19} C; thus, the charge of an electron (negative charge) is -1.602×10^{-19} C. One coulomb of charge equals the charge that can light a 120-watt-bulb for one second. Therefore,

$$\begin{aligned} \mathbf{E}_{mot} &= \frac{\mathbf{f}_m}{q} \\ &= \mathbf{U} \times \mathbf{B}, \end{aligned} \tag{10}$$

where $\mathbf{E}_{mot}$ is the motional electric field seen by the charged particle q, and its direction is perpendicular to both $\mathbf{U}$ and $\mathbf{B}$.

Thus, the voltage Φ_{mot} is defined as the line integral of the corresponding electric field, which is $\mathbf{E}_{mot}$ in this case,

$$\begin{aligned} \phi_{mot}(t) &= -\oint_C \mathbf{E}_{mot} \cdot \mathbf{dl} \\ &= -\oint_C (\mathbf{U} \times \mathbf{B}) \cdot \mathbf{dl}. \end{aligned} \tag{11}$$

This term has been considered in the ideal gyrator.

Finally, the total voltage becomes:

$$\begin{aligned} \phi(t) &= \phi_{tr}(t) + \phi_{mot}(t) \\ &= \int \frac{\partial \mathbf{B}(t)}{\partial t} \cdot \mathbf{dA} - \oint_C (\mathbf{U} \times \mathbf{B}) \cdot \mathbf{dl}. \end{aligned} \tag{12}$$

In the frequency domain, Equation (12) is rewritten as:

$$\begin{aligned} \Phi &= s\Psi - BlU \\ &= sL_eI - BlU, \end{aligned} \tag{13}$$

where $s = j\omega$, L_e is a leakage inductance due to the leakage flux of a self-inductance in the electrical side, $\Psi = L_eI$.

Assuming a static DC magnetic field (B_0), then $s\Psi = 0$, and we find the ideal gyrator definition $\Phi = \Phi_{mot} = -UB_0l$ (Equation (4)). The frequency dependent term shown in Equation (13) ($j\omega\Psi$ and $j\omega L_eI$) is a non-quasistatic (dynamic) term that is not considered in an ideal gyrator. The minus sign for the other term $-UBl$ is related to Lenz's law.

Two types of magnetic fields exist in an electro-mechanical network: one is the DC magnetic field, and the other is the AC magnetic field. In the ideal gyrator formula, only the motional parts (or the DC magnetic field) of the variables (voltage and force) are considered, which dominate an EM system usually. The two modalities in the network (electrical and mechanical) share this DC magnetic field, which is shown in the motional part of each variable. For the non-ideal case, the transduction parts (or AC magnetic field) of variables along with the motional parts must be considered.

For a non-ideal gyrator, we rewrite the mixing system model (from Equation (5)) with the transformer voltage and mechanical impedance (membrane response), Z_m,

$$\Phi(\omega) = -B_0lU(\omega) + j\omega L_eI(\omega) \tag{14}$$

$$F(\omega) = B_0lI(\omega) + Z_mU(\omega) \tag{15}$$

$$Z(\omega) = \frac{-B_0lU(\omega)}{I} + j\omega L_e \tag{16}$$

When the transformer impedance is not small (i.e., $|\omega L_eI(\omega)| \approx |B_0l|$), we have to use a very small constant current, and the question of whether the velocity response in its reasonable operating condition can be reflected in the electrical impedance response might be raised. Therefore, we formulate the relationship between the velocity response and electrical impedance with another approach.

2.2. Velocity Frequency Response with Constant Voltage

From Equation (14),

$$\begin{aligned} \Phi(\omega) &= -B_0 l U(\omega) + j\omega L_e \frac{\Phi(\omega)}{Z(\omega)} \\ U(\omega) &= \frac{\Phi(\omega)}{B_0 l}[\frac{j\omega L_e}{Z(\omega)} - 1] \end{aligned} \tag{17}$$

During a constant voltage frequency sweep (i.e., $\Phi(\omega) = \Phi$),

$$U(\omega) = \frac{\Phi}{B_0 l}[\frac{j\omega L_e}{Z(\omega)} - 1] \tag{18}$$

As ωL_e is the same order as $Z(\omega)$, small changes in the electrical impedance $Z(\omega)$ will be reflected in the velocity response and vice versa.

To illustrate this concept in more detail, we substitute $I(\omega)$ via Equations (15) into (16) to obtain the impedance response,

$$\begin{aligned} Z(\omega) &= \frac{B_0 l U(\omega)}{I(\omega)} + j\omega L_e \\ &= \frac{-(B_0 l)^2 U(\omega)}{F(\omega) - Z_m U(\omega)} + i\omega L_e \end{aligned} \tag{19}$$

Without external force (i.e., no other mixing source in the vicinity or the clamped mechanical system that cannot move), the electrical impedance response in terms of the mechanical impedance is:

$$Z(\omega) = \frac{(B_0 l)^2}{Z_m} + i\omega L_e \tag{20}$$

We also substitute Equations (20) into (17), to obtain the (constant voltage) velocity response in terms of the mechanical impedance,

$$\begin{aligned} U(\omega) &= \frac{\Phi}{B_0 l}[\frac{j\omega L_e}{Z(\omega)} - 1] \\ &= \frac{\Phi}{B_0 l}[\frac{j\omega L_e Z_m}{(B_0 l)^2 + i\omega L_e Z_m} - 1] \\ &= \frac{\Phi}{B_0 l}\frac{-(B_0 l)^2}{(B_0 l)^2 + i\omega L_e Z_m} \end{aligned} \tag{21}$$

When the mechanical impedance is much larger than the gyrator coefficient ($B_0 l$),

$$Z(\omega) \approx \frac{(B_0 l)^2}{Z_m} \tag{22}$$

and

$$U(\omega) \approx \frac{-B_0 l \Phi}{i\omega L_e Z_m} \tag{23}$$

Then, Equation (22) can be rewritten as:

$$Z(\omega) \approx \frac{-i\omega L_e B_0 l}{\Phi} U(\omega). \tag{24}$$

2.3. Mechanical Impedance

From the forced vibration equation, the force due to the thin membrane [35],

$$F_m = \omega \sum_{\phi} \zeta(\omega, \phi, \rho_m, h, c, Y)(\omega_r^2 - \omega^2)U \tag{25}$$

where ζ is the membrane properties function, ρ_m, ϕ is the mode shape of the vibration, h, c, and Y are the membrane density, thickness, viscous solid damping, and Young's modulus, ω_r is the resonant frequencies of the membrane, and F_m is the membrane force.

Assuming quasi-static (considering the large time-scale between vibrations and mixing) conditions, the force on the membrane due to the fluid vibration can be approximated [36,37] by solving the Navier–Stokes equation. The scalar potential flow field, φ, has a solution in the form,

$$\varphi = \frac{U}{i\omega} \sum_{\phi} K_\phi A_{f,\phi} \sinh k(\mu, \omega)(h - z) \tag{26}$$

where k represents the penetration of the vibration, Ki is the amplitude associated with the mode shape, h is the height of the fluid, z is the distance from the membrane, μ is the dynamic viscosity, and A_f is the geometrical area coefficient of the membrane.

The pressure P, therefore, has the following form,

$$\begin{aligned} P &= -\rho_f \ddot{\varphi} & (27) \\ &= -j\omega\rho_f U \sum_{\phi} K_\phi A_{f,\phi} \sinh k(\mu, \omega)(h - z) & (28) \end{aligned}$$

where ρ_f is the density of the fluid.

The force due to the fluid on the membrane, F_f, can be obtained by integrating the pressure across the membrane area, A,

$$\begin{aligned} F_f &= \int_A P|z = 0 dA & \\ &= -j\omega\rho_f U \sinh k(\mu, \omega) h \sum_{\phi} K_\phi \int_A A_{f,\phi} dA & (29) \\ &= U\rho_f \omega h_f(\omega, \mu) A_f & (30) \\ h_f &= -j \sinh k(\mu, \omega) h & (31) \\ A_f &= \sum_{\phi} K_\phi \int_A A_{f,\phi} dA & (32) \end{aligned}$$

where h_f is the equivalent height of the fluid on top of the membrane.

The mechanical impedance can be formulated by the summation of the fluid and membrane force (Equations (25) and (30)),

$$\begin{aligned} F_m + F_f &= Z_m U & (33) \\ Z_m &= \omega[\sum_{\phi} \zeta(\omega, \phi, \rho_m, h, c, Y)(\omega_r^2 - \omega^2) + \rho_f h_f(\omega, \mu) A_f] & (34) \end{aligned}$$

Equation (34) shows that mixing degrees, which can be reflected as a change in fluid density ρ_f and/or equivalent height h_f near the membrane would alter the value of the mechanical impedance. This change would then be reflected in the electrical impedance response, as shown in Kim et al. [17].

2.4. Experimental Setup

For the empirical study of the anti-reciprocal property via the gyrator, we designed a simple electro-mechanical system having a permanent magnet, a voice coil, and a loading chamber. The design process was adopted by Kim et al. [17].

Voice coils are thin and the wires lightweight, and a typical application of the voice coil is a dynamic speaker. With the American Wire Gauge (AWG) standards, AWG 30 (cross-section area = 0.0510 (mm^2), resistance = 338 (Ω/km)) or higher is commonly used in the loudspeaker industry. Note that a higher AWG number indicates a thinner wire. A typical material for the voice coil is copper with an insulation coating (i.e., enamel). The main application of the voice coil is a low-power personal gadget (i.e., an earphone); thus, it can be used as a safe and low-cost application. Such a delicate character of the voice coil is suitable for microfluidic applications.

Once a signal is applied to the coil, the AC magnetic field is generated around the voice coil-loop. Due to the permanent magnet (DC magnetic field), the voice coil moves as it alternates the AC electromagnetic polarities. The coil attached to the polydimethylsiloxane (PDMS) membrane vibrates due to the electromagnetic force at the coil, as depicted in Figure 2. The chamber is used as a fluid container to simulate mechanical loads' variation.

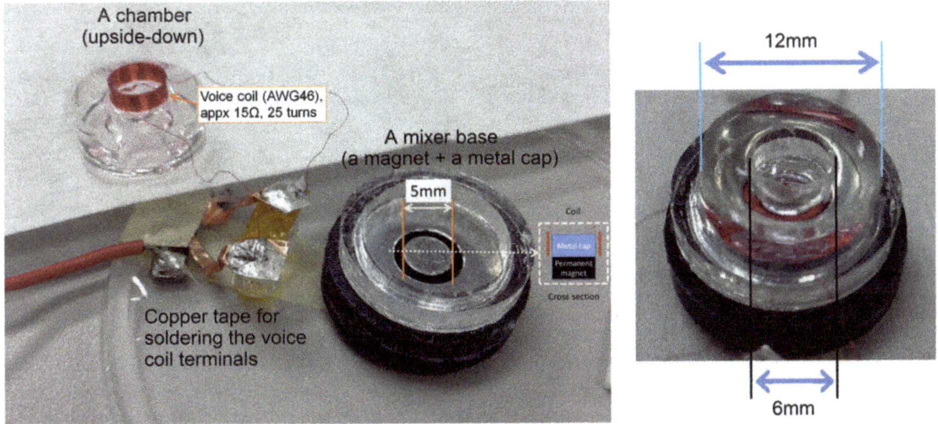

Figure 2. The electromagnetic mixer design. A voice coil is attached to a PDMS-based membrane. A hollow, cylindrical chamber is connected to form a mixer unit. Input signals are driving from the electrical terminals (two ends of the voice coil), and loads vary in mechanical terminals. As shown in the previous study [17], the mechanical loads' variation, including mixing performance, can be captured at the electrical input terminals.

3. Results and Discussion

Based on Ohm's law, the electrical impedance is defined as voltage over current in Ω or V/A, volts per unit current. It can be interpreted as a one-port network's transfer function given an output over an input. The voltage is a potential variable in the electrical field; therefore, in electro-mechanical systems, the velocity (flow in the mechanical side) across a gyrator is linked to the voltage term, as shown in Equation (4) (ideal case). Loads in the mechanical front are receptive to the electrical impedance measured; the mechanical loads' variation should be reflected in the electrical impedance in the electromagnetic systems. The non-ideal case (Equation (13)) contemplates additional AC effects; an EM system is affected by both the induced (AC) and the permanent (DC) magnetic fields. However, when the DC magnetic field governs (i.e., an EM system with a strong magnet and a relatively weak moving coil part), the relationship between the voltage and the velocity shows more similar patterns to each other. In this case, the motional voltage term in Equation (13) dominates the total voltage, Φ, and the electrical impedance response represents the velocity response in a constant current frequency sweep. However, when the transformer impedance becomes more substantial, which is generally the case when we drive the system to its full capability to overcome a sizeable mechanical impedance, we require a constant voltage frequency sweep in order for the velocity response to be represented by the electrical impedance response (see Equation (24)).

Taking a case when the permanent DC magnetic field is influential and the transformer impedance is not negligible, we perform both electrical and mechanical experiments and electrical point impedance and mechanical velocity measurements. Figure 3 represents our experimental materials, setting, and methods.

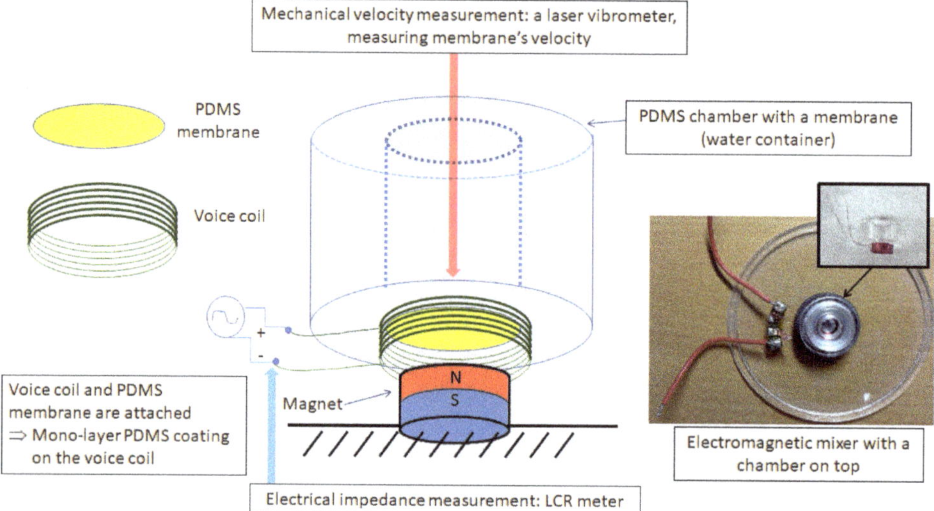

Figure 3. A schematic illustration and a picture to explain the experimental concept. The signal (i.e., sweep frequencies generated by a function generator) is applied to the coil's terminals. Then, due to the permanent magnet, the electromagnetic force drives the coil to move. An LCR meter is used to take the electrical driving point impedance of the system, and a laser vibrometer light is focused on the membrane to measure the membrane's velocity.

3.1. Electrical Impedance Measurement

There are several ways to gauge electrical impedance. One of the methods is to use an LCR meter. For example, the Agilent E4980A Precision LCR Meter was used in this study. Every physically realizable circuit, such as resistors, capacitors, and inductors, has free-loading components. These include undesired resistance in capacitors, capacitance in inductors, and inductance in resistors. Thus, complex impedance representation is required to model a system precisely. The electrical impedance (Z) has a real part and an imaginary part. The Z can be written in rectangular form as resistance and reactance or in polar style as magnitude and phase. With the LCR meter, one may choose a way to analyse the complex impedance of a physical system.

3.2. Mechanical Velocity Measurement

The Polytec CLV3000 3D laser vibrometer (https://www.polytec.com/int/) 3D laser vibrometer has been used for measuring the mechanical velocity of the electromagnetic system. There are four components to drive the laser system: a laser machine, an Data acquisition box (Polytec VIB-E-400 Junction box) , a laser controller, and a management computer with the software.

Figure 4a,b compares electrical impedance and mechanical velocity data obtained from the same electro-mechanical device introduced in Figure 3. Different amounts of water were loaded into the chamber, then the results were compared from both side experiments.

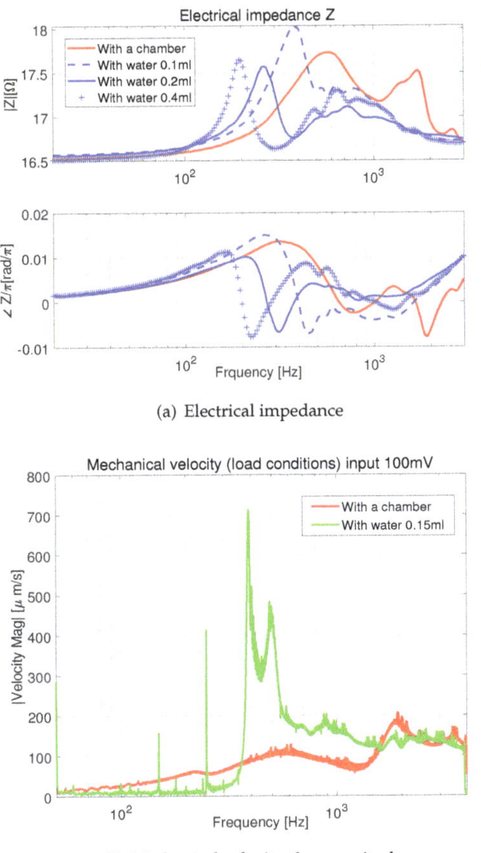

(a) Electrical impedance

(b) Mechanical velocity, the magnitude

Figure 4. The electrical impedance and mechanical velocity measurement results from the electromagnetic mixer introduced in Figure 2. Two loading conditions are examined: an empty chamber or filled with water. For the electrical impedance measurement, three different water volumes were tested to validate the numeric model suggested in Equations (22) and (23).

There were four experimental conditions in the electric impedance data (Figure 4a): a device with a chamber and a device with a chamber loaded with three different amounts of water (change in equivalent fluid height). The same system was tested mechanically. In Figure 4b, the magnitudes of the mechanical velocity (using laser) with two conditions are introduced.

The analogy between the two modalities (electrical and mechanical with the electromagnetic coupling) was discussed in Section 2. Figure 4 carries the empirical evidence; the peak frequency location and overall shape of the first two data of each subfigure corresponded to each other.

A water-filled chamber shifted the damping, as well as the mass of the system. This effect entirely changed the characteristic frequency. The result was reflected well in the electrical impedance data (Figure 4a; the resonance moved to a lower frequency with the mechanical loads' increment); while in the laser data (Figure 4b), the effect was disturbed due to laser light deflection. Maintaining a clear focus was essential for the laser experiments. However, on account of the electromagnetic system's dynamic nature, keeping an excellent focal point of the laser light was laborious as the chamber's membrane was oscillating. Analysing the electrical side data was straightforward by minimizing unnecessary efforts such as filtering out noises and interference. For example, to maximize movement (vibration) with the electromagnetic system, we used 0.15–0.2 mL water loading, and we could drive a 300–400 Hz AC signal, supported by Figure 4a.

As an extension of our previous project, the current study focused on providing theoretical insight into the physical system, our electromagnetic micromixer. Despite the excellent series of results, there are also many exciting challenges for our future work, primarily focusing on mixing technologies, enhancing micromixers' performance, and design aided by physical simulation. Hosseini Kakavandi et al. [38] investigated mass transfer characteristics in micromixers by varying the junctions and channel shapes of the mixers. Their study demonstrated that the T-shaped mixer's junction shape and pit diameter critically affected the mass transfer coefficients as chaotic advection was generated by the modification of the mixing channel shape. Chandan Kumar and Nguyen [39] developed a numerical model of magnetic nanoparticles and fluorescent dye under a nonuniform magnetic field. They performed a parametric analysis of the mass transfer process to scrutinize the magnetic field strength and nano-particle size effects in a magnetofluidic micromixer. Their simulation demonstrated that the core stream spread into the upper sheath stream due to magnetoconvection, and their experimental results supported their model simulation. Their work inspired us to investigate our mixer's mass transfer coefficients based on the current design concept, especially the effect of the voice coil attachment (i.e., position and size concerning the chamber) on the membrane.

4. Conclusion

In electromagnetic systems, the magnetic field $\dot{\mathbf{H}}$ links the electrical and mechanical sides owing to anti-reciprocal characteristics. The $\dot{\mathbf{H}}$ (induced by the conducting current from the coil) is affected by the permanent magnet and changes its polarity (directions). The induced magnetic field and the constant magnetic field define the net force on the coil [34]. Thus, the movement of the coil follows the net force's direction.

In this study, the electro-mechanic (or electromagnetic) two port system was examined. Starting with electromagnetic theories such as Maxwell's equation and the Lorentz force law, we investigated a gyrator's non-ideal formulation. It represented an anti-reciprocal characteristic of the electro-mechanic network. The theory was further explored with the mobility (or dual) analogy" one may choose "the dual analogy with a transformer" or "the impedance analogy with a gyrator" to model an electromagnetic system. An essential benefit of using a gyrator was keeping us from the mystifying dual analogy: it helped an intuitive analysis of the EM network. To our knowledge, this was the first attempt to derive a non-ideal gyrator ever since the gyrator was invented by Tellegen [28]; he suggested the gyrator as the fifth circuit element (along with a resistor, a capacitor, an inductor, and a transformer).

This study expounded on the reason why the electrical impedance data were analogous to the mechanical velocity data. An electromagnetic micromixer was designed and tested to explain the anti-reciprocal nature. Additionally, a benefit from electrical impedance measurement was highlighted. Focusing a laser beam is not easy, especially when the point of focus is the fluid's surface, which may be sensitive to the surrounding environment such as lights, vibration, noise, etc. These undesignated factors can interfere with the laser light to make it challenging to focus the light on a fixed position. The electrical experiment may be used to overcome this problem, which provided more precise data to characterise the electro-mechanic device.

Author Contributions: Conceptualization, N.K. and W.X.C.; numerical simulation, N.K. and W.X.C. ; formal analysis, N.K. and W.X.C.; investigation, N.K. and J.B.A.; resources, S.H.N. and Y.-J.Y.; data creation, N.K. and W.X.C.; writing, original draft preparation, N.K.; writing, review and editing, N.K., Y.-J.Y., S.H.N., W.X.C., and J.B.A.; supervision, S.H.N., Y.-J.Y., and J.B.A.; project administration, S.H.N., Y.-J.Y., and N.K.; funding acquisition, N.K. All authors have read and agreed to the published version of the manuscript.

Funding: This work was supported by the Research Support grant funded by the Newcastle University in Singapore (No. CSEEED5009).

Conflicts of Interest: The authors declare there are no conflicts of interest.

References

1. Lee, C.Y.; Fu, L.M. Recent advances and applications of micromixers. *Sens. Actuators B Chem.* **2018**, *259*, 677–702.
2. Mahmud, F.; Tamrin, K.F.; Sheikh, N.A. A Review of Enhanced Micromixing Techniques in Microfluidics for the Application in Wastewater Analysis. *Adv. Waste Process. Technol.* **2020**. doi:10.1007/978-981-15-4821-5_1.
3. Hessel, V.; Löwe, H.; Schönfeld, F. Micromixers—A review on passive and active mixing principles. *Chem. Eng. Sci.* **2005**, *60*, 2479–2501. doi:https://doi.org/10.1016/j.ces.2004.11.033.
4. Wen, M.; Kim, C.N.; Yan, Y. Mixing of two different electrolyte solutions in electromagnetic rectangular mixers. *J. Hydrodyn. Ser. B* **2016**, *28*, 114–124. doi:https://doi.org/10.1016/S1001-6058(16)60613-3.
5. Wen, C.-Y.; Yeh, C.-P.; Tsai, C.-H.; Lung-Ming, F. Rapid magnetic microfluidic mixer utilizing AC electromagnetic field. *Electrophoresis* **2009**, *30*, 4179–4186. doi:10.1002/elps.200900400.
6. Wen, M.; Kim, C. Mixing features in an electromagnetic rectangular micromixer for electrolyte solutions. *J. Hydrodyn. Ser. B* **2017**, *29*, 668–678.
7. Chen, Y.; Fan, X.; Kim, C.N. A new electromagnetic micromixer for the mixing of two electrolyte solutions. *J. Mech. Sci. Technol.* **2019**, *33*, 5989–5998. doi:https://doi.org/10.1007/s12206-019-1143-y.
8. Pengwang, E.; Rabenorosoa, K.; Rakotondrabe, M.; Andreff, N. Scanning Micromirror Platform Based on MEMS Technology for Medical Application. *Micromachines* **2016**, *7*, 24. doi:10.3390/mi7020024.
9. Liu, R.H.; Lenigk, R.; Druyor, S.R.L.; Yang, J.; Grodzinski, P. Hybridization enhancement using cavitation microstreaming. *Anal. Chem.* **2003**, *75*, 1911–1917.
10. Tsao, T.R.; Moroney, R.M.; Martin, B.A.; White, R.M. Electrochemical detection of localized mixing produced by ultrasonic flexural waves. In Proceedings of the IEEE 1991 Ultrasonics Symposium, Orlando, FL, USA, USA, 8–11 December 1991, Vol. 2, pp. 937–940. doi:10.1109/ultsym.1991.234251.
11. Ahmed, D.; Mao, X.; Juluri, B.; Huang, T. A fast microfluidic mixer based on acoustically driven sidewall-trapped microbubbles. *Microfluid. Nanofluid.* **2009**, *7*, 727–731. doi:10.1007/s10404-009-0444-3.
12. Catarino, S.; Silva, L.; Mendes, P.; Miranda, J.; Lanceros-Mendez, S.; Minas, G. Piezoelectric actuators for acoustic mixing in microfluidic devices—Numerical prediction and experimental validation of heat and mass transport. *Sens. Actuators B Chem.* **2014**, *205*, 206–214. doi:http://dx.doi.org/10.1016/j.snb.2014.08.030.
13. Ozcelik, A.; Ahmed, D.; Xie, Y.; Nama, N.; Qu, Z.; Nawaz, A.A.; Huang, T.J. An Acoustofluidic Micromixer via Bubble Inception and Cavitation from Microchannel Sidewalls. *Anal. Chem.* **2014**, *86*, 5083–5088. doi:10.1021/ac5007798.
14. Woias, P.; Hauser, K.; Yacoub-George, E. An active silicon micromixer for microTAS applications. In *Micro Total Analysis Systems 2000*; Springer: Dordrecht, The Netherland, 2000; pp. 277–282.
15. Yang, Z.; Matsumoto, S.; Goto, H.; Matsumoto, M.; Maeda, R. Ultrasonic micromixer for microfluidic systems. *Sens. Actuators A Phys.* **2001**, *93*, 266–272. doi:10.1016/S0924-4247(01)00654-9.
16. Chan, W.X.; Kim, N.; Ng, S.H.; Yoon, Y.J. Acoustic energy distribution in microfluidics chip via a secondary channel. *Sens. Actuators B Chem.* **2017**, *252*, 359–366. doi:https://doi.org/10.1016/j.snb.2017.05.166.
17. Kim, N.; Chan, W.X.; Ng, S.H.; Yoon, Y. An Acoustic Micromixer Using Low-Powered Voice Coil Actuation. *J. Microelectromech. Syst.* **2018**, *27*, 171–178.
18. Bayareh, M.; Ashani, M.N.; Usefian, A. Active and passive micromixers: A comprehensive review. *Chem. Eng. Process. Process Intensif.* **2020**, *147*, 107771. doi:https://doi.org/10.1016/j.cep.2019.107771.
19. Raza, W.; Hossain, S.; Kim, K.Y. A Review of Passive Micromixers with a Comparative Analysis. *Micromachines* **2020**, *11*, 455. doi:10.3390/mi11050455.
20. Van Valkenburg, M.E. *Introduction to Modern Network Synthesis*; Wiley: New York, NY, USA, 1960.
21. Cheng, S.; Arnold, D.P. Defining the coupling coefficient for electrodynamic transducers. *J. Acoust. Soc. Am.* **2013**, *134*, 3561–3672.

22. Hunt, F.V. *Electroacoustics: The Analysis of Transduction and Its Historical Background*; Harvard University Press: Cambridge, MA, USA, 1954.
23. Pipes, L. *Applied Mathematics for Engineers and Physicists*; International student edition, Mac graw-Hill: New York City, NY, USA, 1958.
24. Weece, R.E.; Allen, J.B. A clinical method for calibration of bone conduction transducers to measure the mastoid impedance. *Hear. Res.* **2010**, *263*, 216–223.
25. Kim, N.; Allen, J.B. Two-port network Analysis and Modeling of a Balanced Armature Receiver. *Hear. Res.* **2013**, *301*, 156–167.
26. Van Valkenburg, M.E. *Network Analysis*; Prentice-Hall: Englewood Cliffs, NJ, USA, 1964.
27. McMillan, E.M. Violation of the reciprocity theorem in linear passive electromechanical systems. *J. Acoust. Soc. Am.* **1946**, *18*, 344–347.
28. Tellegen, B.D.H. The Gyrator: A new electric network element. *Philips Res. Rep.* **1948**, *3*, 81–101.
29. Allen, Jont B. *An Invitation to Mathematical Physics, and Its History*; Springer-Nature: New York City, NY, USA, 2020.
30. Yan, L.; Lehman, B. Better understanding and synthesis of integrated magnetics with simplified gyrator model method. In Proceedings of 2001 IEEE 32nd Annual Power Electronics Specialists Conference (IEEE Cat. No.01CH37230), Vancouver, BC, Canada, 17–21 June 2001, Vol. 1, pp. 433–438.
31. Young, M.A.; Dimitrovski, A.D.; Li, Z.; Liu, Y. Gyrator-Capacitor Approach to Modeling a Continuously Variable Series Reactor. *IEEE Trans. Power Deliv.* **2016**, *31*, 1223–1232.
32. Zhang, H.; Tian, M.; Li, H.; Dong, L.; Sun, L. Improved gyrator-capacitor model considering eddy current and excess losses based on loss separation method. *AIP Adv.* **2020**, *10*, 035309.
33. Ulaby, F.T. *Fundamentals of Applied Electromagnetics*; Prentice-Hall: Upper Saddle River, NJ, USA, 2007.
34. Kim, N.; Allen, J.B. Anti-Reciprocal Motional Impedance and its Properties. *Int. J. Magn. Electromagn.* **2019**, *5*, 1–12. doi:10.35840/2631-5068/6521.
35. Harris, C.M.; Piersol, A.G. *Harris' shock and vibration handbook*; McGraw-Hill: New York, NY, USA, 2002; Vol. 5.
36. Chan, W.X.; Yoon, Y.J. Effects of basilar membrane arch and radial tension on the travelling wave in gerbil cochlea. *Hear. Res.* **2015**, *327*, 136–142.
37. Ma, C.; Shao, M.; Ma, J.; Liu, C.; Gao, K. Fluid structure interaction analysis of flexible beams vibrating in a time-varying fluid domain. *Proc. Inst. Mech. Eng. Part C J. Mech. Eng. Sci.* **2020**, *234*, 1913–1927.
38. Hosseini Kakavandi, F.; Rahimi, M.; Jafari, O.; Azimi, N. Liquid–liquid two-phase mass transfer in T-type micromixers with different junctions and cylindrical pits. *Chem. Eng. Process. Process Intensif.* **2016**, *107*, 58–67. doi:https://doi.org/10.1016/j.cep.2016.06.011.
39. Kumar, C.; Hejazian, M.; From, C.; Saha, S.C.; Sauret, E.; Gu, Y.; Nguyen, N.-T. Modeling of mass transfer enhancement in a magnetofluidic micromixer. *Phys. Fluids* **2019**, *31*, 063603. doi:https://doi.org/10.1063/1.5093498.

Article

Characterization of Soft Tooling Photopolymers and Processes for Micromixing Devices with Variable Cross-Section

J. Israel Martínez-López [1,2,3,*], Héctor Andrés Betancourt Cervantes [1], Luis Donaldo Cuevas Iturbe [1], Elisa Vázquez [1,2], Edisson A. Naula [1], Alejandro Martínez López [3], Héctor R. Siller [4], Christian Mendoza-Buenrostro [1] and Ciro A. Rodríguez [1,2,*]

1 Tecnologico de Monterrey, Escuela de Ingeniería y Ciencias, Monterrey 64849, Mexico; habc-8@hotmail.com (H.A.B.C.); A00831116@itesm.mx (L.D.C.I.); elisa.vazquez@tec.mx (E.V.); A00825462@itesm.mx (E.A.N.); christian.mendoza@tec.mx (C.M.-B.)

2 Laboratorio Nacional de Manufactura Aditiva y Digital (MADiT), Apodaca, Nuevo Leon 66629, Mexico

3 Centro de Investigación Numericalc, 5 de mayo Oriente 912, Monterrey 64000, Mexico; alex@numericalc.org

4 Department of Mechanical Engineering, University of North Texas, 3940 N. Elm. St., Denton, TX 76207, USA; Hector.Siller@unt.edu

* Correspondence: israel.mtz@tec.mx (J.I.M.-L.); ciro.rodriguez@tec.mx (C.A.R.); Tel.: +52-81-8358-2000 (J.I.M.-L.)

Received: 29 September 2020; Accepted: 28 October 2020; Published: 29 October 2020

Abstract: In this paper, we characterized an assortment of photopolymers and stereolithography processes to produce 3D-printed molds and polydimethylsiloxane (PDMS) castings of micromixing devices. Once materials and processes were screened, the validation of the soft tooling approach in microfluidic devices was carried out through a case study. An asymmetric split-and-recombine device with different cross-sections was manufactured and tested under different regime conditions ($10 < Re < 70$). Mixing performances between 3% and 96% were obtained depending on the flow regime and the pitch-to-depth ratio. The study shows that 3D-printed soft tooling can provide other benefits such as multiple cross-sections and other potential layouts on a single mold.

Keywords: micromixers; split-and-recombine; additive manufacturing; surface metrology; asymmetric split-and-recombine (ASAR); stereolithography; surface roughness; soft tooling

1. Introduction

Microfluidic-based devices tend to operate under laminar flow regimes where reagent mixing is a significant challenge [1]. Bringing together two separate fluid streams from opposite directions [2] is a strategy that researchers have employed to enhance mixing in a group of devices identified as Split and Recombine (SAR). While the simplest type of SAR device, constituted by a system where two streams collide downstream (T-mixer), has limited performance, further configurations of this principle have been implemented successfully with more intricate geometries, such as rhomboids [3,4], right angles [5–7], and arcs [8–11], as well as the introduction of pillars [12]. Most SAR micromixer devices have been fabricated using approaches such as (a) soft lithography plus polydimethylsiloxane (PDMS) casting, (b) micromachining, or (c) laser ablation.

The development of micromixing has made great strides toward improved designs with better performance and functionality. Hence, SAR micromixers have evolved to adopt more complex three-dimensional structures that include ridges and modular designs [13–16]. For example, Chen et al. used an array of triangular baffles with three depths to produce transverse movement of fluids toward a cascaded splitting and combination (C-SAR) [17], while Gidde et al. introduced a new design based on rectangular baffles, triple split and recombination (RB-TSAR), and elliptical-based triple split and recombination (EB-TSAR) [18]. Raza and Kim [19] proposed an improved asymmetric split and recombine design based on semi-circular profiles [20] by adding forward and backward-facing steps along the microchannel. They report mixing capabilities of 86% under Reynolds number (Re) below 20.

Recent developments include micromixers based on stacking and folding channels, showing a mixing efficiency beyond 95% [21], as well as good performance under a wide range of Reynolds number ($0.5 < Re < 100$) for Newtonian and non-Newtonian fluids [22]. These complex designs are fabricated via micromilling [21–23]. Other advanced micromixers are based on serpentines with non-rectangular cross-sections [24,25].

The toolkit of materials and manufacturing technologies available for microdevice designers has expanded significantly using additive manufacturing [26–28]. For example, Shallan et al. [29] manufactured a three-dimensional micromixer design through direct 3D printing. The concept is based on a 10× scaled-up version (500 µm) of a three-dimensional mixer based on the Baker Map on a theoretical work developed by Carriere [30]. The design is a paragon of the ideal micromixer device. It represents a three-dimensional projection of the split and recombine concept within all the Cartesian directions and mixing at perpendicular angles. Nonetheless, direct device manufacturing (one-step) is still challenging due to the difficulty of 3D printing internal channels [31].

The rapid or soft tooling approach, initially developed for a low-volume production environment, is a viable alternative when complex microdevices are required [32]. Depending on the application, additive manufacturing can be applied directly or indirectly to build tooling (molds). With a direct tooling approach, a mold or die is created directly through 3D printing [33,34]. Complex micromixers with convoluted features can provide an improved homogeneity of samples and tackle the limitations of the low Reynolds number regime [35].

Stereolithography (SLA) or vat photopolymerization comprises several additive manufacturing processes that rely on the selective curing of resin using a UV light. SLA can provide the UV light through several methods, including a laser, a digital micromirror device (DMD), or a combination of a lamp and alLiquid Crystal Display (LCD) based mask. Compared to most 3D printing processes, SLA provides fine resolution, in a range sufficient to reproduce the designs of intricate three-dimensional micromixers. While SLA brings a wide range of possibilities in the development of novel microfluidic devices [36], the available literature is limited in terms of testing and validation of process capability and reliability.

In the past, our group has shown the potential of soft tooling [37]. The objective of this study is to characterize a wider range of photopolymers and associated stereolithography processes in the context of good manufacturing practices for the development of micromixing devices. Surface and dimensional metrology were carried out for soft tooling (molds) in order to assess the process capability and potential chemical interaction with the cast PDMS. Once the photopolymers and SLA processes were screened, an asymmetric split-and-recombine (ASAR) device was built as a case study.

2. Materials and Methods

2.1. Soft Tooling Material Process Screening

Figure 1 describes the process to develop micromixing devices with complex features followed in this work. The design of molds of microdevices that feature a varying cross-section is introduced toward developing complex devices that are more prone to be adapted in-field. The new design introduces the capability to cast devices with multiple cross-sections on a single step toward developing complex devices that are more prone to be adapted in-field. The steps aim to examine an assortment of materials in three different types of stereolithography-based additive manufacturing processes. The materials were selected considering that the employment of resins had become an affordable benchtop technology available for microdevice designers in their lab or as a service [38–41]. Each material's screening is performed with surface characterization and dimensional metrology of selected features and evaluating the usability and the potential to produce functional devices. Insights of the screening can then be applied for the soft tooling of improved devices.

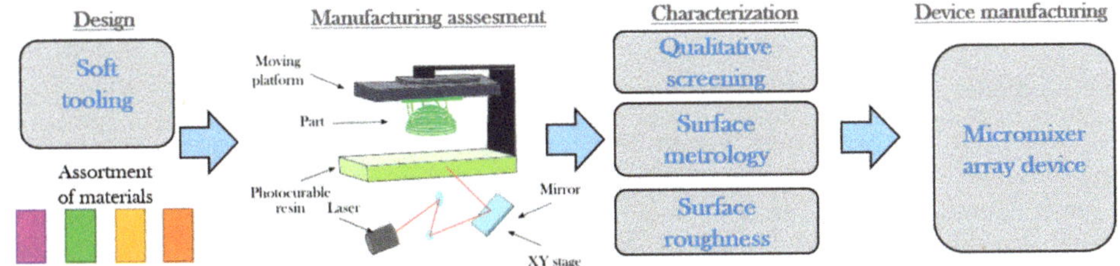

Figure 1. Steps for the screening of soft tooling based on stereolithography.

2.1.1. Screening Mold Geometry

A mold with three main channels and protrusions was designed for material screening (see Figure 2). A depth of 5 mm was selected for the main chamber. Each mold is patterned with three main channels (W_A = 9 mm, W_B = 7 mm, and W_C = 5 mm) and a singular depth or pitch (Δ = 2 mm).

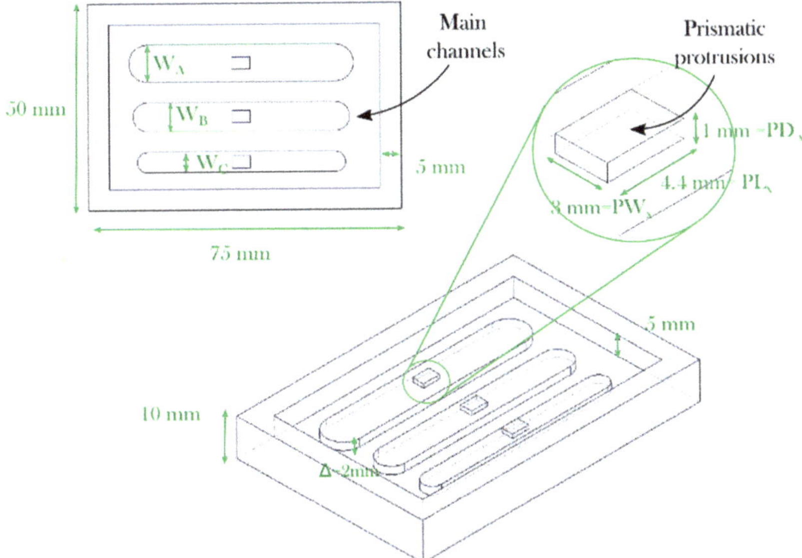

Figure 2. Soft tooling design for photopolymers and process screening.

2.1.2. Qualitative Assessment

The assortment of materials and additive manufacturing techniques were tested experimentally as soft tooling alternatives with the conditions shown below. The test consisted of employing simple manual tools including tweezers, an X-Acto knife, and a Silhouette toolkit that included a hook and a scraper (Silhouette America, West Orem, UT, USA). Once each mold was developed, observations were made to record the usability of the device compared with an aluminum device. The examination of each process included evaluating the difficulty of removing the PDMS elastomer from the device and reviewing the existence of any abnormality observed in the molded pattern after the curation process.

2.1.3. Screening Mold Metrology

For the characterization of the molds, an Alicona Infinite Focus Measurement Machine (Bruker Alicona Headquarters, Graz, Austria) was selected. This device allows for the acquisition of datasets at a high depth of focus, similar to a Scanning Electronic MicroscopeSurface roughness (*Ra*), measured using a 10x optical lens.

2.1.4. Aluminum Mold Manufactured Using a Conventional Subtractive Methodology

A mold was manufactured with conventional subtractive methodology using an aluminum alloy and a 3-axis vertical milling machine center Makino F3 (Makino Inc., MASON, OH, USA) with a maximum spindle speed of 30,000 RPM. A tool holder and a clamp were used for holding the workpiece. A laser measuring system by BLUM (Blum-Novotest Ltd., Staffordshire, UK) was used for tool setting. Three solid tungsten carbide cutting tools were used; a 12.7 mm diameter flat end mill with 4 flutes, a 5 mm diameter flat end mill with two flutes, and a 3 mm flat end- mill with 2 flutes were used for rough, semi-finish, and finish operation. Given the close tolerances achievable through machining, this mold was used as a reference base line.

2.2. *Additive Manufacturing Processes and Materials*

A professional laboratory equipment DLP-SLA Envisiontec Perfactory P3 Mini Multilens, that uses a 60 mm lens system, a work tray of 84 × 63 × 230 mm, and a layer resolution between 15 µm and 150 µm was employed to generate the molds using ABS Flex White, HTM 140, E-Dent 400, ABS Flex Black, and E-Partial. These materials were post-processed in an Otoflash pulse curing chamber of the same supplier (11 W lamp with a wavelength between 300 and 700 nanometers and ten pulses per second).

A benchtop SLA-LF Form 3 additive manufacturing equipment from Formlabs was used employing a 25 µm for the Clear V04 and 50 µm resolution setting for the Flexible V02 resin. Samples were cured using the provider's recommended settings for post-processing the sample accordingly (15 min at 60 °C and 1 h of exposure to UV light).

Additionally, a 3D Systems ProJet 6000 that works with an ultraviolet laser was employed by a service provider [42] to develop devices in Accura SL 5530 resin. The thermoset resin selected for the molds was a high-temperature resistant stereolithography material, Accura SL 5530 (3D Systems, Rock Hill, SC, USA). A post-curing process was performed to the mold by exposing it for 90 min to UV light and baking it at a temperature of 160 °C for 2 h.

2.3. *Microdevice Polydimethylsiloxane (PDMS) Casting*

The standard process for PDMS casting was followed in this work. First, the Sylgard 184 PDMS (Dow-Corning, Midland, Michigan) was poured into a vessel and mixed with a curing agent in a 10:1 ratio by weight. To control this ratio the material was weighed using a calibrated analytical balance (Mettler AT200, Columbus, OH, USA). Then, the mixture was exposed to a vacuum chamber for 10 min to eliminate most of the air bubbles produced before and after pouring the material into the mold. Finally, each mold was placed over a hot plate at 75 °C for polymerization. The temperature was set below the glass transition temperature overnight (12 h) for the process screening and 45 min for the case study. In the latter, shorter times were considered to recreate a more lifelike and time demanding application. For other materials or experimental setup, different times might be required. The PDMS parts were removed using manual tools such as a cutting knife and tweezers.

For the case study, the inlets and outlet ports were punched using a 1.25 mm (internal diameter) Miltex disposable biopsy plunger (Integra Life Sciences, Princeton, NJ, USA). Then, the part was bonded to a 76 × 52 × 2 mm glass slide (VWR International, Radnor, PA, USA). Once the glass slide was cleaned with water and ethanol, the parts were introduced into a plasma cleaner (Harrick Plasma Inc., Ithaca, NY, USA). A vacuum inside the chamber was created for 3 min and the parts were exposed to plasma for 3 min. After three (3) more minutes, the treated surfaces were put in contact for PDMS-glass bonding. A visual inspection of the device was carried out to verify that the device was properly bonding and without bubbles trapped inside.

2.4. Case study: ASAR Micromixer Array

2.4.1. Micromixing Mold Geometry

The mixing performance of the proposed microdevice is based on the one developed by Ansari et al. [17] and used by our research group in the past [43,44]. This version has the added complexity of a variable depth or pitch **Δ** (see Table 1 and Figure 3a). The device is composed of a single channel of 1000µm that splits into two subchannels of uneven width and converges on a channel six times (see Figure 3b,c). A set of micromixer molds were manufactured using the parameters described in Section 2.1 for ABS Flex White Resin. Additional molds were manufactured using ABS Flex Black and Accura SL 5530 resins.

Table 1. Micromixers array employed to assess the manufacturing technology.

Device Type	Array Elements	Microchannel Width (W)	SAR Subchannels (W_1/W_2)	Micromixer Pitch Δ
Asymmetric split and recombine micromixer	6	1000 µm	667/333 µm	I = 100 µm, II = 250 µm, III = 500 µm, IV = 750 µm, V = 900 µm, VI = 1000 µm

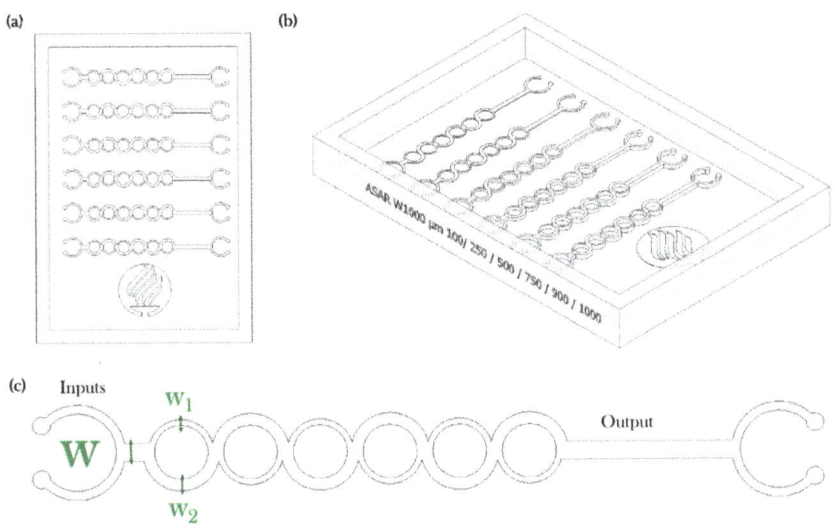

Figure 3. Soft tooling case study: (**a**) Top view, (**b**) Isometric view, (**c**) Detail of micromixer.

2.4.2. Micromixing Performance Evaluation

Reynolds number is conventionally used to characterize the behavior of the flow conditions within microdevices and is defined as the ratio of inertial to viscous forces. Equation (1) represents the Reynolds number (Re) defined as:

$$Re = \frac{inertial\ forces}{viscous\ forces} = \frac{\rho U D_h}{\mu} \quad (1)$$

where μ is the viscosity (Pa s), ρ is the fluid density (kg m^{-3}), U is the average velocity of the flow (m s^{-1}), and D_h is the hydraulic diameter of channel (m), which is defined in Equation (2) as:

$$D_h = \frac{4A}{P} = \frac{4w * \Delta}{2w + 2\Delta} \tag{2}$$

where A and P are the area and the wetted perimeter of the cross-section, which is given by the micromixer width (W) and depth (Δ).

To quantify the mixing behavior, the variance of the liquid species in the micromixer (σ) was calculated. The variance of the species was determined at the cross-sectional area at the output of the micromixer perpendicular to the x-axis. To evaluate the degree of mixing, the variance of the mass fraction of the mixture in a cross-section (σ) that is normal to the flow defined in Equation (3):

$$\sigma = \sqrt{\frac{1}{N}(c_i - \bar{c}_m)^2} \tag{3}$$

where N is the number of sampling points inside the cross-section, c_i is the mass fraction at the sampling point I, and c_m is the optimal mixing mass fraction, which is 0.5 at any cross-sectional plane (ideal mixing). To quantitatively analyze the numerical mixing performance of the micromixer, the mixing index (M) at a cross-sectional plane is shown in Equation (4), which can be defined as:

$$M = 1 - \sqrt{\frac{\sigma^2}{\sigma^2_{max}}} \tag{4}$$

where the mixing efficiency (ranges from 0.00 (0% mixing) to 1.00 (100% mixing). The maximum variance σ_{max} represents a completely unmixed condition.

2.4.3. Micromixing Experimental Setup

An experimental setup for testing mixing efficiency was used for micromixing evaluation. Inlet flows were programmed in a syringe pump (KDS200, Holliston, MA, USA). The micromixer device was evaluated on a range of conditions using the blue dye and distillate water under an Axiovert 200 inverted microscope (Carl Zeiss AG, Oberkochen, Germany). The mixing measures were calculated using intensity profiles. The image processing was performed using the custom software MIQUOD (Mixing Quantification of Devices). The actual flow entering the channels varied considering the hydraulic radius of each of the microchannels, and the Reynolds number at which the micromixer was tested was calculated using the FS3 and FS5 flow meters (Elvesys, Paris, France) connected to the MFS flow reader (Elvesys, Paris, France). The acquisition of 1920 × 1080 pixels images was made using a C930E webcam (Logitech Incorporated, Newark, CA, USA).

The ASAR micromixer array with a variable depth was developed using an Accura SL 5530 resin and tested to assess how relevant the channel pitch for determining the mixing performance. The target areas were equally defined for all the measures (193 × 197 pixels) and calibrated with a channel completely filled with dye and with water before experimentation. The calibration was carried out under the same operating conditions in the experiment (micromixer position and lighting).

3. Results

3.1. Qualitative Assessment of Screened Materials

The qualitative observations derived from the testing of materials for 3D printed mold are shown in Table 2. All the molds were tested under similar conditions. From the assessment it was observed that only the aluminum, E-Partial, ABS Flex white, Clear V04, and Flexible V02 showed the potential for the development of microfluidic devices. In contrast, E-Dent 400, ABS Flex Black, and HTM140 are not recommended for the development of molds. Difficulties for the remotion and swelling have been reported before for untreated bulk and surface of PDMS due to incompatibility with solvents, the absorption of small molecules, or deformations during casting [45–47]. It is possible that these could be some fundamental reasons for the negative observations registered.

Table 2. Qualitative assessment of photopolymers and additive manufacturing processes.

Material	Manufacturing	Tensile Strength/Modulus (MPa) [1]	Observations during Casting
Aluminum 7075	Micromilling	276/ 68900	Excellent surface quality, easy demolding
Clear V04	SLA-LF	65/2800	Excellent surface quality, excellent demolding
Accura SL 5530	SLA	57-63/2854-3130	Good surface quality
E-Partial	SLA-DLP	129/3125	Good surface quality
ABS Flex White	SLA-DLP	65/1772	Good surface quality
Flexible V02	SLA-LF	3.4/8.5	Fair surface quality, easy demolding due flexibility
* E-Dent 400	SLA-DLP	85/2100	Difficulties for demolding
* ABS Flex Black	SLA-DLP	65/1772	Difficulties for demolding
* HTM 140	SLA-DLP	115/3350	PDMS reaction

[1] Information from datasheets. * Note: these resins are not recommended.

PDMS is a kind of silicone. The curing of silicones can be inhibited when in contact with materials that contains sulfur, tin, and nitrogen [48]. In the case of HTM 140, E-Dent 400, and ABS Flex Black, a partial or significant PDMS curing inhibition was found. Therefore, these materials are not recommended as molds. While ABS Flex White resin and ABS Flex Black share the same mechanical properties, it is suspected that the pigment on the latter inhibited PDMS curing. Unfortunately, vendors of 3D printing materials as the ones used in this study do not provide details on the chemical composition and additives. Therefore, it is difficult to provide a detailed analysis as to the reasons behind curing inhibition problems. In addition to the adverse effect of PDMS curing inhibition, the printed mold flexibility plays a significant role in the demanding process of the cast PDMS, as indicated in Table 2, that ranks mold materials from best to worst.

Once this initial qualitative screening was conducted, Clear V04 and Accura SL 5530 were used to produce cast PDMS components. With both of these materials, the plasma treatment was successful when sealing the PDMS component to the microscope slide.

Figure 4a shows a picture of some of the molds manufactured. The materials that are listed with difficulties for demolding either presented bad surface quality or the part could not be removed (See Figure 4b). This problem was most frequently observed between channels B and C and between channel C and the edge of the device, considering that these were the narrowest features.

Compared to the aluminum mold, the polymeric devices were prone to breaking due the force induced by the manual tools. For example, while the Accura SL 5530 resin mold produced a patterned part, one of the walls was shattered during the extraction.

In contrast, compelling results on the usability of the flexible material (Flexible v02) resin were found as the patterned piece could be pulled easily without apparent risks of destroying the device.

3.2. Dimensional and Surface Metrology of Photopolymers and AM Processes

Infinite Focus Microscopy (IFM) has been shown to be capable of capturing images with a lateral resolution down to 400 nm providing 3D data sets with very accurate results [49]. Tables 3 and 4 resume the dimensional and surface metrology of the prismatic protrusions and main channels.

While the deviational error of the aluminum mold using subtractive manufacturing is still a paragon, there are some major advantages on the surface quality of the metallic device compared with the polymeric counterparts, however, such as low surface roughness and a significantly higher bulk modulus (easing the removal of the part from the mold). Low-force stereolithography printing uses a flexible tank and linear illumination to deliver and improve surface quality and print accuracy.

According to the provider, lower print forces allow for support structures to be removed more easily (compared to traditional stereolithography) [50].

Figure 4. Assessment of material compatibility; (**a**) Picture of sample molds); (**b**) Demolding difficulties; (**c**) Polydimethylsiloxane (PDMS) slab removal on a Flexible v02 mold.

Figure 5 Clear V04 and Flexible V02 offered the closest (and lowest) deviational error to the aluminum molds among all the materials. The other materials had an absolute deviational error around between 2.97% and 3.64%. Additional data is available in the Supplementary Material.

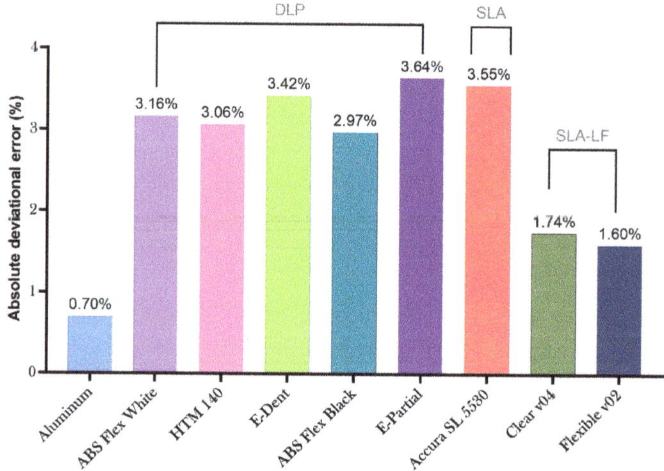

Figure 5. Absolute deviational error of assessed materials.

Table 3. Dimensional and surface metrology of prismatic protrusions.

Feature	Feature	Al	ABS Flex White	HTM 140	E-Dent 400	ABS Flex Black	Accura SL 5530	Clear V04	Flexible V02
PD (mm)	Protrusion depth	1.006 ± 0.002	1.014 ± 0.013	1.026 ± 0.003	0.943 ± 0.002	1.011 ± 0.009	0.890 ± 0.013	1.006 ± 0.011	0.993 ± 0.011
PL (mm)	Protrusion length	4.437 ± 0.002	4.207 ± 0.040	4.217 ± 0.052	4.512 ± 0.024	4.202 ± 0.019	4.450 ± 0.02	4.449 ± 0.041	4.504 ± 0.100
PW (mm)	Protrusion width	3.036 ± 0.002	2.890 ± 0.013	2.887 ± 0.036	3.032 ± 0.012	2.908 ± 0.006	3.11 ± 0.056	3.02 ± 0.009	3.083 ± 0.010
Ra (µm)	Protrusion roughness	0.262 ± 0.085	0.636 ± 0.002	0.526 ± 0.105	0.91 ± 0.170	0.394 ± 0.042	0.303 ± 0.036	0.807 ± 0.098	4.833 ± 0.164

Table 4. Dimensional and surface metrology of main channels.

Feature	Feature	Al	ABS Flex White	HTM 140	E-Dent 400	ABS Flex Black	Accura SL 5530	Clear V04	Flexible V02
W_A (mm)	Channel width A	5.038	4.484	4.758	5.001	4.819	5.035	9.056	9.098
W_B (mm)	Channel width B	7.044	6.634	6.642	6.884	6.714	7.072	7.029	7.15
W_C	Channel width C	9.035	8.574	8.552	8.796	8.642	9.104	5.07	5.056
D_{ABC} (mm)	Channel depth	1.996 ± 0.004	1.966 ± 0.015	2.004 ± 0.011	1.871 ± 0.008	1.954 ± 0.004	2.015 ± 0.005	2.009 ± 0.013	1.970 ± 0.010
Ra (µm)	Channel Roughness	0.225 ± 0.066	0.653 ± 0.032	0.460 ± 0.056	1.095 ± 0.288	0.664 ± 0.120	0.354 ± 0.10	0.839 ± 0.043	4.871 ± 0.214

In terms of the surface roughness of the devices, the values were very similar among the prismatic protrusions and the channels (see Tables 3 and 4). The average surface roughness of each material is shown in Figure 6. The roughest surface was the Flexible v02, and the smoothest molds were produced with the Accura SL 5530 and the HTM 140 resins.

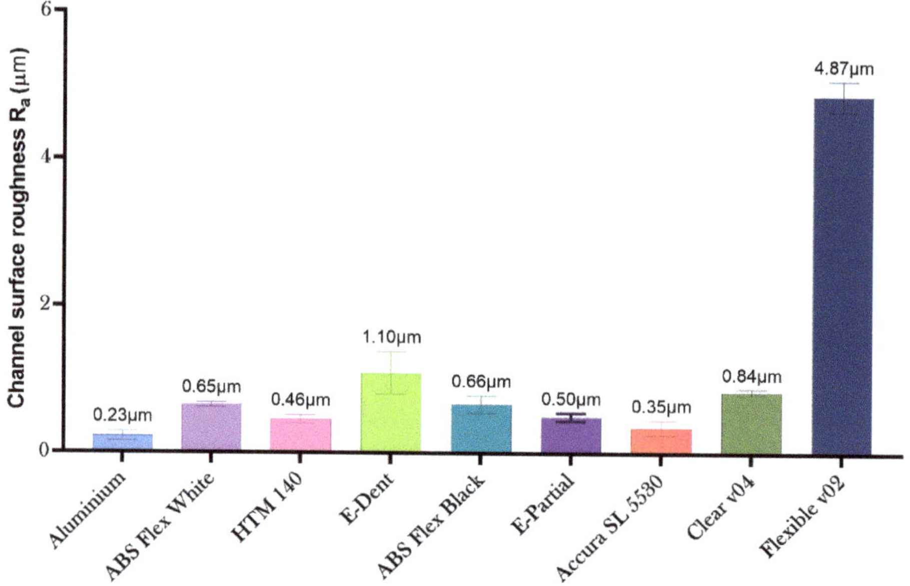

Figure 6. Mean surface roughness of assessed materials.

3.3. *Case Study: Asymmetric Split-and-Recombine (ASAR) Microdevice Surface Metrology*

An additional evaluation of features was done for some of the screened materials to evaluate if the features of a device with finer features could be produced and screened with a similar methodology as the molds made in Section 2.1.1.

Figure 7 shows examples that highlight the advantages of Infinite Focus for evaluating the features compared with other metrology options, such as a stylus-based profilometer. With this methodology, it is possible to obtain details in the micro and meso-scale concurrently in three-dimensional models and determine important functional features such as the width of the subchannels of the ASAR micromixer (Figure 7a).

Overall, the manufacture using the materials was consistent with the data described above. For example, the characterization of the set of ABS Flex White showed to be below 10 μm (see Figure 7b). Figure 7c shows an example of a device manufactured using the proposed methodology and employed for the micromixing performance manufactured with Accura SL 5530 (see Section 3.4). The material was selected because it displayed more potential for usability (see Table 2).

3.4. *Micromixing Performance*

Pictures obtained with the experimental setup described in Section 2.4.3 are shown in Figure 8a as examples of the varying conditions on the degree of mixing, depending on the Reynolds number and the pitch Δ. Figure 8b describes the mixing index (*M*) values quantitatively. As expected, a higher Reynolds number can be associated with a higher degree of mixing overall, considering that in conditions where the inertia forces are more predominant then the viscous forces, the chaotic advection can act to create secondary vortices and enhance mixing downstream.

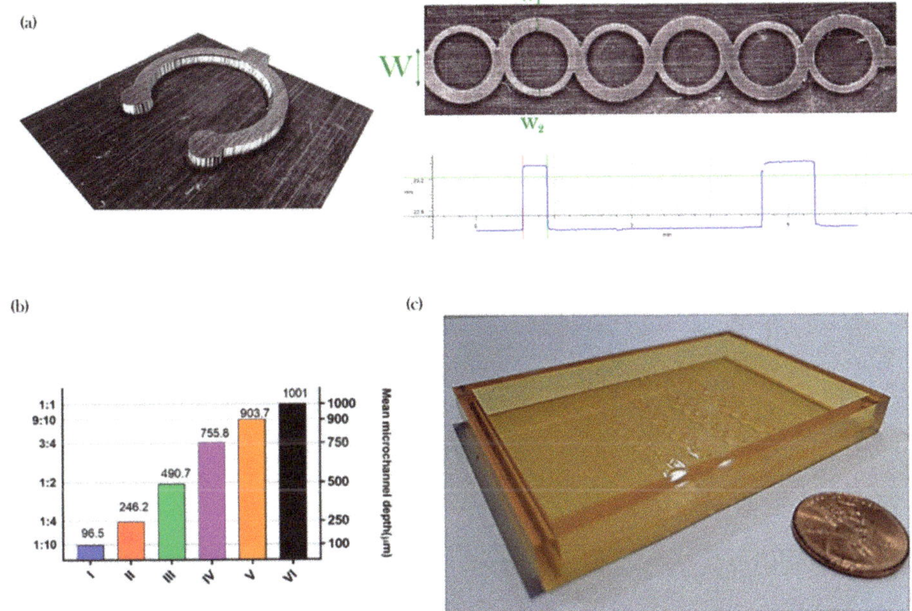

Figure 7. Micromixing device metrology; (**a**) Measurement of a key feature; (**b**) Mean microchannel pitch vs to width ratio and (**c**) Picture of an asymmetric split-and-recombine (ASAR) soft tool mold.

The disposition of the array allowed us to evaluate the performance of the device under similar conditions for different depths and hence different width-to-height ratios (see Figure 8b). Notice that the color of the circle around the picture indicates the corresponding pitch (orange for 100 μm, green for 250 μm, purple for 500 μm, yellow for 750 μm, blue for 900 μm, and pink for 1000 μm). An examination of the data suggests that increasing the microchannel pitch can help increase the degree of mixing. However, the mixing performance improvement between the flow regime of Reynolds number 50 and Reynolds number 70 is limited. During the process, some bubbles were formed on some of the channels ($Re = 70$ and $\Delta = 250$ μm); however, these appeared adjacently to the microdevice walls, and a disruption of the flow mixing due was not observed. The mixing performance showed a dependence on the pitch of the microchannel and the Reynolds number.

The capability to modify the mixing efficiency using a ratio of a single feature to the pitch of the channel has been reported previously [51,52]. Graph (Figure 8b) shows an upward trend between pitch and mixing efficiency. While the ASAR device was developed successfully, there are limitations that offer opportunities for further research to comprehend the phenomena underlying the operation.

3.5. Learned Lessons and Future Work

The applications of the learned lessons can be resumed in the case study as follows:

- It is possible to produce different versions of the same device on a single mold with a single demolding step. Other possible layouts include different devices on a single molding step or an array of a single device.
- Device identifier: engraving symbols on the device can be implemented for identifying the mold among different variations.
- Other potential futures were prospected for future work, as removable wall(s) that could ease the remotion of the PDMS and the capability to dispose placeholders for inlet or outlet pins as part of the mold.

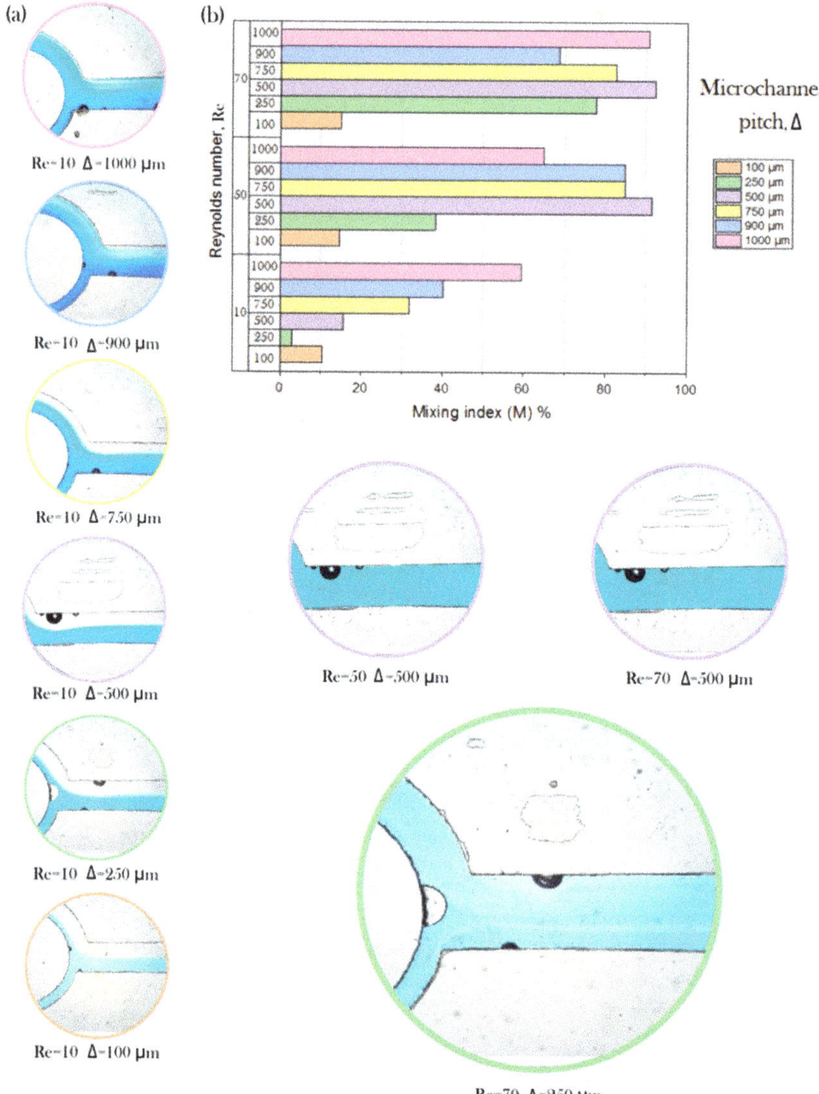

Figure 8. Case study; (**a**) Channel depth assessment comparison using the calculated mixing indexes in a range of 10 < Re < 70; and (**b**) Mixing index M vs Reynolds number vs microchannel pitch (Δ).

4. Conclusions

The concluding remarks of this work are summarized as follows:

- The rapid or soft tooling approach was screened for eight (8) different photopolymers as a viable option for developing complex micromixing devices.
- The experimental data provided valuable insights on acceptable manufacturing practices toward a new generation of devices.
- The novel design of the mold with variable depth was successfully implemented to test different regime conditions within the same device.

- Methodology for the production of an array of micromixers with a variable cross-section was successfully implemented.
- Multiple cross-sections on a single device could be implemented using stereolithography. Other device setups (an array of a single device or different type of devices on a single mold) could be implemented using the methodology presented in this work.
- Surface characterization showed an absolute deviational error within 10 micrometers.
- Stereolithography is a viable option for the development of complex three-dimensional molds for the development of micromixers, but it is necessary to consider the surface-to-surface interaction between the mold and the resin.
- Further studies are required to evaluate the effect of the geometrical features of the ASAR micromixer thoroughly.

Supplementary Materials: The following are available online at http://www.mdpi.com/2072-666X/11/11/970/s1, Supplementary file S1: Dimensional and surface metrology of materials for stereolithography-based AM processes.

Author Contributions: Conceptualization, J.I.M.-L., E.V. and E.A.N.; Data curation, L.D.C.I. and C.M.-B.; Formal analysis, H.A.B.C., L.D.C.I., A.M.L. and C.A.R.; Funding acquisition, J.I.M.-L., E.V., A.M.L., H.R.S, and C.A.R.; Investigation, H.A.B.C., L.D.C.I. and E.A.N.; Methodology, J.I.M.-L., E.V., H.R.S., and C.A.R.; Project administration, J.I.M.-L., E.V. and A.M.L.; Resources, E.V. and H.R.S.; Software, H.A.B.C.; Supervision, J.I.M.-L. and C.M.-B.; Validation, E.V., A.M.L., C.M.-B. and C.A.R.; Visualization, J.I.M.-L.; Writing—original draft, J.I.M.-L.; Writing—review & editing, J.I.M.-L., E.V., A.M.L. and C.A.R. All authors have read and agreed to the published version of the manuscript.

Funding: The CONACyT Basic Scientific Research Grant no. 242634 supported this work. The Research Group of Advanced Manufacturing provided additional support. The authors acknowledge the support from Centro de Investigación Numericalc for providing access to the Formlabs equipment and the technical support from Pavel Celis Retana.

Conflicts of Interest: The authors declare no conflict of interest.

References

1. Gambhire, S.; Patel, N.; Gambhire, G.; Kale, S. A Review on Different Micromixers and its Micromixing within Microchannel. *Int. J. Curr. Eng. Technol.* **2016**, *4*. [CrossRef]
2. Su, Y.; Chen, G.; Yuan, Q. Ideal micromixing performance in packed microchannels. *Chem. Eng. Sci.* **2011**, *66*, 2912–2919. [CrossRef]
3. Chung, C.K.; Shih, T.R. Effect of geometry on fluid mixing of the rhombic micromixers. *Microfluid. Nanofluidics* **2008**, *4*, 419–425. [CrossRef]
4. Hossain, S.; Kim, K.-Y. Mixing Analysis of Passive Micromixer with Unbalanced Three-Split Rhombic Sub-Channels. *Micromachines* **2014**, *5*, 913–928. [CrossRef]
5. Hong, C.-C.; Choi, J.-W.; Ahn, C.H. A Novel In-Plane Passive Micromixer Using Coanda Effect. In *Micro Total Analysis Systems 2001*; Ramsey, J.M., van den Berg, A., Eds.; Springer: Dordrecht, The Netherlands, 2001; pp. 31–33, ISBN 978-94-010-3893-5.
6. Hong, C.-C.; Choi, J.-W.; Ahn, C.H. A novel in-plane passive microfluidic mixer with modified Tesla structures. *Lab. Chip* **2004**, *4*, 109–113. [CrossRef] [PubMed]
7. Hossain, S.; Ansari, M.A.; Husain, A.; Kim, K.-Y. Analysis and optimization of a micromixer with a modified Tesla structure. *Chem. Eng. J.* **2010**, *158*, 305–314. [CrossRef]
8. Sudarsan, A.P.; Ugaz, V.M. Fluid mixing in planar spiral microchannels. *Lab. Chip* **2006**, *6*, 74–82. [CrossRef] [PubMed]
9. Ansari, M.A.; Kim, K.-Y.; Anwar, K.; Kim, S.M. A novel passive micromixer based on unbalanced splits and collisions of fluid streams. *J. Micromech. Microeng.* **2010**, *20*, 055007. [CrossRef]
10. Ansari, M.A.; Kim, K.-Y. Mixing performance of unbalanced split and recombine micomixers with circular and rhombic sub-channels. *Chem. Eng. J.* **2010**, *162*, 760–767. [CrossRef]
11. Gidde, R.R.; Pawar, P.M.; Ronge, B.P.; Misal, N.D.; Kapurkar, R.B.; Parkhe, A.K. Evaluation of the mixing performance in a planar passive micromixer with circular and square mixing chambers. *Microsyst. Technol.* **2018**, *24*, 2599–2610. [CrossRef]

12. Bazaz, S.R.; Warkiani, M.E.; Mehrizi, A.A.; Ghorbani, S.; Vasilescu, S.; Asadnia, M. A hybrid micromixer with planar mixing units. *RSC Adv.* **2018**. [CrossRef]
13. Nimafar, M.; Viktorov, V.; Martinelli, M. Experimental comparative mixing performance of passive micromixers with H-shaped sub-channels. *Chem. Eng. Sci.* **2012**, *76*, 37–44. [CrossRef]
14. Feng, X.; Ren, Y.; Jiang, H. An effective splitting-and-recombination micromixer with self-rotated contact surface for wide {Reynolds} number range applications. *Biomicrofluidics* **2013**, *7*. [CrossRef]
15. SadAbadi, H.; Packirisamy, M.; Wüthrich, R. High performance cascaded PDMS micromixer based on split-and-recombination flows for lab-on-a-chip applications. *RSC Adv.* **2013**, *3*, 7296–7305. [CrossRef]
16. Jian Chen, J.; Ren Lai, Y.; Tang Tsai, R.; Der Lin, J.; Yang Wu, C. Crosswise ridge micromixers with split and recombination helical flows. *Chem. Eng. Sci.* **2011**, *66*, 2164–2176. [CrossRef]
17. Chenouard, N.; Smal, I.; de Chaumont, F.; Maška, M.; Sbalzarini, I.F.; Gong, Y.; Cardinale, J.; Carthel, C.; Coraluppi, S.; Winter, M.; et al. Objective comparison of particle tracking methods. *Nat. Methods* **2014**, *11*, 281–289. [CrossRef]
18. Gidde, R.R.; Pawar, P.M. Flow feature and mixing performance analysis of {RB}-{TSAR} and {EB}-{TSAR} micromixers. *Microsyst. Technol.* **2020**, *26*, 517–530. [CrossRef]
19. Raza, W.; Kim, K.-Y. Unbalanced Split and Recombine Micromixer with Three-Dimensional Steps. *Ind. Eng. Chem. Res.* **2020**, *59*, 3744–3756. [CrossRef]
20. Li, J.; Xia, G.; Li, Y. Numerical and experimental analyses of planar asymmetric split-and-recombine micromixer with dislocation sub-channels. *J. Chem. Technol. Biotechnol.* **2013**, *88*, 1757–1765. [CrossRef]
21. Chen, X.; Shen, J. Numerical analysis of mixing behaviors of two types of {E}-shape micromixers. *Int. J. Heat Mass Transf.* **2017**, *106*, 593–600. [CrossRef]
22. He, M.; Li, W.; Zhang, M.; Zhang, J. Numerical investigation on the efficient mixing of overbridged split-and-recombine micromixer at low {Reynolds} number. *Microsyst. Technol.* **2019**, *25*, 3447–3461. [CrossRef]
23. Guckenberger, D.J.; De Groot, T.E.; Wan, A.M.D.; Beebe, D.J.; Young, E.W.K. Micromilling: A method for ultra-rapid prototyping of plastic microfluidic devices. *Lab. Chip* **2015**. [CrossRef] [PubMed]
24. Clark, J.; Kaufman, M.; Fodor, P.S. Mixing Enhancement in Serpentine Micromixers with a Non-Rectangular Cross-Section. *Micromachines* **2018**, *9*, 107. [CrossRef] [PubMed]
25. Clark, J.A.; Butt, T.A.; Mahajan, G.; Kothapalli, C.R.; Kaufman, M.; Fodor, P.S. Performance and implementation of centrifugal serpentine micromixers with non-rectangular cross-section. *J. Micromechanics Microengineering* **2019**, *29*, 075012. [CrossRef]
26. Au, A.K.; Huynh, W.; Horowitz, L.F.; Folch, A. {3D}-{Printed} {Microfluidics}. *Angew. Chemie Int. Ed.* **2016**, *55*, 3862–3881. [CrossRef]
27. Chen, C.; Mehl, B.T.; Munshi, A.S.; Townsend, A.D.; Spence, D.M.; Martin, R.S. 3D-printed microfluidic devices: Fabrication, advantages and limitations—A mini review. *Anal. Methods* **2016**, *8*, 6005–6012. [CrossRef]
28. Waheed, S.; Cabot, J.M.; Macdonald, N.P.; Lewis, T.; Guijt, R.M.; Paull, B.; Breadmore, M.C. {3D} printed microfluidic devices: Enablers and barriers. *Lab. Chip* **2016**, *16*, 1993–2013. [CrossRef]
29. Shallan, A.I.; Smejkal, P.; Corban, M.; Guijt, R.M.; Breadmore, M.C. Cost-effective three-dimensional printing of visibly transparent microchips within minutes. *Anal. Chem.* **2014**, *86*, 3124–3130. [CrossRef]
30. Carrière, P. On a three-dimensional implementation of the baker's transformation. *Phys. Fluids* **2007**, *19*, 118110. [CrossRef]
31. Robles-Linares, J.; Ramírez-Cedillo, E.; Siller, H.; Rodríguez, C.; Martínez-López, J.; Robles-Linares, J.A.; Ramírez-Cedillo, E.; Siller, H.R.; Rodríguez, C.A.; Martínez-López, J.I. Parametric Modeling of Biomimetic Cortical Bone Microstructure for Additive Manufacturing. *Materials* **2019**, *12*, 913. [CrossRef]
32. Rajaguru, J.; Duke, M.; Au, C.K. Development of rapid tooling by rapid prototyping technology and electroless nickel plating for low-volume production of plastic parts. *Int. J. Adv. Manuf. Technol.* **2015**, *78*, 31–40. [CrossRef]
33. Levy, G.N.; Schindel, R.; Kruth, J.P. Rapid manufacturing and rapid tooling with layer manufacturing (LM) technologies, state of the art and future perspectives. *CIRP Ann.* **2003**, *52*, 589–609. [CrossRef]
34. Udroiu, R.; Braga, I.C. *Polyjet Technology Applications for Rapid Tooling*; EDP Sciences: Ulis, France, 2017; Volume 112.

35. Charmet, J.; Rodrigues, R.; Yildirim, E.; Challa, P.K.; Roberts, B.; Dallmann, R.; Whulanza, Y. Low-Cost microfabrication tool box. *Micromachines* **2020**, *11*, 135. [CrossRef] [PubMed]
36. Au, A.K.; Lee, W.; Folch, A. Mail-order microfluidics: Evaluation of stereolithography for the production of microfluidic devices. *Lab. Chip* **2014**, *14*, 1294–1301. [CrossRef] [PubMed]
37. Martínez-López, J.I.; Mendoza-Buenrostro, C.; Retana, P.C.; Betancourt, H.; Vazquez, E.; Siller, H.R.; Rodriguez, C.A. Manufacturing and Test of Split and Recombine Micromixers with a Variable Cross-Section. *ECS Meet. Abstr.* **2018**, *MA2018-02*, 1303. [CrossRef]
38. Vangunten, M.T.; Walker, U.J.; Do, H.G.; Knust, K.N. 3D-Printed Microfluidics for Hands-On Undergraduate Laboratory Experiments. *J. Chem. Educ.* **2020**, *97*, 178–183. [CrossRef]
39. Nielsen, J.B.; Hanson, R.L.; Almughamsi, H.M.; Pang, C.; Fish, T.R.; Woolley, A.T. Microfluidics: Innovations in materials and their fabrication and functionalization. *Anal. Chem.* **2020**, *92*, 150–168. [CrossRef]
40. Kotz, F.; Risch, P.; Helmer, D.; Rapp, B.E. High-Performance Materials for 3D Printing in Chemical Synthesis Applications. *Adv. Mater.* **2019**, *31*, 1805982. [CrossRef]
41. Castiaux, A.D.; Pinger, C.W.; Hayter, E.A.; Bunn, M.E.; Martin, R.S.; Spence, D.M. PolyJet 3D-Printed Enclosed Microfluidic Channels without Photocurable Supports. *Anal. Chem.* **2019**. [CrossRef]
42. Protolabs Manufacturing. Available online: http://www.protolabs.com (accessed on 9 September 2020).
43. Martínez-López, J.I.; Mojica, M.; Rodríguez, C.A.; Siller, H.R. Xurography as a Rapid Fabrication Alternative for Point-of-Care Devices: Assessment of Passive Micromixers. *Sensors* **2016**, *16*, 705. [CrossRef]
44. Martínez-López, J.I.; Betancourt, H.A.; García-López, E.; Rodriguez, C.A.; Siller, H.R. Rapid Fabrication of Disposable Micromixing Arrays Using Xurography and Laser Ablation. *Micromachines* **2017**, *8*, 144. [CrossRef]
45. Mukhopadhyay, R. When PDMS isn't the best. *Anal. Chem.* **2007**, *79*, 3249–3253. [CrossRef]
46. Dangla, R.; Gallaire, F.; Baroud, C.N. Microchannel deformations due to solvent-induced PDMS swelling. *Lab. Chip* **2010**, *10*, 2972–2978. [CrossRef] [PubMed]
47. Wolf, M.P.; Salieb-Beugelaar, G.B.; Hunziker, P. PDMS with designer functionalities—Properties, modifications strategies, and applications. *Prog. Polym. Sci.* **2018**, *83*, 97–134. [CrossRef]
48. Leow, M.E.L.; Pho, R.W.H. RTV silicone elastomers in hand prosthetics: Properties, applications and techniques. *Prosthet. Orthot. Int.* **1999**, *23*, 169–173. [CrossRef]
49. Schroettner, H.; Schmied, M.; Scherer, S. Comparison of 3D surface reconstruction data from certified depth standards obtained by SEM and an infinite focus measurement machine (IFM). In *Microchimica Acta*; Springer: Berlin/Heidelberg, Germany, 2006; Volume 155, pp. 279–284.
50. Formlabs. Introducing the Form 3 and Form 3L. Available online: https://formlabs.com/blog/introducing-form-3-form-3l-low-force-stereolithography/ (accessed on 2 October 2020).
51. Hossain, S.; Afzal, A.; Kim, K.Y. Shape Optimization of a Three-Dimensional Serpentine Split-and-Recombine Micromixer. *Chem. Eng. Commun.* **2017**, *204*, 548–556. [CrossRef]
52. Raza, W.; Hossain, S.; Kim, K.Y. Effective mixing in a short serpentine split-and-recombination micromixer. *Sensors Actuators B Chem.* **2018**, *258*, 381–392. [CrossRef]

Article

A 3D Printed Jet Mixer for Centrifugal Microfluidic Platforms

Yunxia Wang, Yong Zhang, Zheng Qiao and Wanjun Wang *

Department of Mechanical Engineering, Louisiana State University, Baton Rouge, LA 70803, USA; ywan222@lsu.edu (Y.W.); yzha148@lsu.edu (Y.Z.); zqiao1@lsu.edu (Z.Q.)
* Correspondence: wang@lsu.edu

Received: 12 June 2020; Accepted: 14 July 2020; Published: 17 July 2020

Abstract: Homogeneous mixing of microscopic volume fluids at low Reynolds number is of great significance for a wide range of chemical, biological, and medical applications. An efficient jet mixer with arrays of micronozzles was designed and fabricated using additive manufacturing (three-dimensional (3D) printing) technology for applications in centrifugal microfluidic platforms. The contact surface of miscible liquids was enhanced significantly by impinging plumes from two opposite arrays of micronozzles to improve mixing performance. The mixing efficiency was evaluated and compared with the commonly used Y-shaped micromixer. Effective mixing in the jet mixer was achieved within a very short timescale (3s). This 3D printed jet mixer has great potential to be implemented in applications by being incorporated into multifarious 3D printing devices in microfluidic platforms.

Keywords: three-dimensional (3D) printing; micronozzles; Y-shaped structure; mixing efficiency; histogram and standard deviation

1. Introduction

Centrifugal microfluidic discs, also known as "lab-on-a-disc"(LOD) can eliminate the requirement for an external pump and complicated fluidic interconnections. As a result, many microfluidic centrifugal systems, commonly named as lab-on-CDs, have been reported. In recent years, centrifugal microfluidic technology has found wide applications in the development of microfluidic devices for applications in biomedical fields and drug discovery [1–3]. They also have played a significant role in the agriculture, food engineering, spilled-oil detection and chemical industries due to the potential of becoming automated, disposable, and point-of-care (POC) clinical diagnostic tools [4–8]. In general, microfluidic devices have channels of scale normally less than 1mm to transport and manipulate materials, offering a portable and inexpensive possibility to operations and experiments [9]. Because only microscopic amounts of samples and reagents are consumed in such microfluidic systems, much less waste is generated.

Mixing liquids on the microscale is a challenging task because most mixing processes in microfluidic devices is limited to low Reynolds numbers caused by laminar flows in microchannels. However, for many microfluidic applications including sample dilution, reagent homogenization, chemical reactions, bioassays [10] and medical analysis [11], and materials synthesis [12], mixing is often a crucial step. An efficient mixing process helps to accelerate the rate of reaction, enhance the system sensitivity, and even affect the final product distribution. Therefore, considerable efforts have been made to develop novel micromixers to boost mixing performance in microfluidic devices for more applications in broadly various fields.

In general, micromixers can be classified into two major categories: the active and the passive ones. The active mixers always utilize an external power source and control system to assist in mixing.

In general, the external power sources include vapor pneumatic power [13,14], mechanical pulsation [15], acoustic vibration [16,17], ultrasonic actuation [18], magnetic force [19–23], electro-kinetical force [24–26] and electroosmotic force [27]. The active mixers allow for more complicated processes, which always require an external instrument that adds to the size, weight, complexity and cost of the operating system. In contrast with the active types, the passive mixers generally accomplish mixing by adding geometric obstructions or altering geometries of the flow channels with no external energy sources. In terms of the mixing techniques, passive mixers can be divided into parallel flow-mixers (Y or T-types) [28–31], hydrodynamic focusing flow [32], multiple lamination flow [33,34], split and recombined flow [35], chaotic advection flow [36–38], structured packing flow [39], jet collision flow [40,41], recirculation flow [42], and moving droplet configuration flow-mixers [43]. Passive mixers are considered as an advantageous technique because of their simplicity and ease of fabrication in contrast with their active counterparts.

Jet collision mixing is one of the most efficient passive mixers. It maximizes the interfacial area of fluids by transforming them into plumes. The concept of impingement mixing as a microplume mixer was first investigated in some research in the late 1970s [44]. In the 1990s, the microjet mixer was comprised of micronozzle arrays to flow into a mixing chamber filled with another liquid [45]. This concept has been further optimized into two fluid streams flowing through two opposed arrays of micronozzles [41]. The fabrication of these micromixers is normally done using conventional fabrication technologies such as lithography in cleanroom facilities. In addition, bulky syringe pumps are often employed in experimental setups in those reports [41,44,45]. Three-dimensional printing, because of its fast speed and low cost, and wide choice of materials, has become an attractive tool for fabrication for micro-meso fluidics and mechanical systems. In addition, 3D printed models can be directly applied into testing and even as end products.

The fast-developing additive manufacturing (AM) technology with reasonable fabrication efficiency has provided a powerful tool to fabricate high precision, integrated microfluidic systems. Three-dimensional-printed microfluidics is suitable for a wide range of applications such as immunoassays [46], material science [47], particle analysis [48] and extraction of preterm birth biomarkers [49]. In addition to many other applications, much research in applying 3D printing technologies to make mixers have also been reported in recent years for different applications. The 3D printed mixer with integrated staggered herringbones has been studied to exploit the limits of 3D flow magnetic resonance imaging (MRI) for the analysis of fluid dynamics on a sub-millimeter scale [50]. Using high-definition MultiJet 3D printing to remake five different passive micromixers has also been reported [51]. The indirect 3D printing method was used to fabricate a helical and a cubic electrode mixer and to increase the mass transfer up to 47% the standard flat electrode by some researchers [52]. The 3D printed modular mixing components operate on the basis of splitting and recombining fluid streams to decrease interstream diffusion length [53]. A 3-inlet 3D printed, disposable, high-throughput micromixer for production of therapeutic nanoparticles with syringe pumps has also been studied [54].

One of the commonly used additive manufacturing (AM) technologies, fused deposition modelling (FDM) has emerged as a rapid prototyping approach for various applications in recent years. FDM was first reported to fabricate templates for soft lithography of microfluidic devices with PDMS in 2002 [55]. FDM was demonstrated as the direct fabrication method for microfluidic chemical reaction-ware in 2012 [56]. Compared with other 3D printing technologies such as stereolithography (SLA) and manufacturing using femtosecond laser, FDM is much cheaper and easier to use, and faster [57]. However, it has the drawback of lower resolution, and therefore larger minimum feature sizes in final products.

In this paper, we report a jet collision micromixer (jet mixer) fabricated using a low-cost FDM printer. The micromixer is designed to have two opposed arrays of micronozzle to generate microplumes to enhance the mixing efficiency by both quick diffusion and turbulent mechanisms. By controlling the distance between the neighboring filament lines and also taking on the additive manufacturing nature of FDM, a jet mixer with large arrays of micronozzles (microchannels) was easily generated in a vertical direction. The mixer was installed on a centrifugal fluidic platform for testing. Flow visualization by high-speed camera demonstrated that the mixing process was complete within a few seconds. The mixing efficiency of the jet mixer was experimentally measured by contrastively analyzing with a Y-shaped mixer.

2. Principle and Design of the Jet Mixer

The design of the jet mixer and Y-shaped mixer in this paper is schematically demonstrated in Figure 1. Figure 1a–d were created via using Solidworks (Dassault Systems SolidWorks Corp, Waltham, MA, USA). Figure 1a shows the general layout of the mixer with the sample loading chambers A and B (loading liquid volume 100 μL for each chamber), the mixing chamber, the vent, and the collecting chamber (200 μL). The capillary valves connecting the sample loading chamber and the mixing chamber were designed to have a cross-section of 0.8 mm (width) × 0.5 mm (height). The corresponding cross-sectional view of the jet mixer is shown in Figure 1c. At the beginning of the mixer operation, the sample liquids A and B were preloaded into the sample loading chambers A and B separately. The mixer was then installed on the platform to rotate. When the rotational speed of the platform reached the burst frequency, the capillary valves were turned to open, centrifugal forces helped to propel the liquid samples A and B to flow into the mixing chamber. Figure 1d and e show the more detailed schematic.

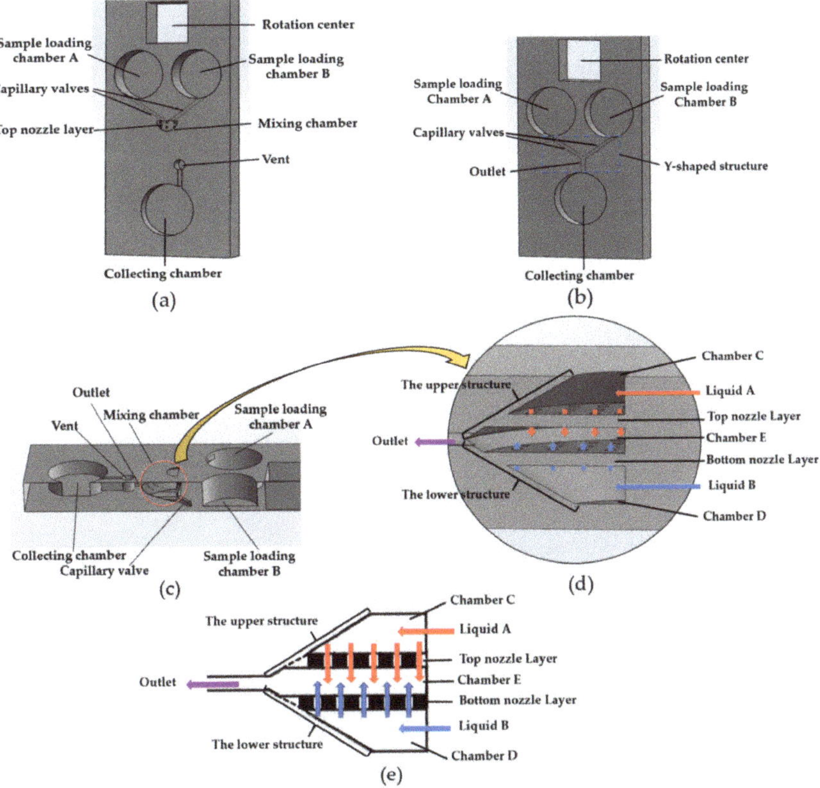

Figure 1. Design of the 3D jet mixer: (**a**) the exploded view of the assembly of the jet mixer to show the components; (**b**) design of the Y-shaped micromixer for comparison with the jet mixer's performance; (**c**) the cross-sectional view on the right plane; (**d**) detailed design of the mixing chamber; (**e**) schematic diagram of the mixing chamber on the right plane.

The design of the mixing chamber: the mixing chamber contains two opposite arrays of micronozzles, which divide this chamber into three subchambers C, D and E. During the mixing process, two different liquid samples (A and B) are propelled into the chambers C and D, respectively, and then flow into chamber E to mix through the two up-down arrays of micronozzles (Figure 1e). Finally, the mixed liquid is delivered to the collecting chamber. The vents are designed for air release.

In fabricating the two arrays of micronozzles, the smaller filament size was created from an extruder primarily by increasing the distance between neighboring filament lines to form micronozzles. In the additive manufacturing process with FDM, the structure was made in a layer by layer fashion. As a result, the micronozzles can be "naturally" generated in a vertical direction. This is the main reason why the entire layout for the mixing chamber is in a vertical direction instead of in the horizontal plane.

For the sake of comparison, a Y-shaped mixer was also designed, built, and tested. The design of a Y-shaped mixer can be fully realized by a 3D FDM printer, to achieve optimal mixing performance. The design of the Y-shaped mixer is schematically shown in Figure 1b. It consists of sample loading chambers A and B, a collecting chamber, and a Y-shaped structure of channels with capillary valves and an outlet. All the chambers of the Y-shaped mixer are in the same plane, which is different from the placement of the chambers in the jet mixer. In the jet mixer, jet collision was designed to happen in a vertical (up-down) direction, not in a right-left direction. This was due to the limitation of FDM printer that can be used to produce arrays of nozzles in a vertical direction. Because chambers C and D are designed to face each other in a vertical direction and are also connected with chambers A and B, respectively, this design limitation requires an increased depth for chambers A and B. To have the same designed volumes of chambers A and B for both the Y-shaped mixer and the jet mixer, the loading chambers A and B for the jet mixer and Y-shaped mixer are designed to have the same volume but different diameters and depths as marked in Figure 1.

The theoretical burst frequencies of the capillary valves for both the Y-shaped mixer and the jet mixer are designed to be the same. The burst frequency f_b of a capillary valve is defined in terms of rotations per minute (RPM), and can be calculated using the following equation:

$$f_b = \sqrt{\frac{P_{cap}}{\rho \Delta R \overline{R}}} \left(\frac{30}{\pi}\right) \tag{1}$$

where, $P_{cap} = 4cos\theta Y/D_h$, ρ is the density of the liquid, ΔR is the difference between the top and bottom of the liquid levels with respect to the rotation center, $\overline{R}$ is the average distance of the liquid to the rotation center, P_{cap} is the capillary pressure, Y is the surface tension, θ is the contact angle, and D_h is the hydraulic diameter. With the design dimensions shown in Table 1, the theoretical burst frequency was estimated to be 473.31 RPM.

Table 1. The dimensions of the common components for the jet mixer and Y-shaped mixer (radius—R, height—H, width—W).

Components	Dimensions in Jet-Mixer	Dimensions In Y-Shaped Mixer
Sample Loading Chamber A	3.37 mm (R) × 5.00 mm (H)	4.77 mm (R) × 2.50 mm (H)
Sample Loading Chamber B	3.37 mm (R) × 5.00 mm (H)	4.77 mm (R) × 2.50 mm (H)
Capillary Valve	0.80 mm (W) × 0.50 mm (H)	0.80 mm (W) × 0.50 mm (H)
Outlet	0.80 mm (W) × 0.50 mm (H)	0.80 mm (W) × 0.50 mm (H)
Collecting Chamber	4.81 mm (R) × 2.75mm (H)	5.05 mm (R) × 2.50 mm (H)

The main difference between the jet mixer and Y-shaped mixer is the extra mixing chamber in the jet mixer. In the Y-shaped mixer, the fluid samples from loading Chambers A and B flow through the Y-shaped channels and enter the collecting chamber directly. The collection chamber also serves as the mixing chamber in the Y-shaped mixer. In comparison, in the design of the jet mixer, the fluid samples from the two loading chambers must flow through two arrays of micronozzles in the mixing chamber before it enters the collecting chamber in which additional mixing happens. The dimensions of the capillary valves (the flow channels to the mixing chambers) for both types of mixers are designed to be the same. The nominal flow resistances are therefore designed to be the same. By assuming the flows in these microchannels are laminar, the nominal flow resistances are calculated to be 1.76×10^9 (Pa·s/m^3).

Figure 2 shows common design options for the arrays of the micronozzles. There are normally two options to arrange the two facing arrays of impinging jets: directly facing as shown in Figure 2a, or to have a fixed shift as shown in Figure 2b. Figure 2a,b are common design principles for jet flow collision, we designed our mixing chamber based on these principles. The cross-section diagram of designs (a) and (b) is shown graphically in Figure 2c,d, respectively. The function of two facing arrays of micronozzles is to transform liquid samples into microjets. The liquid samples are driven by centrifugal force to feed through multiple micronozzles, converting liquids into impinging plumes and mixing in Chamber E.

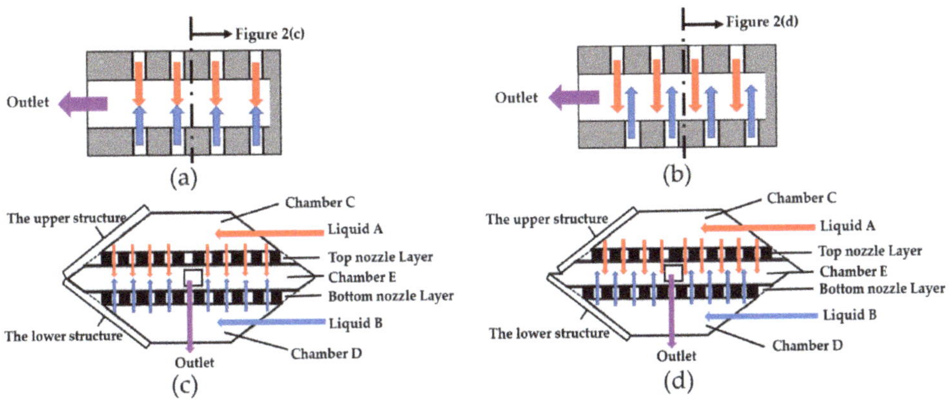

Figure 2. Design options of the jet-impinge mixer: (**a**) Opposite micronozzles directly facing each other; (**b**) Opposite side micronozzles, off-set; (**c**) Cross-section on the top plane of design (**a**); (**d**) Cross-section on the top plane of design (**b**).

As shown in Figure 2c,d, the sample liquids (A and B) are simultaneously introduced into the Chambers C and D in up and down directions in our design. After introduction, the location of the sample liquids become upper and lower plane instead of the same plane. These micro-plumes of flows are introduced oppositely into chamber E, impinging upon each other, either face-to-face as shown in Figure 2c or passing by each other as shown in Figure 2d. The design of Figure 2d obviously has the advantage of maximizing the effective contact surfaces between the impinging plumes, and therefore boosting the diffusion process. The off-set also makes it more likely to produce more vortexes due to flowing across the mixing chamber from one side to the other [41]. It was therefore the preferred design for the mixer.

3. Fabrication of the Jet Mixer

The jet mixer was 3D printed using an FDM 3D printer (Ultimaker 3). Simplify3D software was used to adjust the relevant parameters such as nozzle diameter, extrusion width, primary layer height and layer modifications in achieving the key feature (the micronozzles) and the designed dimension of the jet mixer. The following operational parameters were used in the fabrication process: (1) layer thickness for top and bottom nozzle layer, 0.05 mm; (2) layer thickness for other components, 0.15 mm; (3) nozzle diameter, 400 μm; (4) flow rate, 0.9; (5) print speed, 3600 mm/min; (6) nozzle temperature, 215 °C; (7) heating bed temperature, 55 °C; (8) spacing between neighboring filaments, 600 μm; (9) heated building chamber (25 °C).

White colored PLA (Polylactic Acid, from Dynamism, Chicago, IL, USA) was used as the structural material (and therefore the printer filament) for the jet mixer. We designed the whole model using

Solidworks (Dassault Systems SolidWorks Corp, Waltham, MA, USA). Figure 3 shows the 3D printing process of the jet mixer layer by layer with FDM. It took about one and a half hours to print one prototype jet mixer with the overall size of 100 mm (in length) × 24 mm (in width) × 6 mm (in height). The entire fabrication process was divided into 5 processes in Simplify3D software: Process 1-1, Process 2-2, Process 3-3, Process 4-4 and Process 5-5 in Figure 3. As shown in Figure 3a,b, in Process 1-1, the structure's height increased from 0 mm to 2.5 mm. It consists of the substrate layer (0–1 mm) and chamber D (2.00 mm (lower radius), 5.46 mm (upper radius), 2.00 mm (H)) in the mixing chamber, the sample loading chambers A and B, and the connection capillary valve between sample loading Chamber B and the Chamber B of the mixing chamber. The height of the bottom nozzle layer in the mixing chamber was set to be 0.5 mm, corresponding to the Process 2-2 from 2.5 mm to 3 mm in Figure 3c. In Figure 3d, the Chamber E (6.33 mm (lower radius), 7.20 mm (upper radius), 2 × 0.50 mm (H)) was formed as a hollow structure in the mixing chamber with its height increased from 3 mm to 4 mm as shown in the Process 3-3. Meanwhile, the deposition for the output channel and collecting chamber for collecting mixed fluid were started. Figure 3e shows the top nozzle layer starting to be printed on the hollow Chamber E from 4 mm to 4.5 mm in the Process 4-4. The Process 5-5 is shown in Figure 3f–h, the height of the printed structure increased from 4.5 mm to 6 mm. The fabrication of Chamber C (2.00 mm (lower radius), 5.46 mm (upper radius), 2.00 mm (H)) of the mixing chamber, and the other capillary valve connecting sample loading Chamber A to Chamber C of the mixing chamber started at height 5.0 mm in Figure 3f. Figure 3g depicts the structure of the jet mixer before the cover was made. Figure 3h is the finished 3D FDM-printed jet mixer.

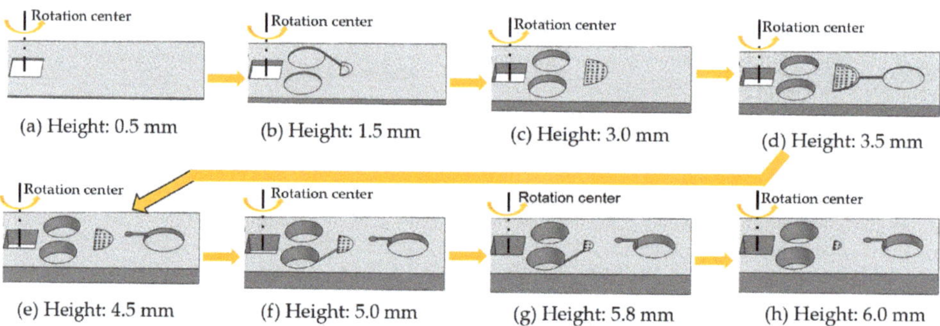

Figure 3. The flowchart of the additive manufacturing process for the jet mixer cartridge: (**a**) the substrate layer (height: 0.0–1.0 mm); (**b**) the mixing Chamber D layer (height: 1.0–2.5 mm); (**c**) the bottom nozzle layer (height: 2.5–3 mm); (**d**) the mixing Chamber E layer (height: 3.0–4.0 mm); (**e**) the top nozzle layer (height: 4.0–4.5 mm); (**f**) the mixing Chamber C layer (height: 4.5–6.0 mm); (**g**) the final geometry of the jet mixer before the cover was made; (**h**) the completed jet mixer. (Process 1-1: 0.0–2.5 mm; Process 2-2: 2.5–3.0 mm; Process 3-3: 3.0–4.0 mm; Process 4-4: 4.0–4.5 mm; Process 5-5: 4.5–6 mm).

The photo images of the 3D printed micronozzles and jet mixer with PLA as structural material are shown in Figure 4, Figure 4a is the photo image of the micronozzles. Figure 4b shows the image of the micronozzles as captured by LED Digital Binocular Compound Microscope (AmScope, Orange County, CA, USA). In 3D software, rectilinear infill patterns with infill angles in +45° and −45° directions were chosen for automatically forming micronozzles during the 3D printing process. The angle formed by two adjacent print layers is therefore 90° as shown in Figure 4b. It can be observed that the largest size of micronozzle was 200 µm by 250 µm. The smallest size was 50 µm by 50 µm. This wide range of size variation was caused by the error of the low-cost FDM printer used to fabricate the mixer. The size of the micronozzles can be adjusted according to design requirements by changing the settings of the extruder, the thickness of each printing layer, the infill, as well as other printer parameters in Simplify3D software. Two different types of sandpaper were then used to polish the 3D printed micromixer. The fabrication of the jet mixer took about one and a half hours to complete. It was time-saving, labor-saving, cost-effective and easy-to-make in comparison with other technologies such as lithography. Finally, transparent polypropylene adhesive tapes (Scotch tape, 3M) were applied to seal the micromixer. The vent holes were punched on the covering tape to prevent air pressure building up. The finished 3D printed jet mixer is shown in Figure 4c. Figure 4d shows a photo image of the 3D printed Y-shaped mixer.

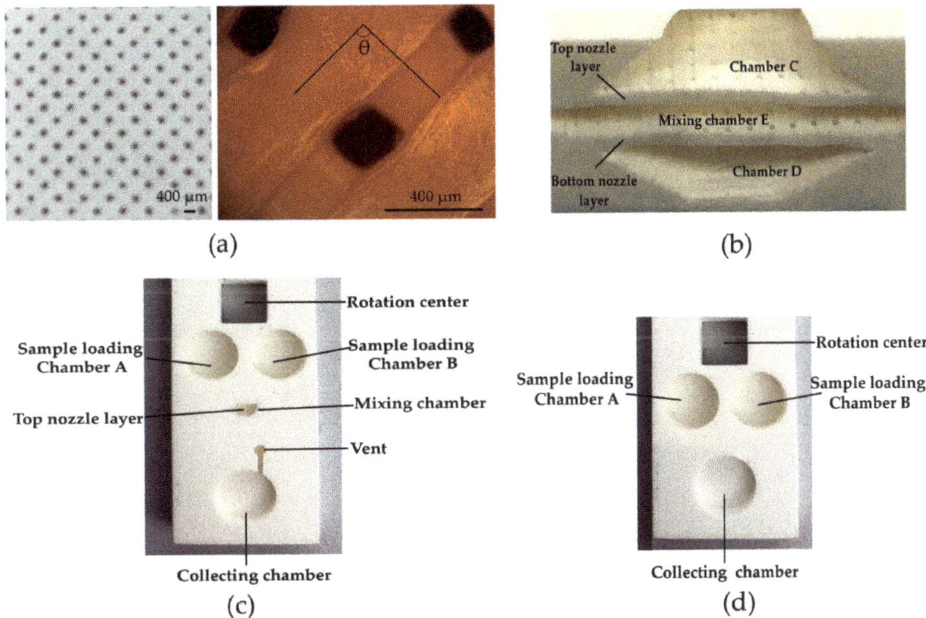

Figure 4. (**a**) A photo image of the 3D printed micronozzles; (**b**) a microscope image of the 3D printed micronozzles, θ = 90°; (**c**) a photo image of the 3D printed jet mixer; (**d**) a photo image of the 3D printed Y-shaped mixer.

In our design of the jet mixer, two arrays of impinging micronozzles were arranged in the vertical or out of plane orientation instead of horizontal or in plane. The main reason for choosing such a design layout is because of the resolution limitation of the 3D printer used in the fabrication. It is difficult to make arrays of micronozzles in plane with the desired resolution, depending on purpose-designed micronozzles in Solidworks. The unique advantage of the fused filament fabrication (FFF) in the FDM printer is that it makes it much easier to form arrays of micronozzles automatically in a vertical direction by manipulating the filament size from the extruder in selecting the infill pattern in the Simplify3D software. It should be noted the line spacing in the FDM printer is normally considered a drawback because it lowers the surface quality of the finished structure. For the fabrication of the jet mixer reported in this paper, it actually helped to simplify the fabrication process.

4. Experimental Results and Discussions

4.1. The Experimental Setup

Figure 5 shows a photo image of the centrifugal platform setup to test the micromixers. The micromixer was mounted on the DC servo motor controlled by a laptop computer. A high-speed camera with frame rate 80 fps (Pixelink, Rochester, NY, USA) was installed above the fluidic system to monitor the mixing processing in real time. A series of experiments were conducted to test the mixing performances of the jet mixer. Experiments with a simple Y-shaped mixer which is one of the most commonly used and extensively studied, were also conducted for comparison.

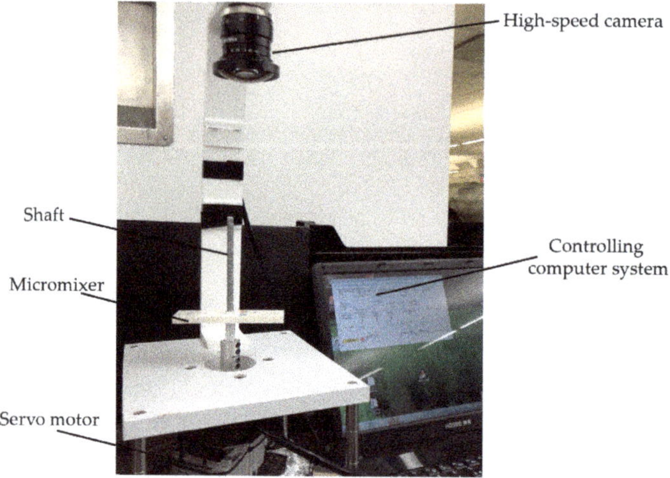

Figure 5. The experimental setup of the centrifugal platform with 3D microfluidic cartridge mounted.

4.2. Mixing Efficiencies and Discussions

In the experiments, 100 μL of red and blue dyed deionized (DI) water were injected into loading Chambers A and B first. Experiments were first run using blue dyed DI water and red dyed DI water because it is easy to observe under a high-speed camera. The mixers were then attached to the rotational platform for tests. The mixers were installed in an orientation so that the loading chambers were located closer to the shaft and the collecting chambers were toward the edge of the platform along the radial direction. The centrifugal platform was driven to rotate in a counter-clockwise direction.

It should be noted that the burst frequencies for the capillary valves in both the Y-shaped and the jet mixer were designed to be the same at 473.37 RPM. However, because of fabrication errors in the microchannels, the burst frequencies were found to be much higher. When the rotational speed of the micromixer was increased to 1000 RPM, the capillary valves were turned open roughly at the same time for both the Y-shaped mixer and the micro-jet one. The red and blue dye samples in the loading chambers A and B were released and flowed through the Y-shaped channels and entered the collecting chamber as shown by the sequential photo images in Figure 6(a1–a7). At the same time, the mixing process for the jet mixer is shown by the sequential photo images of Figure 6(b1–b7). The blue and red dyes flowed into the two facing arrays of micronozzles, and then entered the mixing chamber, and were finally released into the collecting chamber. The mixing process was recorded in real time by the monitoring camera at the rate of two pictures per second.

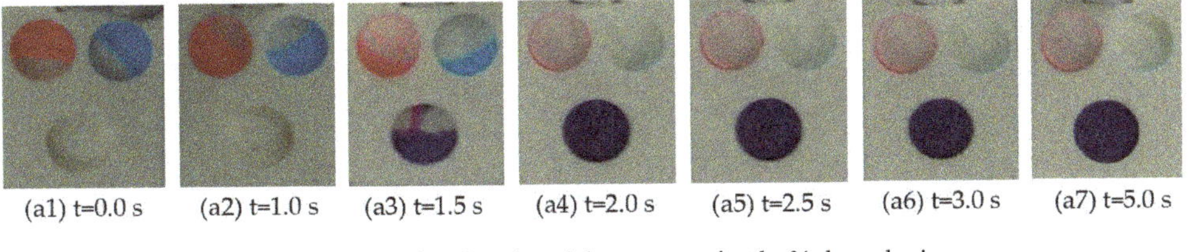

(**a**) Photo images showing the mixing process for the Y-shaped mixer.

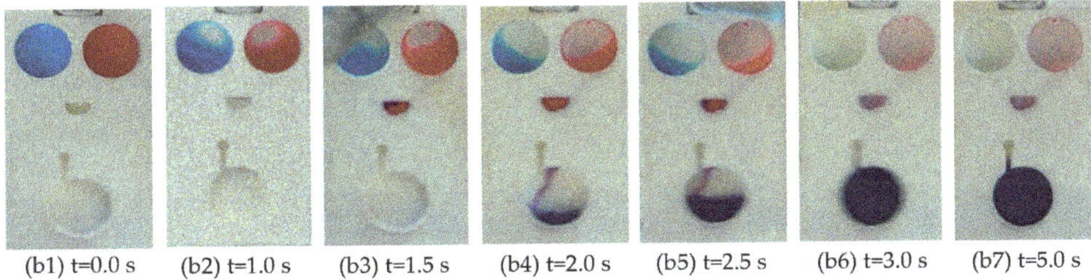

(**b**) Photo images showing the mixing process for the jet mixer.

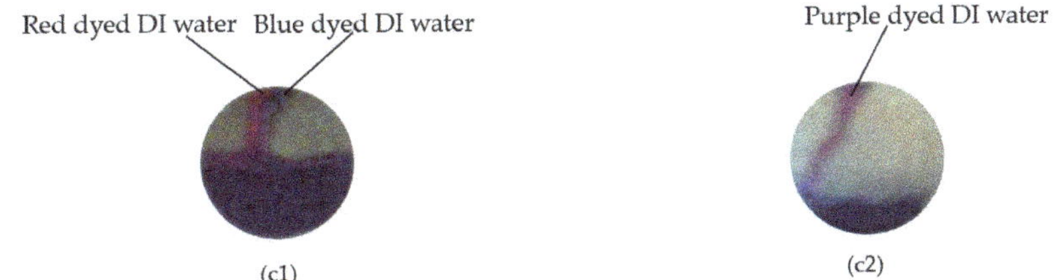

(**c**) The images of mixing fluids entering the collecting chamber for the Y-shaped mixer (c1) and the jet mixer (c2)

Figure 6. Photo images showing the experimental results: (**a**) for the Y-shaped mixer; (**b**) for the jet mixer; (c) the images of fluid entering collecting chambers: (**c1**) for the Y-shaped mixer chambers and (**c2**) for the jet mixer. The rotation of the platform was brought to a complete stop at t = 5s.

At the beginning of the experiments, the two loading chambers were filled with blue and red solutions as shown in Figure 6(a1,b1). In the experiments, the same volumes of the sample fluids, 100 µL, were preloaded into chambers A and B. However, because chambers A and B for the jet mixer have a slightly smaller diameter and larger depth than those of the Y-shaped mixer, the sample fluids completely covered the bottom surfaces of both chambers in the jet mixer and only partially covered the loading chambers for the Y-shaped mixer. This made the photo image Figure 6(a1) seem to complete the mixing for the Y-shaped mixer. For the jet mixer, the mixing time was a little longer, because the flow resistance of the two arrays of micronozzles slowed the flow rate of the fluid samples. As can be observed from Figure 6(b1–b7), it took more than 0.5 s before the blue and red dyes entered the collecting chambers through the arrays of micronozzles. It took about 3 s for the two loading chambers with blue and red dyes to be emptied and the mixed solution to flow into the collecting chamber for the jet mixer. When the mixing was completed after 5 s, the rotation of the platform was stopped. The photo images in Figure 6(a7,b7) were taken in the resting condition of the platform to improve image resolution for more accurate analysis of the mixing effects for both mixers as shown in Figure 7(c1,c2). It should be noted that the collecting chamber in Figure 6(b6) was already empty: because the micronozzles were stained by the dye in the fluids, the photo image looks like there was still some fluid leftover. Figure 6(c1,c2) show the situation of the mixing fluids flowing into the collection chambers for the two types of mixers. The red and blue dyed DI water flowed in parallel into the collecting chamber of the Y-shaped mixer at 1.5 s (Figure 6(c1) corresponding to Figure 6(a3)), indicating that the mixing mainly happened in the collecting chamber due to the vortex. As displayed in Figure 6(c2) corresponding to Figure 6(b4), purple dyed DI water was propelled into the collection chamber of the jet mixer at 2 s, this meant that the red and blue dyed DI water had already been mixed well in the mixing chamber. Injecting into the collection chamber helps to further mix the sample fluid. The time difference for the mixing to complete in Figure 6 is mainly due to the fact that mixing fluids flow through different structures in the Y-shaped mixer and the jet mixer to enter the collecting chambers. Flow resistance from the two arrays of micronozzles in the mixing chamber of the jet mixer slowed the flow rate of the fluid samples.

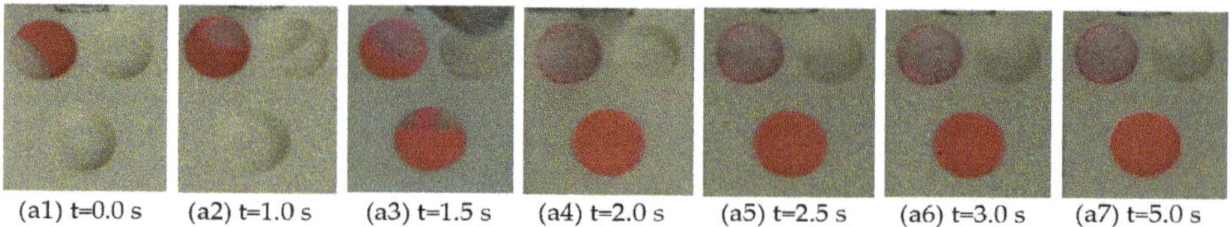

(**a**) Photo images showing the mixing process for the Y-shaped mixer.

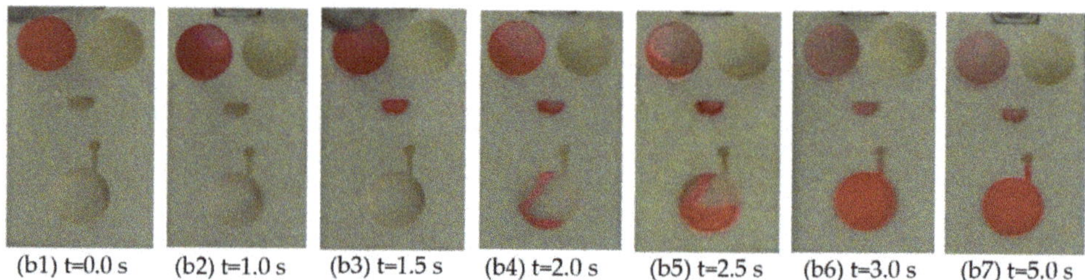

(**b**) Photo images showing the mixing process for the jet mixer.

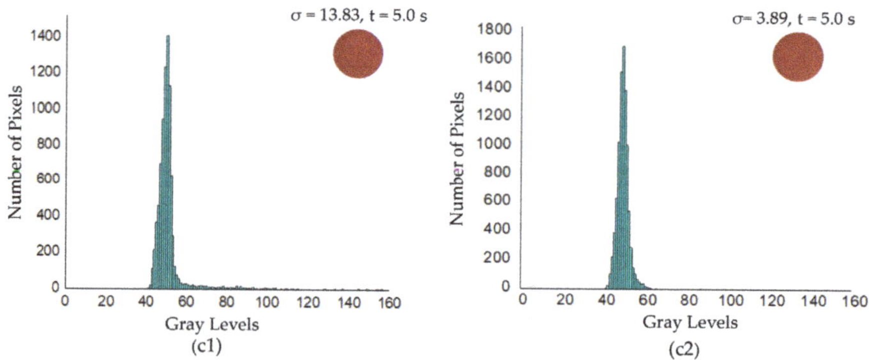

(**c**) Histogram and σ of the area of interest.

Figure 7. Analysis of the mixing performances for the two types of mixers. Photo images showing the experimental results for (**a**) the Y-shaped and (**b**) the jet mixer; (**c**) histogram and σ of the photo image of the mixed fluid samples for (**c1**) the Y-shaped mixer and (**c2**) the jet mixer.

Standard deviation (σ) is a common method in analyzing mixing performance. Greyscale distribution for the colorless DI water and the red dyed DI water are about from 200–255 and 0–50, respectively. Therefore, experiments were also run using colorless DI water and red dyed DI water to check the mixing efficiencies. The whole experimental process was shown in Figure 7a,b. The mixing process of colorless and red dyed DI water under the observation of a high-speed camera was the same as that obtained using red and blue dyed DI water in Figure 6a,b. The images of the final mixed solution in the collecting chamber were processed and analyzed by using the software «Image J» (NIH Image, Bethesda, MD, USA) and «Origin 8.5» (OriginLab, Northampton, MA, USA) to obtain the σ and the distribution of the gray intensity histogram. The histogram and the corresponding σ for both micromixers are shown in Figure 7(c1,c2).

To quantify the effectiveness of the two different types of mixers, the standard deviations σ of the pixel intensity of the images of the mixed solutions in the collecting chambers were utilized to numerically represent the mixing effectiveness. The standard deviation σ, defined by the square root of the variance, was used to measure the dispersion of the data values. The data points which are near the mean value in the histogram with low values of σ indicate higher mixing effect. Oppositely, the data points that have widely spread (with the high σ), indicate lower mixing effect. For the final mixed solutions shown in images in Figure 7(a7,b7), the areas of interest were the region of the collecting chambers, which were digitally cut out for analysis. Each of the images of the collecting chambers was analyzed by employing the software «Image J» and «Origin» for further processing to get a histogram plot, the frequency of the pixel's gray levels of the regions was plotted. The gray intensities of all the pixels were obtained and the σ of each image was obtained. The x-axis of the histogram graph represents the gray levels and the y-axis represents the number of pixels in that specific tone. Figure 7(c1) shows the distribution of pixels for the mixed fluid sample in the collecting chamber for the jet mixer was less dispersed and more concentrated (gray levels from 40 to 60) in contrast with that for the Y-shaped mixers as shown in Figure 7(c2) (gray levels from 40 to 160). Therefore, the histogram distribution displays the jet mixer has a much better mixing efficiency. The change in σ of the mixed liquids for the two mixers is shown in Figure 7(c1,c2). Based on the foregoing analysis, the smaller σ is, the better the mixing efficiency. The σ for both images shown in Figure 7(c1,c2) was 13.83 for the Y-shaped mixer and 3.89 for the jet mixer. Hence, the mixing efficiency of the jet mixer can be considered as 3.56 times of that for the Y-shaped one. Very obviously, the two facing arrays of micronozzles have helped to enhance the mixing efficiency dramatically. The mixing performance of the jet mixer outperformed that of the Y-shaped mixer.

Finally, the experiments were repeated five times for both mixers. The corresponding results are shown in Table 2. The average σ of the jet mixer was found to be 3.56 (3.89, 3.59, 3.24, 3.89, 3.21), in comparison with the average σ of 13.77 for the Y-shaped mixer in five tests (13.53, 13.83, 13.47, 13.88, 14.15). Furthermore, the σ for the five standard deviations of both mixers was 0.25 (Y-shaped mixer) and 0.29 (jet mixer). In addition, the required mixing time for colorless DI water and red dye colored DI water of the jet mixer is shorter than that reported by Esmail et al. [58] using the moment of inertia mixing method in the second mixing cycle. It should be noted that using the image analysis method to estimate mixing performance is limited by the high-speed camera resolution and the irregular shape of the liquid samples in the collecting chamber. However, the σ of the gray intensities can still be a useful estimation for mixing effectiveness. In addition, in the design of traditional Y-shaped mixers, S-shaped flow channels are used to further improve the mixing efficiency. For both the Y-shaped mixer and the jet mixer in this paper, S-shaped channels could also be added for better mixing efficiencies.

Table 2. The standard deviation value of the Y-shaped and the jet mixer under different numbers of the test.

Test #	Standard Deviation of Jet-Mixer	Standard Deviation of Y-Shaped Mixer
1	3.89	13.53
2	3.59	13.83
3	3.24	13.47
4	3.89	13.88
5	3.21	14.15

5. Conclusions

A jet mixer with two facing arrays of micronozzles was designed and successfully fabricated using 3D printing technology. The micronozzles were realized by using Simplify3D software to change the size of filament and easily printed using a low-cost FDM printer. Experiments were conducted to study the mixing efficiencies of the jet mixer and compared with the commonly used Y-shaped mixer. The jet mixer has demonstrated very good mixing efficiency with an average σ of 3.56 for colorless DI water and red dye colored DI water with a mixing time of less than 5 s. The main reason behind the dramatically enhanced mixing efficiency of the jet mixer is the mixing chamber that includes two facing arrays of micronozzles. This unique fabrication method makes it possible to print micro-nozzles via an increased distance between neighboring filament lines. The design significantly increased the effective contact areas of the two liquid samples by impinging plumes from two opposite arrays of micronozzles and significantly enhanced the mixing efficiency. The study also proved that FDM 3D printing technology can be used in the fabrication of the jet mixer and is time-saving, labor-saving and cost-effective. PLA was used as the structural material for demonstration purposes only. Many other materials can be used in FDM and other 3D printing technologies for specific applications and needs. Additionally, 3D printing technology made it very easy to fabricate complete microfluidic CDs without extra equipment installation. It may help to simplify the design, reduce the cost of production, and enable the development of novel fluidic systems for many applications.

Author Contributions: Y.W. investigated the experiments and wrote the original draft; Y.Z. provided methodology and analysis; Z.Q. provided 3D-printing technology support. W.W. administrated the project, reviewed the manuscript, visualized the concept and provided methodology. All authors have read and agreed to the published version of the manuscript.

Funding: This research and the PAC were funded by a grant from the Gulf of Mexico Research Initiative (Grant # AWD 001115).

Acknowledgments: This research was made possible by a grant from the Gulf of Mexico Research Initiative. Data are publicly available through the Gulf of Mexico Research Initiative Information & Data Cooperative (GRIIDC) at https://data.gulfresearchinitiative.org. (https://data.gulfresearchinitiative.org/data/R6.x831.000:0003, https://data.gulfresearchinitiative.org/data/R6.x831.000:0002).

Conflicts of Interest: The authors declare no conflict of interest.

References

1. Cheng, Y.; Zhang, X.; Cao, Y.; Tian, C.; Li, Y.; Wang, M.; Zhao, Y.; Zhao, G. Centrifugal microfluidics for ultra-rapid fabrication of versatile hydrogel microcarriers. *Appl. Mater. Today* **2018**, *13*, 116–125. [CrossRef]
2. Phaneuf, C.R.; Seamon, K.J.; Eckles, T.P.; Sinha, A.; Schoeniger, J.S.; Harmon, B.; Meagher, R.J.; Abhyankar, V.V.; Koh, C.Y. Ultrasensitive multi-species detection of CRISPR-Cas9 by a portable centrifugal microfluidic platform. *Anal. Methods* **2019**, *11*, 559–565. [CrossRef] [PubMed]
3. Gao, Z.; Chen, Z.; Deng, J.; Li, X.; Qu, Y.; Xu, L.; Luo, Y.; Lu, Y.; Liu, T.; Zhao, W.; et al. Measurement of carcinoembryonic antigen in clinical serum samples using a centrifugal microfluidic device. *Micromachines* **2018**, *9*, 470. [CrossRef] [PubMed]
4. Zhang, Y.; Xiang, J.; Wang, Y.; Qiao, Z.; Wang, W. A 3D printed centrifugal microfluidic platform for spilled oil enrichment and detection based on solid phase extraction (SPE). *Sens. Actuators B Chem.* **2019**, *296*, 126603. [CrossRef]

5. Ould, A.; Moctar, E.; Aubry, N.; Batton, J. Electro-hydrodynamic micro-fluidic mixer. *Lab Chip* **2003**, *3*, 273–280. [CrossRef]
6. Chen, L.; Chen, S.H.; Russell, D.H. An experimental study of the solvent-dependent self-assembly/disassembly and conformer preferences of gramicidin A. *Anal. Chem.* **2013**, *85*, 7826–7833. [CrossRef]
7. Neethirajan, S.; Kobayashi, I.M.; Nakajima, M.; Wu, D.; Nandagopal, S.; Lin, F. Microfluidics for food, agriculture and biosystems industries. *Lab Chip.* **2011**, *11*, 1574–1586. [CrossRef]
8. Zhang, Y.; Xiang, J.; Wang, Y.; Qiao, Z.; Wang, W. A novel gravity valve and its application in a 3D printed centrifugal fluidic-system for solid phase extraction (SPE). In *Microfluidics, BioMEMS, and Medical Microsystems XVIII*; International Society for Optics and Photonics: San Francisco, CA, USA, 2020. [CrossRef]
9. Sanders, G.H.W.; Manz, A. Chip-based microsystems for genomic and proteomic analysis. *TrAC Trends Anal. Chem.* **2000**, *19*, 364–378. [CrossRef]
10. El-Ali, J.; Sorger, P.K.; Jensen, K.F. Cells on chips. *Nature* **2006**, *442*, 403–411. [CrossRef]
11. Stott, S.L.; Hsu, C.H.; Tsukrov, D.I.; Yu, M.; Miyamoto, D.T.; Waltman, B.A.; Rothenberg, S.M.; Shah, A.M.; Smas, M.E.; Korir, G.K.; et al. Isolation of circulating tumor cells using a microvortex-generating herringbone-chip. *Proc. Natl. Acad. Sci. USA* **2010**, *107*, 18392–18397. [CrossRef]
12. Haswell, S.J.; Watts, P. Green chemistry: Synthesis in micro reactors. *Green Chem.* **2003**, *5*, 240–249. [CrossRef]
13. Xu, L.; Lee, H.; Jetta, D.; Oh, K.W. Vacuum-driven power-free microfluidics utilizing the gas solubility or permeability of polydimethylsiloxane (PDMS). *Lab Chip* **2015**, *15*, 3962–3979. [CrossRef]
14. Tsai, J.H.; Lin, L. Active microfluidic mixer and gas bubble filter driven by thermal bubble micropump. *Sens. Actuators A Phys.* **2002**, *97–98*, 665–671. [CrossRef]
15. Bottausci, F.; Mezić, I.; Meinhart, C.D.; Cardonne, C. Mixing in the shear superposition micromixer: Three-dimensional analysis. *Philos. Trans. R. Soc. A Math. Phys. Eng. Sci.* **2004**, *362*, 1001–1018. [CrossRef] [PubMed]
16. Yang, Z.; Matsumoto, S.; Goto, H.; Matsumoto, M.; Maeda, R. Ultrasonic micromixer for microfluidic systems. *Sens. Actuators A Phys.* **2001**, *93*, 266–272. [CrossRef]
17. Zhen, Y.; Goto, H.; Matsumoto, M.; Maeda, R. Active micromixer for microfluidic systems using lead-zirconate-titanate(PZT)-generated ultrasonic vibration. *Electrophoresis* **2000**, *21*, 116–119. [CrossRef]
18. Yaralioglu, G.G.; Wygant, I.O.; Marentis, T.C.; Khuri-Yakub, B.T. Ultrasonic mixing in microfluidic channels using integrated transducers. *Anal. Chem.* **2004**, *76*, 3694–3698. [CrossRef] [PubMed]
19. Lu, L.H.; Ryu, K.S.; Liu, C. A magnetic microstirrer and array for microfluidic mixing. *J. Microelectromechanic. Syst.* **2002**, *11*, 462–469. [CrossRef]
20. Rida, A.; Gijs, M.A.M. Manipulation of self-assembled structures of magnetic beads for microfluidic mixing and assaying. *Anal. Chem.* **2004**, *76*, 6239–6246. [CrossRef] [PubMed]
21. Ryu, K.S.; Shaikh, K.; Goluch, E.; Fan, Z.; Liu, C. Micro magnetic stir-bar mixer integrated with parylene microfluidic channels. *Lab Chip* **2004**, *4*, 608–613. [CrossRef] [PubMed]
22. Schiele, I.; Huber, J.; Hillerich, B.; Kozlowski, F. Surface-micromachined electrostatic microrelay. *Sens. Actuators A Phys.* **1998**, *66*, 345–354. [CrossRef]
23. Mensing, G.A.; Pearce, T.M.; Graham, M.D.; Beebe, D.J. An externally driven magnetic microstirrer. *Philosoph. Transact. Royal Soc. Lond. Series A: Math. Phys. Eng. Sci.* **2019**, *362*, 1059–1068. [CrossRef]
24. Lin, C.H.; Fu, L.M.; Chien, Y.S. Microfluidic T-form mixer utilizing switching electroosmotic flow. *Anal. Chem.* **2004**, *76*, 5265–5272. [CrossRef] [PubMed]
25. Wang, S.C.; Lai, Y.W.; Ben, Y.; Chang, H.C. Microfluidic mixing by dc and ac nonlinear electrokinetic vortex flows. *Ind. Eng. Chem. Res.* **2004**, *43*, 2902–2911. [CrossRef]
26. Lee, C.Y.; Lee, G.B.; Fu, L.M.; Lee, K.H.; Yang, R.J. Electrokinetically driven active micro-mixers utilizing zeta potential variation induced by field effect. *J. Micromech. Microeng.* **2004**, *14*, 1390–1398. [CrossRef]
27. Biddiss, E.; Erickson, D.; Li, D. Heterogeneous surface charge enhanced micro-mixer for electrokinetic flows. In Proceedings of the Second International Conference on Microchannels and Minichannels, Rochester, NY, USA, 17–19 June 2004; Volume 76, pp. 869–874. [CrossRef]
28. Wu, Z.; Nguyen, N.T.; Huang, X. Nonlinear diffusive mixing in microchannels: Theory and experiments. *J. Micromech. Microeng.* **2004**, *14*, 604–611. [CrossRef]
29. Wong, S.H.; Ward, M.C.L.; Wharton, C.W. Micro T-mixer as a rapid mixing micromixer. *Sens. Actuators B Chem.* **2004**, *100*, 359–379. [CrossRef]

30. Bökenkamp, D.; Desai, A.; Yang, X.; Tai, Y.C.; Marzluff, E.M.; Mayo, S.L. Microfabricated silicon mixers for submillisecond quench-flow analysis. *Anal. Chem.* **1998**, *70*, 232–236. [CrossRef]
31. Gobby, D.; Angeli, P.; Gavriilidis, A. Mixing characteristics of T-type microfluidic mixers. *J. Micromech. Microeng.* **2001**, *11*, 126. [CrossRef]
32. Knight, J.B.; Vishwanath, A.; Brody, J.P.; Austin, R.H. Hydrodynamic focusing on a silicon chip: Mixing nanoliters in microseconds. *Phys. Rev. Lett.* **1998**, *80*, 3863–3866. [CrossRef]
33. Löb, P.; Drese, K.S.; Hessel, V.; Hardt, S.; Hofmann, C.; Löwe, H.; Schenk, R.; Schönfeld, F.; Werner, B. Steering of liquid mixing speed in interdigital micro mixers—From very fast to deliberately slow mixing. *Chem. Eng. Technol.* **2004**, *27*, 340–345. [CrossRef]
34. Hardt, S.; Schönfeld, F. Laminar mixing in different interdigital micromixers: II. Numerical simulations. *AIChE J.* **2003**, *49*, 578–584. [CrossRef]
35. Schönfeld, F.; Hessel, V.; Hofmann, C. An optimised split-and-recombine micro-mixer with uniform "chaotic" mixing. *Lab Chip* **2004**, *4*, 65–69. [CrossRef]
36. Jen, C.P.; Wu, C.Y.; Lin, Y.C.; Wu, C.Y. Design and simulation of the micromixer with chaotic advection in twisted microchannels. *Lab Chip* **2003**, *3*, 77–81. [CrossRef]
37. Stroock, A.D.; Dertinger, S.K.W.; Ajdari, A.; Mezić, I.; Stone, H.A.; Whitesides, G.M. Chaotic mixer for microchannels. *Science* **2002**, *295*, 647–651. [CrossRef]
38. Liu, R.H.; Stremler, M.A.; Sharp, K.V.; Olsen, M.G.; Santiago, J.G.; Adrian, R.J.; Aref, H.; Beebe, D.J. Passive mixing in a three-dimensional serpentine microchannel. *J. Microelectromech. Syst.* **2000**, *9*, 190–197. [CrossRef]
39. Bertsch, A.; Heimgartner, S.; Cousseau, P.; Renaud, P. Static micromixers based on large-scale industrial mixer geometry. *Lab Chip.* **2001**, *1*, 56–60. [CrossRef]
40. Holger, L.; Guan, H.; Wu, C.-J.; Tu, S.-T. Three-dimensional numerical study of flow structures of impinging jets in Y-typed micro-mixers. *J. Nonlinear Sci. Numer. Simul.* **2007**, *8*, 425–434.
41. Yang, R.; Williams, J.D.; Wang, W. A rapid micro-mixer/reactor based on arrays of spatially impinging micro-jets. *J. Micromech. Microeng.* **2004**, *14*, 1345–1351. [CrossRef]
42. Mengeaud, V.; Josserand, J.; Girault, H.H. Mixing processes in a zigzag microchannel: Finite element simulations and optical study. *Anal. Chem.* **2002**, *74*, 4279–4286. [CrossRef]
43. Paik, P.; Pamula, V.K.; Fair, R.B. Rapid droplet mixers for digital microfluidic systems. *Lab Chip* **2003**, *3*, 253–259. [CrossRef] [PubMed]
44. Gooch, J.W. Liquid injection molding. *Encycl. Dict. Polym.* **2011**, *77*, 431. [CrossRef]
45. Miyake, R.; Lammerink, T.S.J.; Elwenspoek, M.; Fluitman, J.H.J. Micro mixer with fast diffusion. *IEEE Micro Electr. Mech. Syst.* **1993**, 248–253. [CrossRef]
46. Kadimisetty, K.; Spak, A.P.; Bhalerao, K.S.; Sharafeldin, M.; Mosa, I.M.; Lee, N.H.; Rusling, J.F. Automated 4-sample protein immunoassays using 3D-printed microfluidics. *Anal. Methods* **2018**, *10*, 4000–4006. [CrossRef]
47. Alizadehgiashi, M.; Gevorkian, A.; Tebbe, M.; Seo, M.; Prince, E.; Kumacheva, E. 3D-printed microfluidic devices for materials science. *Adv. Mater. Technol.* **2018**, *3*, 1800068. [CrossRef]
48. Hampson, S.M.; Rowe, W.; Christie, S.D.; Platt, M. 3D printed microfluidic device with integrated optical sensing for particle analysis. *Sens. Actuators B Chem.* **2018**, *256*, 1030–1037. [CrossRef]
49. Parker, E.K.; Nielsen, A.V.; Beauchamp, M.J.; Almughamsi, H.M.; Nielsen, J.B.; Sonker, M.; Gong, H.; Nordin, G.P.; Woolley, A.T. 3D printed microfluidic devices with immunoaffinity monoliths for extraction of preterm birth biomarkers. *Anal. Bioanal. Chem.* **2019**, *411*, 5405–5413. [CrossRef]
50. Wiese, M.; Benders, S.; Blümich, B.; Wessling, M. 3D MRI velocimetry of non-transparent 3D-printed staggered herringbone mixers. *Chem. Eng. J.* **2018**, *343*, 54–60. [CrossRef]
51. Enders, A.; Siller, I.G.; Urmann, K.; Hoffmann, M.R.; Bahnemann, J. 3D printed microfluidic mixers—A comparative study on mixing unit performances. *Small* **2019**, *15*, 1804326. [CrossRef]
52. Hereijgers, J.; Schalck, J.; Lölsberg, J.; Wessling, M.; Breugelmans, T. Indirect 3D printed electrode mixers. *Chem. Electr. Chem.* **2019**, *6*, 378–382. [CrossRef]
53. Bhargava, K.C.; Ermagan, R.; Thompson, B.; Friedman, A.; Malmstadt, N. Modular, discrete micromixer elements fabricated by 3D printing. *Micromachines* **2017**, *8*, 137. [CrossRef]
54. Bohr, A.; Boetker, J.; Wang, Y.; Jensen, H.; Rantanen, J.; Beck-Broichsitter, M. High-throughput fabrication of nanocomplexes using 3D-printed micromixers. *J. Pharm. Sci.* **2017**, *106*, 835–842. [CrossRef] [PubMed]

55. McDonald, J.C.; Chabinyc, M.L.; Metallo, S.J.; Anderson, J.R.; Stroock, A.D.; Whitesides, G.M. Prototyping of microfluidic devices in poly (dimethylsiloxane) using solid-object printing. *Anal. Chem.* **2002**, *74*, 1537–1545. [CrossRef] [PubMed]
56. Symes, M.D.; Kitson, P.J.; Yan, J.; Richmond, C.J.; Cooper, G.J.; Bowman, R.W.; Vilbrandt, T.; Cronin, L. Integrated 3D-printed reactionware for chemical synthesis and analysis. *Nat. Chem.* **2012**, *4*, 349–354. [CrossRef]
57. Rupal, B.S.; Garcia, E.A.; Ayranci, C.; Qureshi, A.J. 3D printed 3d-microfluidics: Recent developments and design challenges. *J. Integr. Des. Proc. Sci.* **2018**, *1*, 5–20. [CrossRef]
58. Pishbin, E.; Eghbal, M.; Fakhari, S.; Kazemzadeh, A.; Navidbakhsh, M. The effect of mo ment of inertia on the liquids in centrifugal microfluidics. *Micromachines* **2016**, *7*, 215. [CrossRef] [PubMed]

Article

Technical Model of Micro Electrical Discharge Machining (EDM) Milling Suitable for Bottom Grooved Micromixer Design Optimization

Izidor Sabotin [1], Gianluca Tristo [2] and Joško Valentinčič [1,3,*]

1 Faculty of Mechanical Engineering, University of Ljubljana, Aškerčeva 6, 1000 Ljubljana, Slovenia; izidor.sabotin@fs.uni-lj.si

2 Department of Industrial Engineering, University of Padua, Via Gradenigo, 6/a, 35131 Padua, Italy; gianluca.tristo@unipd.it

3 Chair of Micro Process Engineering and Technology (COMPETE), University of Ljubljana, Večna pot 113, 1000 Ljubljana, Slovenia

* Correspondence: jv@fs.uni-lj.si; Tel.: +386-1-477-1-730

Received: 20 May 2020; Accepted: 14 June 2020; Published: 16 June 2020

Abstract: In this paper, development of a technical model of micro Electrical Discharge Machining in milling configuration (EDM milling) is presented. The input to the model is a parametrically presented feature geometry and the output is a feature machining time. To model key factors influencing feature machining time, an experimental campaign by machining various microgrooves into corrosive resistant steel was executed. The following parameters were investigated: electrode dressing time, material removal rate, electrode wear, electrode wear control time and machining strategy. The technology data and knowledge base were constructed using data obtained experimentally. The model is applicable for groove-like features, commonly applied in bottom grooved micromixers (BGMs), with widths from 40 to 120 μm and depths up to 100 μm. The optimization of a BGM geometry is presented as a case study of the model usage. The mixing performances of various micromixer designs, compliant with micro EDM milling technology, were evaluated using computational fluid dynamics modelling. The results show that slanted groove micromixer is a favourable design to be implemented when micro EDM milling technology is applied. The presented technical model provides an efficient design optimization tool and, thus, aims to be used by a microfluidic design engineer.

Keywords: micromachining; micro EDM milling; empirical modelling; micromixer; design for manufacturing; computational fluid dynamics

1. Introduction

Micromachining plays a crucial role in bringing new chemical, medical, optical, automotive and semiconductor applications and products to market [1,2]. The industry strives for shorter cycle times to minimize the manufacturing costs. The research field that largely benefited from microengineering technologies (MET) development is microreactor technology, which exploits microstructured devices for realization of (bio)chemical processes [3,4].

A microreactor is commonly understood to be a continuous flow reactor which utilizes microchannels and similar micro features in order to manipulate flow of reactants. The basic configuration of a microreactor consists of a micromixer, a reaction unit and a separator. The micromixer, as a crucial functional part of the microreactor with the task of mixing reactants, usually represents the most challenging geometry to be machined.

One of the established micromixer designs is the bottom grooved micromixer (BGM), first introduced by Stroock et al [5] in 2002. The working principle of the BGM is based on the grooves embedded in the bottom of the microchannel which laterally transport fluid fractions and cause lateral circulation. This circulation motion promotes mixing by stretching and folding the fluid along the microchannel [5,6]. Generally, there are two basic designs of BGMs, namely, slanted groove micromixer (SGM) and staggered herringbone micromixer (SHM) (Figure 1). Grooves in SGM design are inclined at an angle with respect to the axis along the channel, whereas grooves in SHMs are the shape of a herringbone pattern. Many publications dealing with BGMs geometry optimization to enhance mixing exist, the most notable being by: Yang et al. [7], who applied the Taguchi method; Aubin et al. [8], who analysed BGM design using spatial data statistics and the maximum striation thickness parameter; Ansari and Kim [9], who applied the radial basis neural network method; Lynn and Dandy [6], who investigated helical flows; Williams et al. [10], who developed the analytical model of mixing; Cortes-Quiroz et al. [11], who used design of experiments, the function approximation technique and the multi-objective genetic algorithm; and Hossain et al. [12], who investigated grooves embedded in the top and bottom walls. The topic of BGM design optimization is highly relevant up to the present time, as demonstrated in very recent publications [13,14]. Thus, optimal design of a BGM has yet to be discovered with respect to specificities of its application and taking into account applied micromanufacturing technology and substrate material.

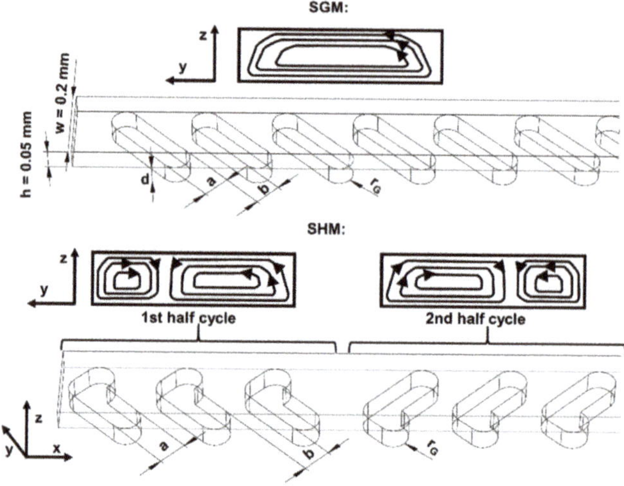

Figure 1. In the upper part, a slanted groove micromixer (SGM) geometry with schematic presentation of path lines in the channel cross-section is depicted. In the lower part, a staggered herringbone micromixer (SHM) geometry with schematic presentation of path lines in the cross-section for each half-cycle is depicted. Denotations: h—channel height, w—channel width, d—groove depth, a—groove width, b—ridge width, and r_G—groove corner rounding radius.

Traditionally, BGMs [5,10,15] and microfluidic platforms with similar geometrical features such as micro heat exchangers [16,17] and fuel cell components [18] are machined in silicon, thus, utilizing microfabrication technologies (i.e., various lithographic processes). On the other hand, metallic materials are gaining substantial importance in microfluidic devices and micro-electro-mechanical systems (MEMS) for their ability to withstand high pressures, harsh environmental conditions and chemically reactive conditions [19]. In this respect, stainless steel substrates offer significant advantages over silicon: excellent mechanical, electrical, thermal and chemical properties, and they are easy to clean, thus, they are reusable mostly due to material robustness [20–22]. Accordingly, micro EDM milling is a versatile and well established micromachining process which is readily capable of machining geometries applied in aforementioned applications [23].

Micro EDM milling is a thermal (or energy-assisted) process for contactless material removal of electrically conductive materials where a rotating cylindrical electrode removes material layer by layer with kinematics similar to those of conventional milling. Material is removed due to local melting and evaporation caused by a sequence of electrical discharges occurring in an electrically insulated gap between the tool electrode and the workpiece. Technology advantage stems from being able to machine complex shapes with high precision regardless of material hardness.

To date, a generally valid physical model of EDM material removal does not exist and there is a lack of a direct link between theoretical models of material removal mechanisms and industrial applications [24]. In the literature, many micro EDM milling modelling approaches were executed, such as mathematical modelling [25], cellular automata approach [26,27], non-uniform rational basis spline (NURBS) surface warping [28] and many others thoroughly reviewed in [29]. These modelling approaches yield valuable information in the context of understanding the underlying physical principles of the technology and can be applied for its advancements. However, in the view of process selection and industrial applicability, simple technical models can yield enough information to the designing engineer for the purpose of product design optimization in line with the design for manufacturing (DFM) paradigm [30]. By incorporating a few simple empirical equations encompassing the main technology characteristics, the technical model can be used as a tool by the design engineer, who does not need a thorough in-depth knowledge about the technology, for designing a functional product with lower machining costs.

For this reason, a simple technical model of micro EDM milling is constructed and presented in this paper. Main relations specific to micro EDM milling technology were obtained using empirical modelling. The output of the model is the total machining time of a single feature to be machined, which is determined by the contribution of duration of three operations, namely electrode dressing time, electrode wear control time and actual feature erosion time. The feature geometries are parametrically presented as an input to the technical model. One of the aims of this paper is to further familiarize the microfluidic research community with the micro EDM milling technology and its aspects on designing a microfluidic device. Demonstration of the technical model usage is presented through optimization of a BGM design. The developed technical model presents a simple platform, the applicability of which could be further expanded to a wider range of microfeatures by upgrading the technology data and knowledge base (DKB). At present, its applicability is confined to groove microfeatures which are commonly applied in BGMs; however, similar repetitive geometries are applied in various microfluidic applications [17,31,32]. The novelty of this work is in construction of an easy to understand technical model that can be used by a non-expert for microfluidic design optimization. Additionally, the results of intertwining the technical and computational fluid dynamics modelling give novel insights on the influence of geometrical parameters on the mixing performance of the BGM.

2. Materials and Methods

2.1. Investigation Workflow

The workflow of this investigation is presented in Figure 2. The mixing efficiency of the BGM is critically dependent on the geometry of one groove, which is defined by its width and depth, its basic shape, namely slanted (SG) or staggered herringbone (SH), and their number and orientation in a mixer configuration (see also Section 2.2). The defined single groove design presents the main input to the micro EDM milling technical model. Output of the technical model is the overall machining time t_{TOT} for the input geometry and, by defining a time constraint, results in the number of grooves (N_G) for the whole BGM configuration that can be machined in the given time. The machining time of the main microchannel is included in the time constraint as well. The technical model is supported by the micro EDM milling DKB, which was constructed based on the experimental approach (see Section 2.3). After determining N_G, permutations of the groove sequence layout were manually determined, resulting in BGM configurations with a different number of half-cycles (N_{HC}) and number of grooves in one half-cycle (N_{GH}). Mixing performances of adequate designs were determined by simulations using the numerical finite element software COMSOL (5.0, COMSOL, Inc., Burlington, MA, USA).

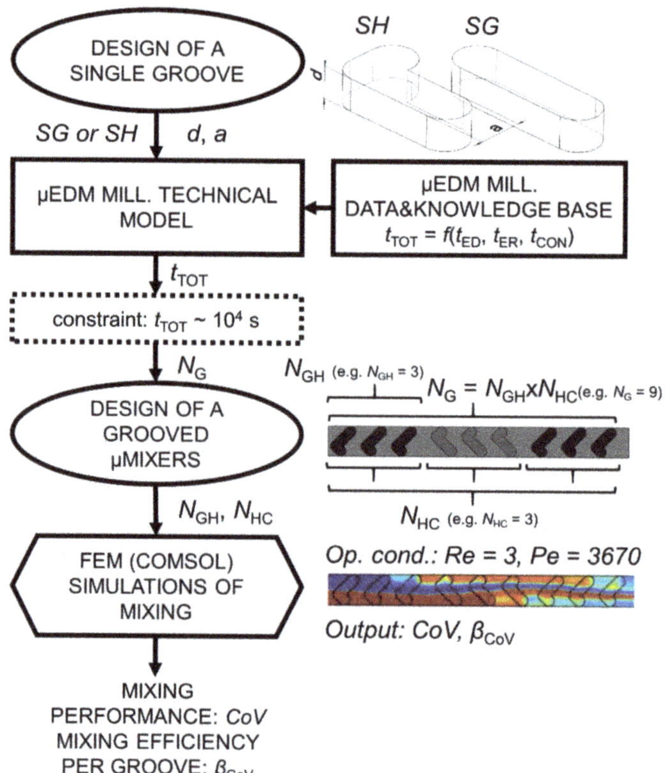

Figure 2. Investigation workflow. Denotations: *SH*—staggered herringbone groove design, *SG*—slanted groove design, *d*—groove depth, *a*—groove width, t_{ED}—electrode dressing time, t_{ER}—erosion time, t_{CON}—electrode wear control time, t_{TOT}—overall machining time, N_G—number of grooves in a configuration, N_{GH}—number of grooves in a half-cycle, N_{HC}—number of half-cycles, *Re*—Reynolds number, *Pe*—Peclet number, *CoV*—mixing coefficient of variance, and β_{CoV}—mixing efficiency per groove.

2.2. BGM Geometry and Mixing Simulation Tool

For the purpose of this investigation, the microchannel is oriented along the x-axis and its cross-section is fixed to width $w = 200$ µm and height $h = 50$ µm (Figure 1). This cross-section dimension is typical for generic microfluidic chips [33] and, due to its low aspect ratio ($h/w = 0.25$), it narrows the residence time distribution and reduces axial dispersion of the fluid species [34]. The grooves are inclined at a 45° angle to the x-axis and asymmetry of the SH grooves apex position was fixed to 1/3 x w, since this was found to be the optimal groove geometry [5–7]. Groove width a varied between 48 µm and 106 µm and groove depth d between 40 and 75 µm. The maximum aspect ratio of the grooves (d/a) incorporated in micromixer configurations was limited to 0.85 due to the possibility of air entrapment within deeper grooves [35] and to avoid dead volumes. The minimum groove width of 48 µm was selected since it is a typical width for micromixers produced by soft lithography and can be stably machined by micro EDM milling with a ~40 µm thick electrode. The upper boundary for groove width a is defined by the width of the channel w, since wider grooves machined with a circular electrode would be more similar to a circular blind hole than a groove. The rounding of the groove corners r_G corresponds to the smallest electrode diameter used for its machining and r_G varied between 25 and 53 μm. Ridge length b between consecutive grooves, being one of the least influential geometric parameters [7,36], was fixed to 70 μm. The geometrical parameters of simulated groove designs are gathered in Table 1.

Table 1. Geometrical parameters of simulated micromixer designs.

Main Channel Cross-Section (μm)	Groove Inclination Angle	SH Groove Apex Asymmetry	Groove Width a (μm)	Groove Depth d (μm)	Groove Aspect Ratio	Ridge Length b (μm)	Radius of the Groove Corners r_G (μm)
$w \times h = 200 \times 50$	45°	$1/3 \times w$	48–106	40–75	0.38–0.85	70	25–53

Mixing performances of BGM configurations compliant with micro EDM milling technology were evaluated by computational fluid dynamics (CFD) modelling using COMSOL Multiphysics 5.0 software. The software implements the finite element method to numerically solve governing equations. Since the applied simulation software settings are described in detail in [36], only a condensed description is given here. The flow field at steady state was calculated by solving Navier–Stokes (NS) equations for an incompressible fluid, and mass transport of dissolved species was numerically calculated by solving convection–diffusion (CD) equations. Since numerical simulations of mixing are prone to artificial diffusion of species [6,37], the simulation tool setup was verified against experimentally obtained results presented in [5,10], where confocal microscopy was applied to obtain precise spatial mixing profiles. Both quantitative (coefficient of variance) and qualitative (concentration profiles in the channel cross-section) results showed excellent agreement up to the Péclet number (Pe) of 6250, thus, we considered numerical model setup to be adequate.

Mixing performance was evaluated by applying coefficient of variance CoV as recommended in [38], defined by:

$$CoV = \frac{\sqrt{1/N \sum_{n=1}^{N} (c_n - \bar{c})^2}}{\bar{c}} \quad (1)$$

where CoV denotes the coefficient of variance in the yz-cross-section plane at the end of the channel, c_n is the concentration at a point in the cross-section plane, N is the number of concentration points in the plane (exported in a uniform grid with 2.5 μm spacing resulting in $N = 1600$), and $\bar{c}$ denotes complete mixing. A value of CoV closer to 0 corresponds to better mixing performance.

To take into account the geometrical parameter of the number of grooves N_G per BGM configuration, an additional parameter interpreted as mixing efficiency per groove was estimated as:

$$\beta_{CoV} = 1/(CoV \cdot N_G) \quad (2)$$

where a higher β_{CoV} value corresponds to a higher mixing efficiency of BGM configuration.

All further simulations were performed using the following conditions: the working fluid was water at 20 °C, the solute diffusion coefficient $D = 10^{-9}$ m^2/s, the average inflow velocity $v = 37.8$ mm/s corresponding to the Reynolds number of 3 and $Pe = 3670$, the boundary condition at outflow set to 0 Pa (pressure, no viscous stress), no-slip condition at walls, fluid concentration in one half of the channel of 1 mol/m^3 (color coded red) and the other half of 0 mol/m^3 (color coded blue). Data visualizations and post-processing of simulation results were done using associated functions in COMSOL and MATLAB (R2019b, MathWorks Inc., MA, USA).

2.3. Micro EDM Milling

The technical model is based on experimental data obtained on Sarix SX-200 micro EDM milling machine (SARIX SA, Sant'Antonino, Switzerland), thus, the set-up parameters, also referred to as the machining regime, are specific to the used machine (further elaborated in Section 2.3.2).

Input parameters encompass the workpiece, electrode, machining strategy, dielectric fluid and machining parameters. The latter define the settings of the pulse generator and the electrode feeding system. For clearer presentation of the micro EDM milling working principle, a diagram is shown in Figure 3.

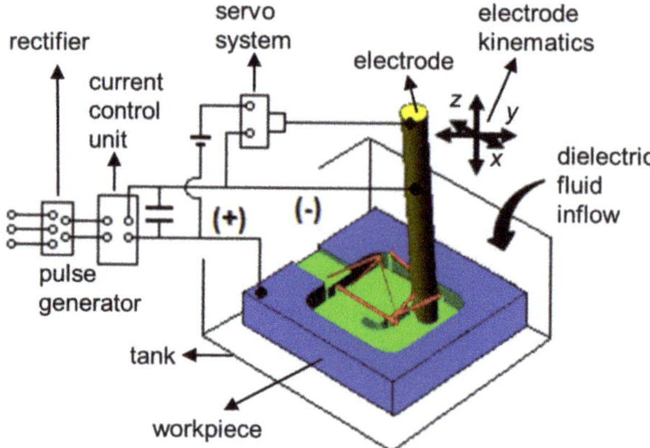

Figure 3. Schematic presentation of the micro EDM milling machine.

Thin cylindrical tungsten carbide rods with initial diameter d_{E0} of 300 µm were used as tool electrodes. The machine was equipped with a wire dressing unit (ARIANNE) and a laser scan micrometer Mitutoyo LSM-500s (Mitutoyo Corporation, Kawasaki, Kanagawa, Japan), which monitors the reduced diameter (d_E) of dressed electrodes (Figure 4a). By means of mentioned integrated components, the initial electrode is grinded down to the desired diameter, which depends on the width of the geometry to be machined and the spark gap. Slotting and pocketing machining strategies were considered (Figure 4c,d). Hydrocarbon oil was used as a dielectric fluid (HEDMA 111). All experiments to acquire micro EDM milling DKB were performed on corrosive resistant steel X2CrNiMo17-12-2 (Acroni Ltd., Jesenice, Slovenia), which is commonly used in chemical and pharmaceutical industry. Prior to machining, the steel workpieces were ground on both sides in order to establish parallel surfaces. In the next step, the surfaces were hand polished to achieve a mirror-like surface finish (roughness of $Ra \sim 0.05$ µm, measured with MARSURF PS 10 profilometer (Mahr GmbH, Göttingen, Germany). Due to imperfect polishing, minor surface scratches can be noticed on the confocal microscope images of the grooves (Figure 4c,d). However, these scratches were small enough and have insignificant influence on the obtained machining results.

Although the technical model built within this investigation is based on the specific machine and material used, the generalized findings can also be applied when using other micro EDM milling systems in the view of part design optimization for manufacturing.

In general, the aim of all machining processes is to produce parts with the desired geometrical accuracy and surface integrity. In micro EDM milling, accuracy greatly depends on compensation of the tool electrode wear [39–42]. SX-200 implements the linear tool wear compensation method (LCM) [43]. Electrode wear is often measured with linear wear l_F (Figure 4b), which defines the coefficient of linear wear ϑ_F by dividing the l_F with erosion time t_{ER} [44].

In micro EDM milling, the machining is done in a sequence of 3 machining operations:

1. op_1: dressing of the electrode using on the machine integrated wire dressing unit. Time needed for this operation is denoted as t_{ED} and is a function of:$t_{ED} = f(d_E(\text{geometry, gap(regime)}), \text{electrode working length}(\vartheta_F, \text{regime}))$;
2. op_2: actual erosion time $t_{ER} = f(d_E, \text{regime, strategy}, MRR, \text{volume of geometry})$;
3. op_3: time for electrode wear control $t_{CON} = f$ (depth of the geometry, ϑ_F).

To achieve the desired tolerances of the machined geometries (in our case, within ±4 μm), all three operations are executed every time when a new feature is being machined. Thus, the feature machining time is a sum of all three operations.

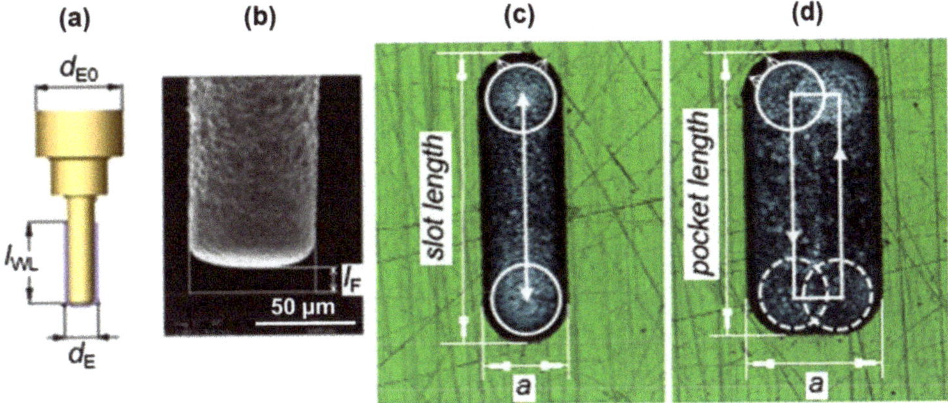

Figure 4. (**a**) Schematic presentation of the dressed electrode used for micro EDM milling with its main features: d_{E0}—diameter of primary electrode, d_E—electrode working diameter, l_{WL}—electrode working length. (**b**) A scanning electron microscope (FEI Quanta 400, FEI, Hillsboro, Ore., USA) image of a worn electrode where l_F denotes linear wear. (**c**) A confocal microscope image (Sensofar PLμ, NEOX, Sensofar Medical, Terrassa, Spain) of a groove (*a* = 90 μm) presenting slotting strategy for machining. The white circles represent the cross-section of the electrode and the arrows at the circle's circumference denote the sparking gap. (**d**) Presentation of pocketing strategy for machining of a pocket groove in 2 passes (*a* = 140 μm).

2.3.1. Dressing of the Tool Electrode

Electrode dressing time t_{ED} was investigated with respect to electrode nominal diameter d_E and electrode working length l_{WL} (see Figure 4a). These two parameters, in combination with starting electrode diameter d_{E0}, determine the shape and dressing time of the tool electrode. The electrode l_{WL} defines the maximum depth of the slot that could be potentially machined considering no electrode wear. Default factory settings for the on-machine wire dressing unit were used. Investigated d_E range was from 40 to 120 μm. The electrode diameter tolerance was set to ±4 μm. In the first part of experiments, l_{WL} was fixed at 150 μm. For investigation of the influence of l_{WL} on t_{ED}, the length of l_{WL} was fixed to 8 × d_E, which is the maximum length ratio recommended by the machine tool manufacturer. For each d_E, electrode dressing was repeated 4 times to estimate the electrode dressing time t_{ED}. Altogether, 80 experiments on electrode dressing were performed.

2.3.2. Material Removal Rate and Erosion Time

Due to required low roughness (*Ra* ~ 0.5 μm), the finishing machining regime suggested by the Esprit CAD/CAM system suitable for stainless steel was used (Table 2). The same machining regime is suggested for electrode diameters between 50 and 120 μm. For d_E = 40 μm, the system suggests finer settings with lower pulse current and voltage. Machining parameters applied for the erosion of the main channel (d_E = 183 μm) are also presented.

Table 2. Machining parameters setup particular to Sarix SX-200 EDM milling machine for corresponding d_E.

Parameters	d_E = 40 µm	d_E = 50–120 µm	d_E = 183 µm
E (index)	13	13	105
(+,-)	-	-	-
t_P (indeks)	2	2	6
f_P (kHz)	180	180	100
i_P (index)	80	100	65
u (V)	77	90	100
z_{1L} (µm)	0.7	0.7	0.7
$gain$ (index)	100	100	80
g_i (index)	74	74	80
n_{CON}	every 10 µm	every 10 µm	every 10 µm

Denotation of parameters: energy index E, polarity, duration of discharge pulse t_p, discharge frequency f_p, discharge current i_p, ignition voltage u, gain index $gain$, gap index g_i, machining layer depth z_{1L}, and number of controls of electrode axial wear n_{CON}.

Several sets of slots and pockets were machined to determine the influence of d_E and machining regimes on material removal rate (*MRR*) and electrode wear. Each feature was machined by an electrode producing total width of the feature in one pass (slotting strategy), except in the case of pockets:

- slots: length of 310 µm, d = 100 µm, d_E = 40–120 µm, 2 repetitions (Figure 4c);
- pockets: length of 310 µm, d = 100 µm, d_E = 40–120 µm, 2 passes with 40% overlap, 2 repetitions (Figure 4d);
- *SG* slots: d = 80 µm, d_E = 50, 75, 100 µm, 3 repetitions (Figure 5a);
- *SH* slots: d = 80 µm, d_E = 40–120 µm, 3 repetitions (Figure 5b,c);
- *SH* slots: d = 40, 120 µm, d_E = 40, 70, 100 µm, 1 repetition;
- *SH* groove—2-electrode strategy: d = 100 µm, d_{E1} = 92 µm, d_{E2} = 44 µm, 2 repetitions (Figure 6).

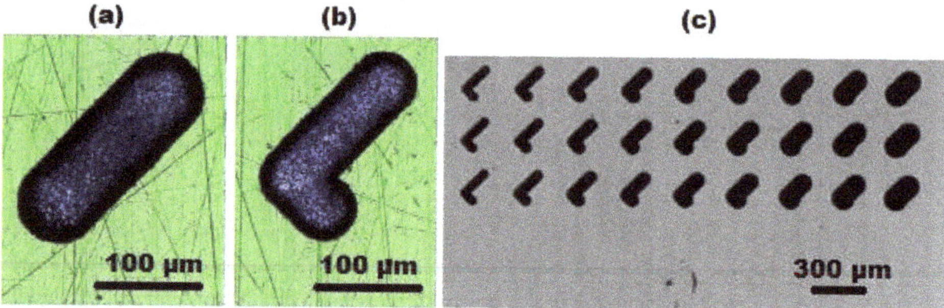

Figure 5. Microscope images of *SG* and *SH* slots, which can be implemented in the w = 200 µm wide microchannel. (**a**) *SG* slot (d_E = 75 µm, d = 80 µm) and (**b**) *SH* slot (d_E = 60 µm, d = 80 µm). (**c**) *SH* slots (d_E = 40–120 µm, d = 80 µm).

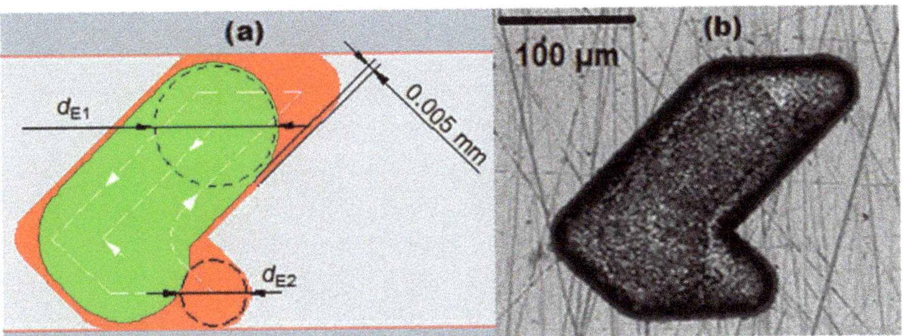

Figure 6. (**a**) Depiction of the 2-electrode strategy for machining of the SH groove. 5 µm band is left by larger electrode d_{E1}, thus, its diameter is determined by: $d_{E1} = a - 2 \times l_{GAP} - 2 \times 5$ µm. (**b**) Confocal microscope image of the SH groove machined with d_{E1} = 92 µm and d_{E2} = 44 µm (d = 100 µm, a = 106 µm, r_G = 25 µm).

This resulted in 76 machined grooves. Removed workpiece volume (V_W) was determined by measuring groove dimensions with the Sensofar PLµ NEOX confocal microscope (Sensofar Medical, Terrassa, Spain). *MRR* was determined as V_W/t_{ER}. In all experiments stable machining was observed, thus suitable machining regimes were used. The results were added to DKB.

When groove corner roundings are relatively narrow, a 2-electrode strategy is employed: a wider electrode is applied to remove a larger portion of groove volume and, afterwards, thinner electrode removes material around the edges (Figure 6). The 2-electrode strategy is significantly faster compared to machining the whole groove volume with only one thinner electrode (1-electrode pocketing strategy). The technical model considers both strategies and selects the most appropriate one for the given geometry.

At the end of machining, linear wear l_F was measured by the touch method. The touch method was implemented on the used machine and consists of traversing the electrode to the reference measuring point. Then, the electrode is gradually approaching the reference point in the z-axis. Upon touch, detected by a short circuit signal, the difference in readout of the z-coordinate compared to the previous iteration determines the electrode linear wear. For further information about the touch method, see [45]. Volumetric wear ratio ϑ was calculated as [46]:

$$\vartheta = \frac{V_E}{V_W} = \frac{l_F \pi d_E^2}{4V_W}, \tag{3}$$

where V_E denotes volumetric wear of the electrode.

2.3.3. Electrode Wear Control

Compensation of the tool electrode wear during erosion is crucial for the achieved machining accuracy. For this reason, LCM is implemented on the machine, where linear electrode wear is intermittently monitored during erosion after selecting the number of layers by the touch method. Consequently, linear electrode wear is compensated during machining with a gradual movement of the electrode tip position in the z-axis. The results of preliminary experiments (groove length = 310 µm, d = 100 µm) showed that the control after erosion of every 10 µm in depth is adequate for achieving tolerances in the z-axis within ±4 µm. Individual control time (t_{CON}) comprises of touch method measurement duration and electrode traveling to and from the reference measuring point.

2.4. Implementation of the Technical Model

The technical model was programmed in MATLAB and is schematically presented in Figure 7. The input to the model is parametrically defined BGM geometry, basically determined by the variant (SG or SH) and dimensions of a single groove (step *I*). Based on the groove corner rounding radius (r_G), 2-electrode strategy is selected if r_G is smaller than $a/2$ (Figure 7, step *II*). In step *III*, groove volume V_W is calculated. According to groove width a, electrode working diameter d_E is determined by taking into account spark gap (l_{GAP}) consistent with the applied machining regime. Needed electrode working length l_{WL} is determined by the absolute depth of the groove (i.e., sum of h and d) and corresponding tool wear. In step *IV*, durations of all three machining operations are calculated with respect to relevant relations defined in DKB. In step *V*, the machining time of a single groove is multiplied by the number of grooves in a BGM configuration.

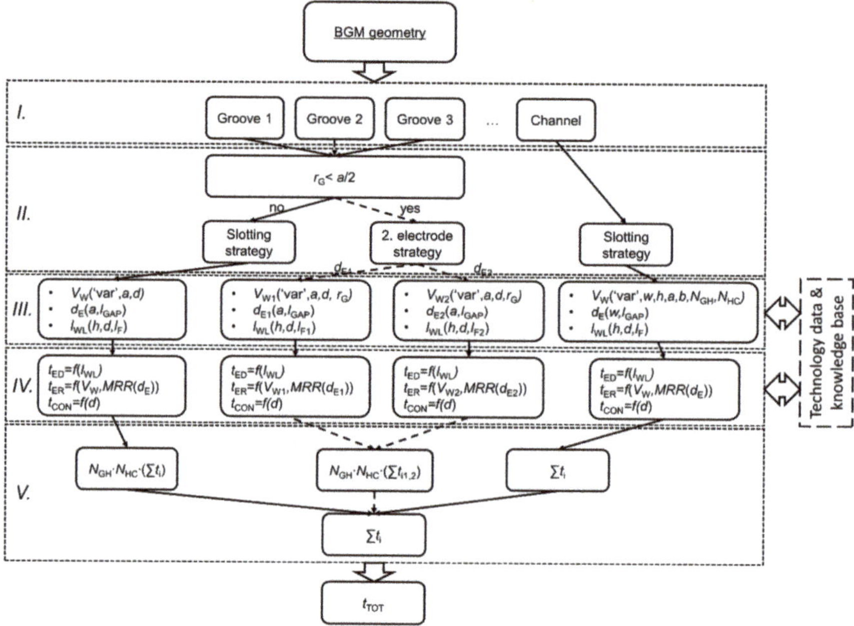

Figure 7. Detailed scheme of the implemented technical model where 'var' in dashed box *III.* denotes either SG or SH design.

Microchannel machining time is calculated separately and, at fixed width w and height h, depends only on its length, which is determined by the BGM configuration. The sum of machining times of all features (i.e., grooves and microchannel) defines total machining time t_{TOT}.

2.5. Simulated BGM Designs

Overall, 23 different BGM designs compliant with the dimensional boundaries given in Table 1. were simulated to estimate their mixing performances. A particular BGM design was considered on the criteria of having overall machining time t_{TOT} of approximately 10^4 s, and that the number of half-cycle N_{HC} at SH configurations corresponded to an integer. Relaxation of the t_{TOT} constraint was made for SH configurations where 2-electrode strategy was applied. Since this strategy requires longer machining time per groove and mixing performance of the SH design is significantly enhanced with the increase of number of half-cycles, the t_{TOT} criterion was extended up to 1.5×10^4 s, so that at least

9 grooves were machined (resulting in configuration with N_{HC} and N_G of 3). All configurations are parametrically presented later in Section 3.2 (Table 2).

3. Results and Discussion

3.1. Micro EDM Milling DKB Construction

As explained in Section 2.3.1, two sets of experiments were performed to model the relationship defining electrode dressing time t_{ED}: first set with constant l_{WL} and the second set with l_{WL} being $8 \times d_E$. The results show that t_{ED} depends on the electrode working length l_{WL} and obtained accuracy of the electrode diameter in the first step of the dressing. As evident from Figure 8, the on-machine factory implemented electrode dressing procedure takes about the same amount of time when set to l_{WL} = 150 μm regardless of the set d_E: t_{ED} = 182 ± 1.5 s. Thus, for the electrodes dressed in the first iteration, t_{ED} is influenced only by the electrode working length l_{WL}.

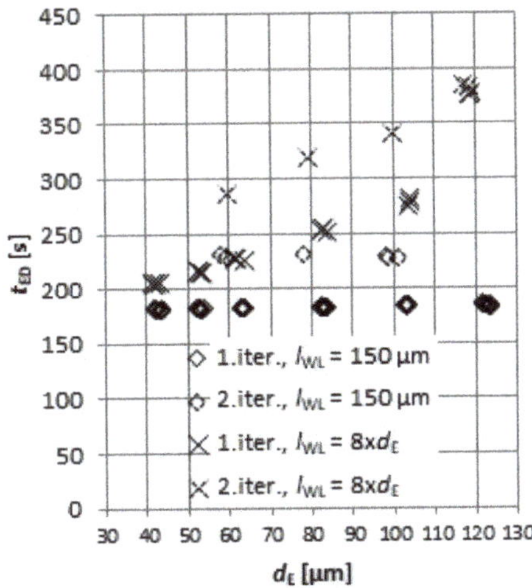

Figure 8. Electrode dressing times t_{ED} for various electrode diameters d_E and electrode working lengths l_{WL}.

It should be noted that, in minor cases, the electrode is dressed in second iteration. Namely, if the laser scan measurement of the d_E after the first iteration of dressing is out of set tolerances, the second iteration of electrode dressing is executed, but this happened randomly and rarely (only in 16% of the cases) and, thus, it is not included in the technical model. t_{ED} operation was therefore modelled with linear function (Figure 9a) by applying linear regression resulting in equation

$$t_{ED} = 0.145 l_{WL} + 160 \ [s], \tag{4}$$

where the units for l_{WL} are in μm and a correlation coefficient of $R^2 = 0.99$ was obtained.

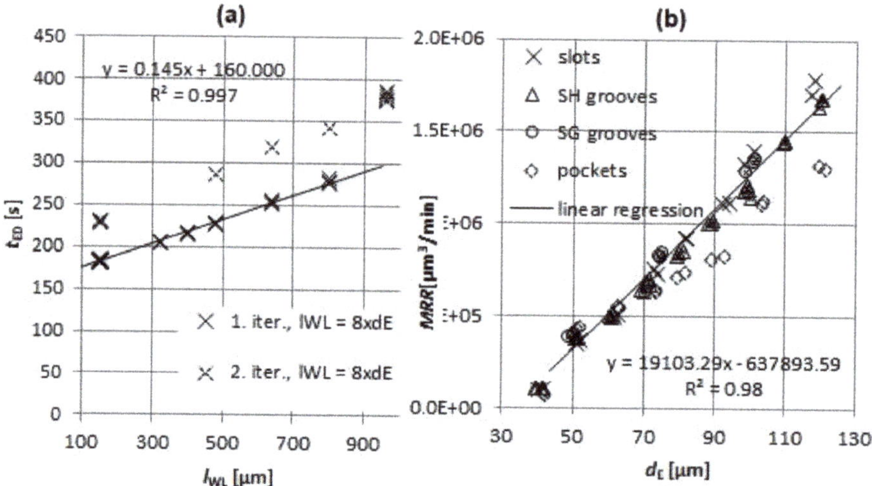

Figure 9. (**a**) Electrode dressing times t_{ED} for various electrode working lengths ($l_{WL} = 8 \times d_E$). Linear regression was applied to the data set, where the electrodes are dressed in the first iteration. (**b**) Material removal rate (*MRR*) with respect to electrode diameter d_E. For the linear regression, only slotting experiments (slots, SH and SG grooves) were considered.

Investigation of *MRR* relation with respect to d_E exhibits a linear trend (Figure 9b). For micro EDM milling, it is known that material removal per discharge mainly depends on pulse discharge energy [47]. Since, for most electrode diameters (with exception of $d_E = 40$ μm), the same machining regime was used, the *MRR* increased due to the increase of electrode face surface rendering favourable spark conditions. Thus, *MRR*(d_E) was modelled by applying linear regression resulting in equation

$$MRR = 19103d_E - 637893 \left[\mu m^3/min\right], \tag{5}$$

where the units for d_E are in μm and correlation coefficient of $R^2 = 0.98$ was obtained. A plausible explanation for rising linear trend of *MRR*(d_E) is in linear increase of the effective eroding surface of the electrode which is, at constant machining layer depth z_{1L}, defined by $(d_E + 2l_{GAP})\cdot z_{1L}$ as stated in [43]. With the increase of the active area, the discharges are more sparsely distributed thus, rendering favourable spark conditions. Average gap at groove machining was measured to be $l_{GAP} = 3 \pm 0.3$ μm.

MRR for pocketing strategy is lower due to the overlap which reduces effective eroding surface of the electrode. Thus, the slotting strategy is favourable in terms of machining BGM designs.

Electrode tool wear was monitored during machining. It influences machining time t_{TOT} via adding a compensation length to l_{WL}. It can be noted (Figure 10a) that for a fixed groove depth and machining *regime*, the volumetric wear ratio ϑ linearly decreases with increase of d_E. This is due to a higher wear resistance of larger electrodes, since wear is correlated to electrode thermal properties. Thus, larger surface area of the electrode tip, which grows proportionally with d_E^2, conducts heat in the electrode bulk and, due to larger electrode volume, the wear caused by melting is reduced. At the same time, discharges are less densely distributed due to the bigger eroding surface. Simultaneously, when slotting with larger electrodes, more dielectric fluid surrounds the electrode within the groove and acts as a coolant. Obtained results are consistent with the findings in [46], which states that larger electrodes are more wear resistant due to their bulk volume and correlated manifestation of thermal erosion caused by sparks.

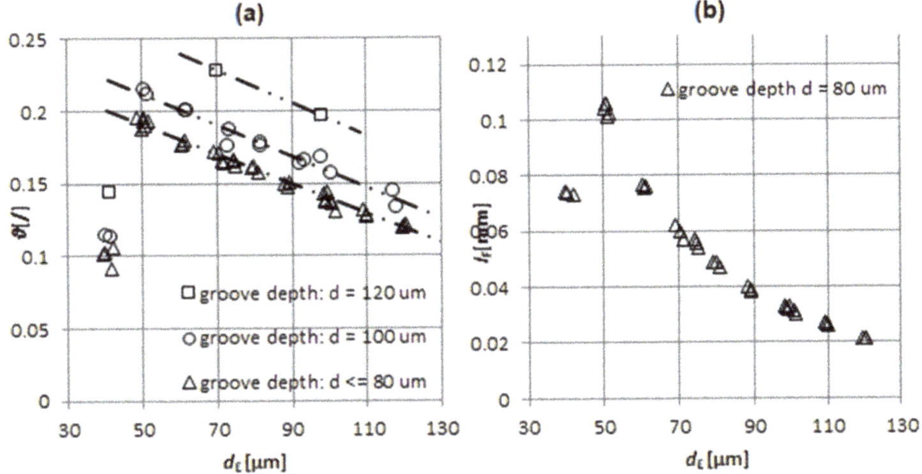

Figure 10. (**a**) Volumetric wear ratio ϑ with respect to electrode working diameter for grooves of different depths. (**b**) Linear wear l_F with respect to electrode working diameter for groove depths of d = 80 µm.

In the case of machined grooves deeper than 80 µm, ϑ increases by offsetting a linear trend (Figure 10a). A plausible explanation of this observation would be in worse flushing conditions occurring in deeper grooves. This effects electrode wear in two ways. Firstly, the debris within the spark gap start to accumulate, which cause arcing and, thus, promote electrode wear. Secondly, the cooling effect of the dielectric is reduced. Figure 10a also indicates the influence of machining *regime* with lower pulse current and voltage used for $d_E = 0.04$. Volumetric wear ration ϑ and linear wear l_F (Figure 10b) significantly decrease due to less aggressive machining regime.

Stemming from practice and considering the simplicity of the technical model, adding a safety factor of 100% to the highest linear wear recorded ($l_F = 103$ µm at $d_E = 50$ µm, $d = 80$ µm) was introduced to calculation of electrode working length

$$l_{WL} = structure\ depth + 200\ \mu m, \tag{6}$$

where structure depth in the case of a micromixer machining represents a sum of main channel height h and groove depth d.

Analysis of 60 measurements of the time required for control of the electrode linear wear resulted in an average control time $\bar{t}_{1CON}$ of 22.6 s with a standard deviation of 0.9 s. With respect to reasoning described in Section 2.3.3, t_{CON} was determined as:

$$t_{CON} = \frac{d\ [\mu m]}{10\ [\mu m]} \bar{t}_{1CON}\ [s]. \tag{7}$$

The technical model was verified by machining of test grooves with varying depths and widths. The model prediction of the machining time was within 8% of the measured machining time. With regards to the stochastic nature of the EDM process, we consider that the technical model is adequate for its purpose.

To give a clearer presentation on how the technical model algorithm works, a short manual demonstration of machining time calculation for a single slanted groove is demonstrated. For parametrical input parameters $a = 88$ µm, $r_G = a/2$, d = 40 µm, $w = 200$ µm and $h = 50$ µm the algorithm works as follows:

a. based on the premise $r_G == a/2$ single electrode slotting strategy is selected;
b. volume to be removed is calculated as: $V_W = \left(\frac{\pi a^2}{4} + \sqrt{2} \cdot a \cdot (w - a)\right) \times d = 8 \times 10^5 \mu m^3$;
c. electrode diameter is calculated as: $d_E = a - 2 \cdot l_{GAP} = 82$ μm;
d. electrode working length (Equation (6)): $l_{WL} = f(h,d) = 290$ μm;
e. time of electrode dressing (Equation (4)): $t_{ED} = f(l_{WL} = 290 \mu m) = 202$ s;
f. material removal rate (Equation (5)): $MRR = f(d_E = 82 \mu m) = 9.3 \times 10^5 \mu m^3/min$;
g. erosion time: $t_{ER} = V_W/MRR = 51$ s;
h. control time (Equation (7)): $t_{CON} = f\ (d = 40 \mu m) = 90$ s;
i. one SG groove machining time: $t_{1groove} = t_{ED} + t_{ER} + t_{CON} = 343$ s;

The key to automize the machining calculation time is in deriving a parametrical equation for feature volume. At fixed input parameters defining BGMs, the groove's volume calculation becomes a simple geometrical problem.

3.2. Micromixer Design for Micro EDM Milling

Altogether, 23 different BGM configurations compliant with micro EDM milling technology were simulated in order to determine their mixing performance. Geometrical parameters, machining times obtained using the technical model and mixing performances determined by simulations are gathered in Table 3.

Table 3. Geometrical definition of BGM configurations and simulated mixing performances. Denotations: r_G—groove corner rounding (a/2 denotes 1-electrode strategy), N_{HG}—number of grooves in a half-cycle, N_{HC}—number of half-cycles, $t_{1groove}$—machining time of 1 groove, t_{GR}—machining time of all grooves in a configuration, t_{CH}—channel machining time, t_{TOT}—overall machining time, CoV—mixing coefficient of variance , β_{CoV}—mixing efficiency per groove.

N°	Var.	a (μm)	d (μm)	r_G (μm)	N_{GH}	N_{HC}	$t_{1groove}$ (s)	t_{GR} (s)	t_{CH} (s)	T_{TOT} (s)	CoV	β_{CoV}
#1	SGM	48	40	a/2	12	1	478	5741	4069	9810	0.57	0.15
#2	SGM	71	40	a/2	15	1	357	5348	4587	9934	0.15	0.44
#3	SGM	88	40	a/2	15	1	342	5133	4784	9917	0.14	0.48
#4	SGM	106	40	a/2	15	1	335	5019	4981	10000	0.15	0.44
#5	SGM	71	60	a/2	13	1	437	5676	4376	10052	0.07	1.10
#6	SGM	88	70	a/2	12	1	474	5690	4429	10119	0.05	1.67
#7	SGM	106	75	a/2	12	1	467	5603	4587	10189	0.06	1.39
#8	SHM	48	40	a/2	2	6	477	5718	4209	9927	0.32	0.26
#9	SHM	48	40	a/2	3	4	477	5718	4139	9857	0.37	0.23
#10	SHM	48	40	a/2	4	3	477	5718	4104	9822	0.47	0.18
#11	SHM	48	40	a/2	6	2	477	5718	4069	9787	0.47	0.18
#12	SHM	71	40	a/2	3	5	355	5331	4692	10023	0.15	0.44
#13	SHM	71	40	a/2	5	3	355	5331	4622	9953	0.25	0.27
#14	SHM	71	60	a/2	3	4	435	5220	4341	9561	0.19	0.44
#15	SHM	71	60	a/2	6	2	435	5220	4271	9491	0.28	0.30
#16	SHM	88	40	25	3	3	711	6395	4109	10503	0.52	0.21
#17	SHM	88	40	25	2	5	711	7105	4297	11402	0.34	0.29
#18	SHM	106	40	25	3	3	728	6548	4227	10774	0.54	0.21
#19	SHM	106	40	25	2	5	728	7275	4429	11704	0.41	0.24
#20	SHM	88	70	25	3	3	994	8948	4109	13056	0.25	0.44
#21	SHM	88	70	25	2	5	994	9942	4297	14239	0.16	0.63
#22	SHM	106	75	25	3	3	1043	9388	4227	13615	0.26	0.43
#23	SHM	106	75	25	2	5	1043	10431	4429	14860	0.18	0.56

It can be noted that the highest number of grooves $N_G = 15$ can be machined within the time constraint for groove widths between 71 to 106 μm and groove depth of 40 μm. Among the 1-electrode strategy per groove designs (#1-15), the lowest $N_G = 12$ is achieved for the narrowest and shallowest of grooves despite removing the smallest volume of material. The reason lies in a very low MRR when machining with electrodes of small diameter (e.g., $d_E = 40$ μm). When machining the widest and deepest grooves (e.g., #7) also 12 grooves can be machined in nearly the same time despite removing 3.5 times larger volume. Grooves with 2-electrode strategy need significantly more time to be machined.

The interpretation of simulated mixing results is further discussed through the following configurations (visualized in Figure 11): #1 as worst performing configuration overall; #6 as best performing configuration overall; #12 as best performing SHM configuration; #14 as second best performing SHM configuration; #21 as best 2-electrode strategy SHM configuration.

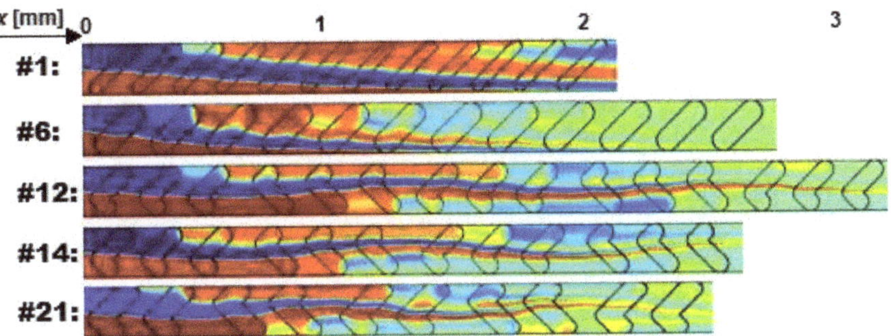

Figure 11. Results of mixing simulations for different BGM designs represented by a concentration surface in the mid-plane of the main channel. Operation conditions are defined by $Re = 3$ and $Pe = 3670$. Green colour denotes complete mixing.

Low mixing performance of configuration #1 is mainly a consequence of low number of too narrow grooves, which are inefficient in lateral transportation of fluid. Additionally, it is known from the literature that narrow and shallow grooves in SG configuration perform worse compared to equivalent SH designs (#8-11) [5,15,48]. However, at adequate groove width to depth ratio the results show that in some cases SG outperform SH configurations, as demonstrated by the most efficient design #6 ($CoV = 0.05$, $\beta_{CoV} = 1.67$). Considering the fixed cross-section of the main channel, this groove design is close to optimal and on the higher end of groove aspect ratios (0.8) investigated in this research. It should be mentioned that SGM realizations reported in the literature were fabricated by soft lithography (e.g., [5,15,49]) and thus, have sharp groove corners. From the perspective of fluid flow pattern, rounded corners seem to induce additional perturbation, which further promotes mixing [36].

SHM configurations #12 and #14 differ in groove depth and number of grooves. Having the same mixing efficiency per groove ($\beta_{CoV} = 0.44$) shows that groove depth is a critical geometrical parameter. In agreement with assumptions of [6] it is evident that more half-cycles are favourable for all investigated SH configurations (Table 2, #8-23). Similar conclusions as for 1-electrode strategy SHM designs can be made for 2-electrode strategy configurations; namely, deeper grooves and more half-cycles improve mixing performance. The benefit of smaller groove corner roundings is also evident, since configuration #21 renders highest $\beta_{CoV} = 0.63$ among all SHM designs. This indicates that mixing mechanism in SHMs is impacted by the geometry of SH grooves' shorter arm.

3.3. Simulation Runs of Technical Model

To show the relations between time durations of particular machining operation for BGM geometry configurations highlighted in Section 3.2, relevant pie-charts are presented in Figure 12. Pie-chart for one groove machining of configuration #1 demonstrates the time distribution when using the thinnest electrode ($d_E = 40$ µm) considered in this investigation. Electrode dressing and erosion time contribute 81% to overall time and control time due to shallow depth only 19%. Using larger electrode ($d_E = 80$ µm) for configuration #6 the share of erosion time significantly drops (18%) due to higher related *MRR*. The electrode control time increases due to bigger depth (38%).

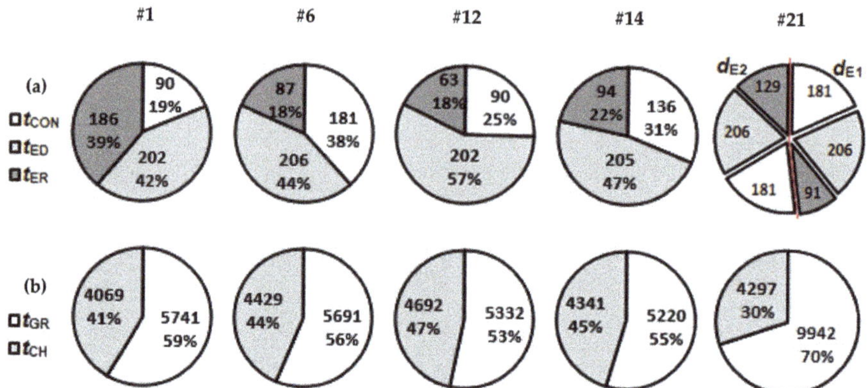

Figure 12. Time required for specific machining operation regarding denoted BGM configurations. (**a**) One groove machining times. For 2-electrode strategy (#21) groove machining times are separately presented with respect to used electrode (d_{E1} = 72 μm and d_{E2} = 44 μm). (**b**) Machining time of all grooves (t_{GR}) in a configuration versus corresponding channel machining time (t_{CH}). Time is given in seconds.

The comparison of one groove machining times for configuration #12 and #14 shows the influence of applying tool electrode of the same diameter (d_E = 65 μm) to different depths of the grooves. The electrode dressing times are similar (202 s and 205 s respectively), although the share of electrode dressing time t_{ED} significantly drops from 57% to 47% on the account of erosion t_{ER} and control time t_{CON} when machining deeper grooves (#14). Machining operation durations for the 2-electrode strategy are presented through configuration #21. It is evident that, despite machining significantly less volume with the thinner electrode (d_E = 44 μm, which results in corner rounding of r_G = 25 μm), it contributes 52% to overall 1 groove machining time.

From the above observations, the following design suggestions related to efficient use of micro EDM milling technology can be drawn. Firstly, for narrow and shallow geometries the *MRR* optimization is crucial for reducing the groove machining times. At wider and deeper grooves efforts to reduce electrode control t_{CON} and dressing time t_{ED} is the strategy to be pursued (e.g., Figure 12 #6, #14). In this respect, novel electrode wear control strategies such as on-line discharge counting [50], in-situ adaptive control [43] and strategy based on scanning area [41] promise significant reduction of a feature machining duration due to reducing t_{CON}.

For low aspect ratio grooves (e.g., Figure 12 #12) the electrode dressing time represents more than half of machining time per groove. Thus, it would be beneficial to reduce t_{ED} by optimization of dressing procedure. As for 2-electrode strategies it can be stated that narrow groove corner roundings should be avoided if possible, as usage of smallest electrode diameters dominantly determines the feature machining time.

3.4. General Applicability of the Technical Model

The presented technical model is specific to used Sarix micro EDM milling machine tool, manufacturer of which is considered leading in the field. However, some findings can be generalized and are expected to be representative regardless of the type of machine tool used. In this context relations determining *MRR* can be generalized as long as similar discharge energies are applied. Namely, unit material removal per discharge and resulting surface finish chiefly depend on the discharge energy. Thus, for the range of d_E investigated in this research by applying discharge energies for finishing would result in similar characteristics. By applying the same reasoning, we expect that presented electrode wear characteristics could also be generalized, since again discharge energy level is the dominant influencing parameter. However, as demonstrated by relatively lower electrode wear at thinnest electrode used (e.g., $d_E = 40$ μm) less aggressive *regime* reduces the wear effects. On the other hand, the contribution of electrode wear to electrode dressing time, as defined in our investigation, is rather small.

Two of the most common microelectrode dressing procedures are based on the sacrificial block technique and wire EDM grinding (as used in this investigation), the latter being more sophisticated and precise solution. In order to implement sacrificial block technique into the technical model adequate experimental campaign would need to be carried out. We expect that derived relation for t_{ED} would significantly differ from relation obtained in this investigation (demonstrated by Equation (4)).

With respect to calculation of t_{CON}, the linear tool wear compensation method is most commonly implemented on machine tools, thus, similar relations as provided in this investigation are expected. However, duration of $\bar{t}_{1CON}$ may vary between the systems.

Built DKB is also specific to applied workpiece material. We expect that for materials in the steels family the relations should remain similar. This may not be the case when using materials with significantly different thermo-physical properties (e.g., conductive ceramics).

As regards to technical model suitability for wider range of geometries additional exhaustive experimental campaigns should be executed. It is evident that applied machining *regime* significantly depends on the tool electrode diameter, especially if larger d_E (e.g., $d_E > 0.12$ mm) are to be used. Additionally, pocketing strategy could be implemented upon further investigation. At this point it should be stated, that presented technical model usage is limited to groove-like repetitive geometries commonly found in BGM designs. However, the presented technical model framework could be used for the purpose of expanding its applicability by upgrading the supporting DKB.

4. Conclusions

In this paper, a simple empirically based technical model of micro EDM milling is constructed. Its applicability is confined to groove-like microfeatures within a given dimensional interval, but it could be expanded by upgrading the technology DKB using presented methodology. The accuracy of the model lies within 8%, when comparing calculated and measured machining times. It represents an easy to use platform which can be applied for product design optimization and process selection. Demonstration of its usage was presented through optimization of a bottom grooved micromixer (BGM) design. Some general design guidelines particular to micro EDM milling technology are derived:

- When machining narrow slots ($a < 60$ μm) optimization of *MRR* is crucial for reduction of feature machining time.
- At machining wider ($a > 60$ μm) and deeper ($d > 40$ μm) slots reducing electrode control and dressing time is sensible optimization strategy.
- For lower aspect ratio slots ($d/a < 0.6$) electrode dressing time contributes dominantly to feature machining time.
- Narrow groove corner roundings should be avoided if possible, despite applying a 2-electrode strategy, due to corresponding very low $MRR(d_{E2})$.
- Predicted future implementations of concurrent electrode wear estimations could eliminate electrode wear control time and thus, significantly shorten feature machining time. In the case of grooves applied in this investigation from 19% up to 40%.

The mixing performances of 23 micromixers, which are comparable in terms of machining duration, were analysed. Findings are summarised below:

- Investigated BGM designs are readily applicable in metal substrates, thus demonstrating the capabilities of micro EDM milling technology, which in terms of obtainable feature aspect ratios surpasses soft lithography methods.
- The best BGM configuration in terms of mixing performance (CoV and β_{CoV}) was SGM configuration with groove dimension of $a = 88$ μm and $d = 70$ μm, representing a favourable groove geometry.

- SG micromixers with round groove corners achieve better mixing performance. In SH micromixers smaller corner roundings (i.e. sharper corners) perform better.
- More half-cycles for fixed number of grooves improves mixing performance of SH micromixers.
- SGM design is more suitable than SHM for micro EDM milling and gives better mixing performance.
- Lower mixing performances of SHMs may be attributed to reduced effectiveness of the shorter arm due to groove corner roundings which are inherently connected with micro EDM milling technology.

The content of the paper also aims at familiarization of micro EDM milling technology to wider microfluidic research community.

Author Contributions: Conceptualization, I.S. and J.V.; data curation, I.S. and G.T.; formal analysis, I.S. and G.T.; funding acquisition, J.V.; investigation, I.S., G.T. and J.V.; methodology, I.S., G.T. and J.V.; project administration, J.V.; resources, J.V.; software, I.S.; supervision, J.V.; Validation, I.S. and G.T.; visualization, I.S. and G.T.; writing—original draft, I.S.; writing—review & editing, I.S. and J.V. All authors have read and agreed to the published version of the manuscript.

Funding: This work was supported by the Slovenian Research Agency (ARRS) (Grant No. P2-0248 (B)) and EU EC H2020 funded Era Chair of Micro Process Engineering and Technology – COMPETE (Grant No. 811040).

Acknowledgments: Authors would also like to thank Laboratory of Biocybernetics (Faculty of Electrical Engineering, University of Ljubljana) for providing resources to run simulations.

Conflicts of Interest: The authors declare no conflict of interest.

References

1. Alting, L.; Kimura, F.; Hansen, H.N.; Bissacco, G. Micro engineering. *CIRP Ann. Manuf. Technol.* **2003**, *52*, 635–657. [CrossRef]
2. Masuzawa, T. State of the art of micromachining. *CIRP Ann. Manuf. Technol.* **2000**, *49*, 473–488. [CrossRef]
3. Hessel, V.; Lowe, H.; Muller, A.; Kolb, G. *Chemical Micro Process Engineering: Processing and Plants*, 1st ed.; Wiley-VCH: Weinheim, Germany, 2005; ISBN 978-3-527-30998-6.
4. Cvjetko, M.; Žnidaršič-Plazl, P. Ionic Liquids within Microfluidic Devices. In *Ionic Liquids: Theory, Properties, New Approaches*; InTech: London, UK, 2011.
5. Stroock, A.D.; Dertinger, S.K.W.; Ajdari, A.; Mezic, I.; Stone, H.A.; Whitesides, G.M. Chaotic Mixer for Microchannels. *Science* **2002**, *295*, 647–651. [CrossRef] [PubMed]
6. Lynn, N.S.; Dandy, D.S. Geometrical optimization of helical flow in grooved micromixers. *Lab Chip* **2007**, *7*, 580–587. [CrossRef]
7. Yang, J.T.; Huang, K.J.; Lin, Y.C. Geometric effects on fluid mixing in passive grooved micromixers. *Lab Chip* **2005**, *5*, 1140–1147. [CrossRef]
8. Aubin, J.; Fletcher, D.F.; Xuereb, C. Design of micromixers using CFD modelling. *Chem. Eng. Sci.* **2005**, *60*, 2503–2516. [CrossRef]
9. Ansari, M.A.; Kim, K.-Y. Application of the Radial Basis Neural Network to Optimization of a Micromixer. *Chem. Eng. Technol.* **2007**, *30*, 962–966. [CrossRef]
10. Williams, M.S.; Longmuir, K.J.; Yager, P. A practical guide to the staggered herringbone mixer. *Lab Chip* **2008**, *8*, 1121. [CrossRef]
11. Cortes-Quiroz, C.A.; Zangeneh, M.; Goto, A. On multi-objective optimization of geometry of staggered herringbone micromixer. *Microfluid. Nanofluidics* **2009**, *7*, 29–43. [CrossRef]
12. Hossain, S.; Husain, A.; Kim, K.Y. Optimization of micromixer with staggered herringbone grooves on top and bottom walls. *Eng. Appl. Comput. Fluid Mech.* **2011**, *5*, 506–516. [CrossRef]
13. Yoshimura, M.; Shimoyama, K.; Misaka, T.; Obayashi, S. Optimization of passive grooved micromixers based on genetic algorithm and graph theory. *Microfluid. Nanofluidics* **2019**, *23*, 1–21. [CrossRef]
14. Rhoades, T.; Kothapalli, C.R.; Fodor, P.S. Mixing Optimization in Grooved Serpentine Microchannels. *Micromachines* **2020**, *11*, 61. [CrossRef] [PubMed]
15. Ianovska, M.A.; Mulder, P.P.M.F.A.; Verpoorte, E. Development of small-volume, microfluidic chaotic mixers for future application in two-dimensional liquid chromatography. *RSC Adv.* **2017**, *7*, 9090–9099. [CrossRef]

16. Sitar, A.; Golobic, I. Heat transfer enhancement of self-rewetting aqueous n-butanol solutions boiling in microchannels. *Int. J. Heat Mass Transf.* **2015**, *81*, 198–206. [CrossRef]
17. Marschewski, J.; Brechbühler, R.; Jung, S.; Ruch, P.; Michel, B.; Poulikakos, D. Significant heat transfer enhancement in microchannels with herringbone-inspired microstructures. *Int. J. Heat Mass Transf.* **2016**, *95*, 755–764. [CrossRef]
18. Hahn, R.; Wagner, S.; Schmitz, A.; Reichl, H. Development of a planar micro fuel cell with thin film and micro patterning technologies. *J. Power Sources* **2004**, *131*, 73–78. [CrossRef]
19. Brousseau, E.B.; Dimov, S.S.; Pham, D.T. Some recent advances in multi-material micro- and nano-manufacturing. *Int. J. Adv. Manuf. Technol.* **2010**, *47*, 161–180. [CrossRef]
20. Hung, J.C.; Yang, T.C.; Li, K.C. Studies on the fabrication of metallic bipolar plates—Using micro electrical discharge machining milling. *J. Power Sources* **2011**, *196*, 2070–2074. [CrossRef]
21. Hessel, V.; Hardt, S.; Löwe, H.; Schönfeld, F. Laminar mixing in different interdigital micromixers: I. Experimental characterization. *AIChE J.* **2003**, *49*, 566–577. [CrossRef]
22. Pennemann, H.; Watts, P.; Haswell, S.J.; Hessel, V.; Löwe, H. Benchmarking of Microreactor Applications. *Org. Process Res. Dev.* **2004**, *8*, 422–439. [CrossRef]
23. Liu, K.; Lauwers, B.; Reynaerts, D. Process capabilities of Micro-EDM and its applications. *Int. J. Adv. Manuf. Technol.* **2010**, *47*, 11–19. [CrossRef]
24. Kunieda, M.; Lauwers, B.; Rajurkar, K.P.; Schumacher, B.M. Advancing EDM through fundamental insight into the process. *CIRP Ann. Manuf. Technol.* **2005**, *54*, 64–87. [CrossRef]
25. Yu, Z.Y.; Kozak, J.; Rajurkar, K.P. Modelling and simulation of micro EDM process. *CIRP Ann. Manuf. Technol.* **2003**, *52*, 143–146. [CrossRef]
26. Jeong, Y.H.; Min, B.-K. Geometry prediction of EDM-drilled holes and tool electrode shapes of micro-EDM process using simulation. *Int. J. Mach. Tools Manuf.* **2007**, *47*, 1817–1826. [CrossRef]
27. Heo, S.; Jeong, Y.H.; Min, B.-K.; Lee, S.J. Virtual EDM simulator: Three-dimensional geometric simulation of micro-EDM milling processes. *Int. J. Mach. Tools Manuf.* **2009**, *49*, 1029–1034. [CrossRef]
28. Surleraux, A.; Pernot, J.-P.; Elkaseer, A.; Bigot, S. Iterative surface warping to shape craters in micro-EDM simulation. *Eng. Comput.* **2016**, *32*, 517–531. [CrossRef]
29. Hinduja, S.; Kunieda, M. Modelling of ECM and EDM processes. *CIRP Ann. Manuf. Technol.* **2013**, *62*, 775–797. [CrossRef]
30. Lovatt, A.M.; Shercliff, H.R. Manufacturing process selection in engineering design. Part 1: The role of process selection. *Mater. Des.* **1998**, *19*, 205–215. [CrossRef]
31. Lynn, N.S.; Bocková, M.; Adam, P.; Homola, J. Biosensor Enhancement Using Grooved Micromixers: Part II, Experimental Studies. *Anal. Chem.* **2015**, *87*, 5524–5530. [CrossRef]
32. Lehmann, M.; Wallbank, A.M.; Dennis, K.A.; Wufsus, A.R.; Davis, K.M.; Rana, K.; Neeves, K.B. On-chip recalcification of citrated whole blood using a microfluidic herringbone mixer. *Biomicrofluidics* **2015**, *9*, 064106. [CrossRef]
33. Žnidaršič Plazl, P.; Plazl, I. Modelling and experimental studies on lipase-catalyzed isoamyl acetate synthesis in a microreactor. *Process Biochem.* **2009**, *44*, 1115–1121. [CrossRef]
34. Aubin, J.; Prat, L.; Xuereb, C.; Gourdon, C. Effect of microchannel aspect ratio on residence time distributions and the axial dispersion coefficient. *Chem. Eng. Process. Process Intensif.* **2009**, *48*, 554–559. [CrossRef]
35. Valentinčič, J.; Sabotin, I.; Tristo, G.; Bissacco, G.; Bigot, S. Tooling and Microinjection Moulding of Bottom Grooved Micromixers. In Proceedings of the 10th International Conference on Multi-Material Micro Manufacture, San Sebastian, Spain, 10 October 2013.
36. Sabotin, I.; Tristo, G.; Junkar, M.; Valentinčič, J. Two-step design protocol for patterned groove micromixers. *Chem. Eng. Res. Des.* **2013**, *91*, 778–788. [CrossRef]
37. Okuducu, M.; Aral, M. Performance Analysis and Numerical Evaluation of Mixing in 3-D T-Shape Passive Micromixers. *Micromachines* **2018**, *9*, 210. [CrossRef]
38. Kukuková, A.; Noël, B.; Kresta, S.M.; Aubin, J. Impact of sampling method and scale on the measurement of mixing and the coefficient of variance. *AIChE J.* **2008**, *54*, 3068–3083. [CrossRef]
39. Yu, Z.Y.; Masuzawa, T.; Fujino, M. Micro-EDM for three-dimensional cavities—Development of uniform wear method. *CIRP Ann. Manuf. Technol.* **1998**, *47*, 169–172. [CrossRef]
40. Nguyen, M.D.; Wong, Y.S.; Rahman, M. Profile error compensation in high precision 3D micro-EDM milling. *Precis. Eng.* **2013**, *37*, 399–407. [CrossRef]

41. Li, J.Z.; Xiao, L.; Wang, H.; Yu, H.L.; Yu, Z.Y. Tool wear compensation in 3D micro EDM based on the scanned area. *Precis. Eng.* **2013**, *37*, 753–757. [CrossRef]
42. Tong, H.; Zhang, L.; Li, Y. Algorithms and machining experiments to reduce depth errors in servo scanning 3D micro EDM. *Precis. Eng.* **2014**, *38*, 538–547. [CrossRef]
43. Wang, J.; Qian, J.; Ferraris, E.; Reynaerts, D. In-situ process monitoring and adaptive control for precision micro-EDM cavity milling. *Precis. Eng.* **2017**, *47*, 261–275. [CrossRef]
44. Bleys, P.; Kruth, J.P.; Lauwers, B. Sensing and compensation of tool wear in milling EDM. *J. Mater. Process. Technol.* **2004**, *149*, 139–146. [CrossRef]
45. Tristo, G.; Balcon, M.; Carmignato, S.; Bissacco, G. Validation of on-machine microfeatures volume measurement using micro EDM milling tool electrode as touch probe. In Proceedings of the 13th International Conference of the European Society for Precision Engineering and Nanotechnology, Berlin, Germany, 27 May–31 May 2013; EUSPEN: Delft, The Netherlands, 2013.
46. Tsai, Y.-Y.; Masuzawa, T. An index to evaluate the wear resistance of the electrode in micro-EDM. *J. Mater. Process. Technol.* **2004**, *149*, 304–309. [CrossRef]
47. Bissacco, G.; Hansen, H.N.; Tristo, G.; Valentinčič, J. Feasibility of wear compensation in micro EDM milling based on discharge counting and discharge population characterization. *CIRP Ann. Manuf. Technol.* **2011**, *60*, 231–234. [CrossRef]
48. Du, Y.; Zhang, Z.; Yim, C.; Lin, M.; Cao, X. Evaluation of Floor-grooved Micromixers using Concentration-channel Length Profiles. *Micromachines* **2010**, *1*, 19–33. [CrossRef]
49. Kim, D.S.; Lee, S.W.; Kwon, T.H.; Lee, S.S. A barrier embedded chaotic micromixer. *J. Micromech. Microeng.* **2004**, *14*, 798–805. [CrossRef]
50. Bissacco, G.; Tristo, G.; Hansen, H.N.; Valentincic, J. Reliability of electrode wear compensation based on material removal per discharge in micro EDM milling. *CIRP Ann. Manuf. Technol.* **2013**, *62*, 179–182. [CrossRef]

Article

Mechanical Characterisation and Analysis of a Passive Micro Heat Exchanger

Francisco-Javier Granados-Ortiz * and Joaquín Ortega-Casanova

Department of Mechanical, Thermal and Fluid Engineering, School of Industrial Engineering, University of Málaga, 29071 Málaga, Spain; jortega@uma.es

* Correspondence: frangranados@live.com

Received: 29 May 2020; Accepted: 7 July 2020; Published: 9 July 2020

Abstract: Heat exchangers are widely used in many mechanical, electronic, and bioengineering applications at macro and microscale. Among these, the use of heat exchangers consisting of a single fluid passing through a set of geometries at different temperatures and two flows in T-shape channels have been extensively studied. However, the application of heat exchangers for thermal mixing over a geometry leading to vortex shedding has not been investigated. This numerical work aims to analyse and characterise a heat exchanger for microscale application, which consists of two laminar fluids at different temperature that impinge orthogonally onto a rectangular structure and generate vortex shedding mechanics that enhance thermal mixing. This work is novel in various aspects. This is the first work of its kind on heat transfer between two fluids (same fluid, different temperature) enhanced by vortex shedding mechanics. Additionally, this research fully characterise the underlying vortex mechanics by accounting all geometry and flow regime parameters (longitudinal aspect ratio, blockage ratio and Reynolds number), opposite to the existing works in the literature, which usually vary and analyse blockage ratio or longitudinal aspect ratio only. A relevant advantage of this heat exchanger is that represents a low-Reynolds passive device, not requiring additional energy nor moving elements to enhance thermal mixing. This allows its use especially at microscale, for instance in biomedical/biomechanical and microelectronic applications.

Keywords: micro heat exchanger; vortex shedding; thermal mixing; computational fluid dynamics (CFD); thermal engineering

---

1. Introduction

Heat exchangers are present in many mechanical, biomechanical, and electronic engineering applications such as automobile refrigeration [1], air conditioning systems [2], powerplants [3], cooling of microelectronics [4], blood warming [5], or pressure ventilators [6,7]. Depending on the application, heat exchangers involving working fluids may aim at cooling/heating a fluid–fluid or fluid-solid system at different temperatures. The present investigation examines a fluid–fluid heat exchanger, which is influenced by an adiabatic fluid-structure interaction.

According to the transfer process, fluid–fluid heat exchangers can be classified into two groups. Indirect contact heat exchangers deal with heating/cooling by using a solid separation media between the two fluids. These are often called surface heat exchangers [8]. On the other hand, direct contact heat exchangers are systems where fluids have no physical separation, flowing within the same space [8]. An important point to consider in heat exchangers with thermal mixing of fluids is the ability to promote heat transfer at a low energy cost. The higher the mixing between the two flows at different temperature, the more efficient the thermal mixing is. Therefore, to find configurations that enhance mixing is the main objective for this type of heat exchangers. However, enhancing thermal mixing in fluids passing through channels is not cost-free. If moving elements are used (stirrers, heaving&pitching elements, etc.), one has to think of the energy needed for these elements to work. These devices that incorporate elements that consume electrical energy are called active devices, and these are a complex to include at microscale application. On the other hand, there is an alternative of using static mechanical devices, which do not require any additional energy consumption nor design of any moving element or structure at microscale. These are the namely passive devices. Examples of this type of device are the use of grooved channels or placing elements inside the channel, as in the present work. Although these new elements in the design do not need additional input energy, the pumping power P_p is likely to be increased because of pressure loss is increased as a consequence of the drag force.

Several examples of the application of microscale heat exchangers and micromixers can be found in the literature. The main advantage of microdevices is that these offer a rapid, portable, simple, and low-cost tool [9]. For instance, Huang et al. [10] developed an experimental work for a microchannel heat transfer, where it was concluded that reentrant cavities are an ideal approach to improve the performance. In their work and review was outlined that to include elements to perturb the flow enhances heat transfer in microchannels. Pressure drop and heat transfer were analysed via simulation in Reference [11], where a single flow passed through a microchannel heat sink of a microelectronic device. In Reference [12] the thermal mixing of two fluids at different temperature was analysed theoretically, experimentally and numerically in a T-shaped microchannel at a very low Reynolds, and different volume flow-rate ratios were tested. Their results showed that the T-shape had two differentiated regions of heat exchange in terms of behaviour: the T-shape junction and the mixing channel flow. A very similar work in T-shape microchannels was developed in Reference [13] studying different temperatures and higher Reynolds numbers. A drawback in the use of T-shapes, F-shapes, etc. is that these increase dramatically the pumping power required to overcome such large pressure drop with respect to a straight microchannel, as shown by other authors [14,15]. In Reference [16] a gas-to-gas micro heat exchanger design with grooved channels was analysed experimentally. In their work, it was observed that when the volume flow rate is low, the impact of heat transfer by conduction was dominant within the wall and fins. During the literature review, the authors did not find any application of thermal mixing via micro heat exchangers enhanced by flow detachment from cylindrical/prism structures.

The micro heat exchanger simulated (2D simulation), analysed and characterised for different configurations is depicted in Figure 1, where all geometric parameters are dimensional. This device consists of two flows at different temperatures (cold and hot, T_1 and T_2, respectively) passing through a microchannel with a rectangular cylinder structure (pillar) of width h and length l, located at the centreline. The microchannel has width H and total length L. Consequently, the cylinder-to-microchannel width ratio (blockage ratio) is $BR = \frac{h}{H}$; and the longitudinal cylinder aspect ratio, $AR = \frac{l}{h}$. The pillar structure is centered at the origin of the coordinates system, whose face nearest to the microchannel inlet is located at L_u distance, and the face nearest to the microchannel outflow is positioned at a distance $L - (L_u + l)$ from it. As a consequence of this geometry, when the flow impinges on the structure for a certain geometrical configuration and flow regime, vortex shedding takes place, which leads to a thermal mixing between the hot and cold fluids.

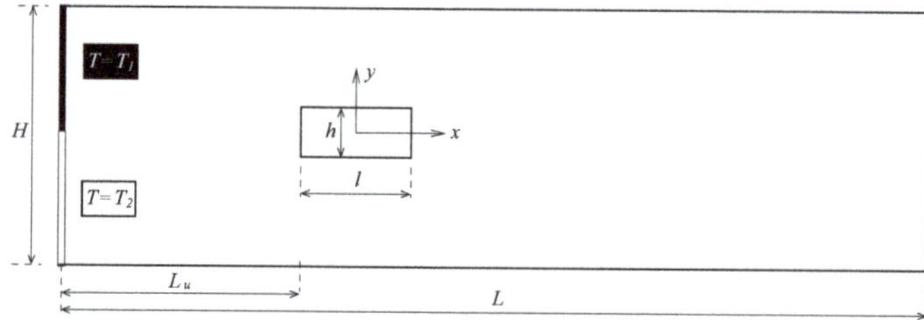

Figure 1. Sketch of the microchannel geometry with a rectangular structure positioning in the centreline.

Vortex shedding generated downstream cylinders has been extensively studied in the literature for more than a century. The first study on the problem dates back to 1907 to the pioneering work developed by Mallock [17]. Few years later, Benard continued on investigating the phenomenon [18,19]. Theodore von Karman probably did the most complete research on the topic back then, as shown in Reference [20]. Since these studies, many more researchers have investigated the mechanics behind this popular problem. Outlining the past-present history of the study of vortex shedding, one can find that in the last 5 years an impressive growing body of +16.400 research papers and patents have been developed on vortex shedding by cylinders. This highlights the importance the phenomenon still has in many engineering applications such as aircraft industry, heat exchangers, building engineering, civil engineering, environmental engineering or automobile industry, among many others.

Regarding the use of vortex shedding in heat transfer applications of laminar flows in channels, the interaction between a single flow and a cylinder at different temperatures is the most frequent application. This is opposite to chemical mixing applications, such as for instance Reference [21], where the molecular diffusion in a high Schmidt number flow was studied in the onset of the vortex shedding for a fixed value of BR and a fixed value of Re, and varying AR. Regarding the popular heat transfer interaction between cylinders in channels with laminar flows, in Reference [22] the convective heat transfer of an unsteady laminar flow (Re varied between 10 and 200) over a square cylinder in a channel (with a fixed $BR = 1/8 = 0.125$ and $AR = 1$) was analysed numerically. In this work, correlation models are found for the Nusselt-Reynolds number dependence, which is a regular practice in thermal engineering studies. It is relevant to outline that this type of model is often referred to as correlation in the engineering literature, but the correlation term will be used in this work joint with the term "model" as label to avoid any confusion with statistical correlation. However, there are no correlation models for other parameters such as pressure drop in any work in the literature of single-object vortex shedding in channels. Notwithstanding, one can find several works in the literature related to arrays of circular pins in microchannels, which are a good reference. For instance, Brunschwiler et al. [23] analysed heat-removal and pressure drop for arrays of in-line and staggered (circular) pins in chip stacks and plain parallel-wall microchannels, for which correlations were obtained. In their work, vortex shedding downstream was not addressed, and only weakly commented in terms of boundary-layer separation. Prasher et al. [24] analysed staggered arrays of circular pins with low pillar height aspect ratios in a microplate, and their experimental work was aimed at obtaining correlations for the Nusselt number and friction factor (which is related to the pressure drop) by including geometric and flow regime dependencies. Koşar et al. [25] also carried out an experimental investigation for low pillar height aspect ratio circular pin arrays. The experimental study was focused on determining correlation models for the pressure drop and friction factor. Focusing back again on single-object cases in channels, similarly to Reference [22], in Reference [26] the effect of channel confinement on the same vortex shedding heat transfer problem was analysed for $Re = 50, 100$ and 150 at different blockage ratio percentages and fixed $AR = 1$, providing a comparison with the unconfined case. Similarly, in Reference [27] the effect of only varying the blockage ratio BR at different values for Re ranging between 62 and 300 was developed. As in most works in the literature, Turki et al. [27] studied variations on a $AR = 1$ cylinder only, which was a limitation to generalise the multivariate nature of the problem to account also the AR effect. Moreover, as in other works found in the literature, this work stated different limits for the Reynolds number (Re_{cr}) above which the flow is unsteady. However, no correlation models as rule of thumb are given to know in advance whether a configuration may or may not lead to vortex shedding for the considered ranges. Surprisingly, only a few works were found in the literature related to the analysis of different AR values for laminar flows. Examples can be Reference [28], where the blockage ratio BR was fixed and AR was varied from 0.15 up to 4, for Re between 100 and 250 for a confined flow solved numerically; or Reference [29], where BR was fixed and different AR were ranged between 0.5 and 2, with Re between 50 and 200.

Unconfined flows have been also widely studied in the literature. For instance, Kelkar et al. [30] analysed a laminar flow which exchanged heat with a single heated square cylinder with Reynolds number Re ranged between 50 and 200. In such work was surprisingly stated that the heat transfer around the object is not pretty much enhanced with respect to steady cases when vortex shedding took place. This is opposite to our thermal mixing between two fluids. Different shapes have been also tested in the literature, not only the popular round or squared pillars. One can find, for instance, Reference [31], where a triangular cylinder shape was placed in a laminar regime channel to study the impact on heat transfer from the bottom side of the channel. Other similar applications of flow detachment in heat transfer can be found in the literature, such as Reference [32] where thermal-fluid-structure interaction was studied in a channel. In their research vortex shedding by means of flexible structure flags was used to promote thermal mixing between the cold flow and heated channel walls. Shi et al. [33] used a cylinder with a flexible plate to enhance heat transfer in a channel flow with heated walls by means of Vortex Induced Vibration (VIV). With this approach, a disruption of the thermal boundary layer by vortex interaction with the walls was aimed, and the use of vortex shedding mechanisms achieved an impressive heat transfer enhancement of up to a 90% with respect to a plain channel. In Reference [34] again a cold flow passed through a channel with flaps at the top and bottom of the heated channel walls. The flexible heated flaps were placed both symmetrically and asymmetrically, in order to study the kinematics of the flaps and to analyse the impact on the mixing promotion by vortex shedding. The vibrating flaps produced instabilities which strongly promoted mixing, and thence, thermal efficiency.

Summarising the literature review and research gaps found, some limitations have been identified. Apart from the absence of works on heat transfer between two microfluids at different temperature in the vortex shedding problem, it has been noticed an important gap in the characterisation of single-object vortex shedding in channels: there are no correlation models for relevant parameters such as the pressure related parameters (pressure drop, drag forces or the pumping power) nor the critical Reynolds (above which the flow is unsteady and von Karman streets do appear). Additionally, those works which intended to characterise and analyse the underlying mechanics of the problem did only focus efforts in indicating the limit values of one geometric factor at a time (and actually, only a couple of works investigated variations in AR), with poor generalisation. In the present investigation, a total amount of 80 simulations are developed (for $Re \in [120, 200]$, $AR \in [0.125, 1]$ and $BR \in [0.2, 0.5]$). All the lacks described are addressed by providing correlation models between the pumping power, drag coefficient and critical Reynolds number Re_{cr}, by including all the geometric and regime parameters that govern the mechanics, as well as analysing the performance of the heat exchanger in thermal mixing.

The paper is divided into different sections as follows. Section 2 introduces the governing equations, defines the most relevant parameters and describes the numerical considerations for the computational simulation. In Section 3 the dependence of the critical Reynolds number with the geometric parameters, and the relations between other influential parameters of the micro heat exchanger set-up are analysed. Finally, in Section 4 the most relevant conclusions are given.

2. Computational Geometry and Numerical Approach

2.1. Governing Equations and Parameters

As shown in Figure 1, the inlet consists of two fluids (same fluid properties) at different temperature T_1 and T_2, whose velocity parabolic profile corresponds to a fully developed laminar flow. Their temperature T_i is made dimensionless as θ_i by

$$\theta_i = \frac{T_i - T_1}{T_2 - T_1} = \frac{T_i - T_1}{\Delta T}, \tag{1}$$

thus the inlet boundary condition is set as:

$$\theta = \theta_1 = 0 \text{ at } x = -(L_u + l/2),\ 0 < y \leq H/2,$$

$$\theta = \theta_2 = 1 \text{ at } x = -(L_u + l/2), \; -H/2 \leq y \leq 0.$$

The thermal diffusivity of the fluids has been input to a very low value. Since the Prandtl number $Pr = \nu/\alpha_t$ is fixed at a high value of 10^4, the thermal diffusivity of the flows is $\alpha_t = \nu \cdot 10^{-4}$, with ν the fluid kinematic viscosity. The reason of simulating the problem at such a low α_t value is that the heat exchange will be thus dominated by the vortex convective mechanics leading to fluid mixing. That is to say, by the designed geometry and flow regime. If α_t is chosen at a high value, then the heat exchanger performance would be less attributed to mixing mechanics and more attributed to fluid properties. The flow simulation is unsteady, 2D and the flow is incompressible. The pressure, p, and velocity, $\mathbf{v} = (u,v)$, fields are then governed by the Navier-Stokes equations in the dimensionless form:

$$\nabla \cdot \mathbf{v} = 0, \tag{2}$$

$$\frac{\partial \mathbf{v}}{\partial t} + (\mathbf{v} \cdot \nabla)\mathbf{v} = -\nabla p + \frac{1}{Re}\nabla^2 \mathbf{v}, \tag{3}$$

whereas the mixing is governed by energy equation, written as

$$\frac{\partial \theta}{\partial t} + (\mathbf{v} \cdot \nabla)\theta = \frac{1}{Re\,Pr}\nabla^2 \theta, \tag{4}$$

where the viscous dissipation term has been neglected, due to the negligible value in comparison with the convection term. To transform all geometric and mechanical parameters into dimensionless quantities, the characteristic length, velocity, pressure and time used in the present work are H, U, ρU^2 and H/U, respectively, with H the channel width, U the mean inlet velocity and ρ the density of the fluid. The Reynolds number defined in the equations above is $Re = UH/\nu$. Thus, the geometric parameters are $L = 5H$ for the microchannel length, and $L_u = H$ for the pillar positioning respect to the inlet. The blockage ratio of the microchannel is defined as $BR = h/H$, and the aspect ratio is $AR = l/h$. When the regime is above a critical Reynolds value, $Re \geq Re_{cr}$, as a consequence of the oscillatory behaviour, the forces on the pillar structure are also oscillating. These forces are drag and lift, whose coefficients in dimensionless notation can be written as

$$Cl = \frac{F_y}{\frac{1}{2}\rho U^2 h}, \; Cd = \frac{F_x}{\frac{1}{2}\rho U^2 h}, \tag{5}$$

where Cl and Cd are the lift and drag coefficients, respectively, and h is the pillar characteristic length, because in flows around objects this is the most appropriate length. The frequency of oscillation f is quantified by the dimensionless Strouhal number, $St = fh/U$. Since fluctuating quantities are complicated to analyse, it is useful to compute the time-averaged values, which vary according to its oscillation period St^{-1}. Thus, a time-averaged arbitrary quantity M can be computed as

$$\langle M \rangle = \frac{1}{St_M^{-1}} \int_{t_0}^{t_0 + St_M^{-1}} M(s)\, ds, \tag{6}$$

where t_0 stands for an initial time reference.

Forces on the rectangular pillar structure have a strong impact on the power requirements of the heat exchanger. The higher the drag force, the higher the pressure loss across the microchannel. This means that the pumping power required for the predicted performance would be also high. Such pressure loss can be modelled as a dimensionless pressure difference between the inlet and outlet of the microchannel. Thus, the dimensionless pumping power, denoted by P_p, can be calculated as

$$P_p = 2\Delta p\, q, \tag{7}$$

which is expressed in dimensionless form using $1/2\rho U^3 H$ in the nondimensionalisation (the $1/2$ is the reason of the 2 factor), and Δp and q stands for the dimensionless pressure drop and volume-flow rate, respectively. According to the characteristic quantities adopted in this work, the dimensionless volume flow-rate is $q = 1$ (made dimensionless with UH), so the pumping power can be finally expressed as $P_p = 2\Delta p$. In other works such as References [26,35] the pumping power is calculated as a function of the drag coefficient and Reynolds number, which is equivalent.

Since thermal mixing is the scope of the micro heat exchanger system, another important parameter is the assessment of the mixing quality. This feature must be evaluated at the outlet section, being defined as the thermal mixing efficiency η, in %, as the temperature deviation with respect to the average temperature (maximum standard deviation value possible) as done in previous literature:

$$\eta = \left(1 - \frac{\langle\sigma\rangle}{\frac{\theta_2-\theta_1}{2}}\right) \times 100, \tag{8}$$

where $\langle\sigma\rangle$ stands for the time average of the standard deviation of the dimensionless temperature θ at the microchannel exit. Since $0 \leq \theta \leq 1$, $\langle\sigma\rangle = 0$ means that there is no deviation in temperature, thus the time-average temperature is unperturbed because the thermal mixing is perfect ($\eta = 100\%$). On the contrary, if the thermal mixing is as poor as there is no heat transfer between the microfluids at different temperature, the standard deviation with respect to the mean value is at its maximum value, $\langle\sigma\rangle = 0.5$. Thus, in this case $\eta = 0\%$. Since very high values of the Prandtl number are used, the thermal mixing values are expected to be low. This means that if heat transfer is enhanced with the thermal mixing mechanism in this unfavourable case, for lower Prandtl fluids (for instance, air or water) the thermal mixing would be notably higher. Finally, both mechanical- and thermal-related properties, P_p and η, can be combined into one parameter to define the thermal mixing energy cost (ϕ) as:

$$\phi = \frac{P_p}{\eta}, \tag{9}$$

which is the required pumping power to generate a 1% of thermal mixing efficiency. Mixing devices with high values of ϕ (high cost) are undesired, since this means high pumping power is required for a given mixing efficiency.

2.2. Numerical Aspects

The numerical investigation consists of a 2D computational mesh, whose governing equations are solved with the CFD software ANSYS-FLUENT 18.2. ANSYS-FLUENT allows the use of a pressure-based formulation for the incompressible flow, with a second-order spatial discretisation; and SIMPLE (Semi-Implicit Method for Pressure-Linked Equations) algorithm to deal with the pressure-velocity coupling. The boundary conditions imposed onto the numerical model in ANSYS-FLUENT (please see Reference [36] for further details on each boundary condition set-up) are:

- *Velocity-inlet*: For the laminar flow conditions, fully developed parabolic velocity profiles are imposed onto the simulation at the microchannel entrance, depending on each Reynolds number. For each fluid, a temperature is imposed (cold and hot), as aforementioned in Equation (1). Density and viscosity of both fluids is the same (related to each Reynolds number Re simulated), and a high Prandtl number of $Pr = 10^4$ is also considered.
- *Pressure-outlet*: The pressure is imposed onto the outflow of the microchannel, imposing an atmospheric pressure. Transported quantities (let denote them by M) have gradients fixed to zero value: $\partial M/\partial x = 0$.
- *Wall*: A fixed zero heat flux (adiabatic surfaces) is imposed onto the upper and bottom walls of the microchannel, as well as onto the pillar structure: $\partial T(x_s, y_s; t)/\partial n = 0$, where n stands for the coordinate normal to the considered wall surface and (x_s, y_s) is the exact position at the wall

surface. The no-slip condition is also imposed in the velocity boundary condition on the wall: $\mathbf{v}(x_s, y_s) = 0$.

In addition to this set-up, the time discretisation step Δt has been selected such that the CFL number does not exceed the unity value: $u_{max}\Delta t/ds < 1$. In the definition of the CFL number, since the mesh is uniform and has the same size in both x and y directions, thus $\Delta x \equiv \Delta y \equiv ds$. The u_{max} term is the maximum velocity value in the computational domain.

In terms of validation of the 2D computational mesh, a mesh convergence analysis is a must. In this investigation, the Grid Convergence Index (GCI) developed by Roache [37] has been used. This method is a useful approach to measure the discretisation uncertainty based on the popular Richardson extrapolation. The objective of the GCI is to find an approximation for the exactness in the computation of a quantity of interest, obtained by different CFD grid refinements in a consistent and uniform basis. The method requires the computation in at least three different meshes, whose difference in results is contrasted one-by-one in increasing level of refinement. In GCI analysis, the mesh is halved in most works in the literature (which would mean a mesh refinement factor of 2), possibly in order to perform a reasonable systematic grid size reduction. Nevertheless, a refinement factor greater than 1.3 is recommended by Roache. An error estimation between grids is calculated by means of a generalised Richardson extrapolation, and a safety factor (recommended to be between 1.25 and 3 value in the literature) is applied to generate the grid uncertainty estimates. For further details, please see Reference [37].

In this analysis, three different uniform structured meshes (with cell size $ds_j = 0.0125$ H, 0.025 H and 0.05 H, namely by their indexes from fine to coarse as j = 1, 2, 3) have been tested for four different values of the blockage ratio ($BR = 0.125$, 0.2, 0.25 and 0.33), a fixed Reynolds number ($Re = 100$) and a fixed aspect ratio $AR = 1$ (pillar of square section). The GCI results are presented in Table 1 as percentage for the Strouhal number of Cl (St_{Cl}, which is calculated from the frequency of the lift coefficient Cl time series) and $\langle Cd \rangle$. Additionally, the numerical results have been validated against experimental data, as shown in Figure 2, where also the Richardson extrapolated data (numerical value inferred if cell size tends to zero value) is shown.

Table 1. Mesh convergence study (GCI).

			St_{Cl}	$\langle Cd \rangle$
Re	BR	**Grid:** j	$GCI_{j+1,j}$	$GCI_{j+1,j}$
100	0.125	1	1.0%	2.1%
		2	2.9%	4.6%
100	0.2	1	0.5%	0.6%
		2	2.6%	1.8%
100	0.25	1	1.9%	0.5%
		2	4.5%	4.1%
100	0.33	1	1.0%	0.6%
		2	4.1%	4.2%

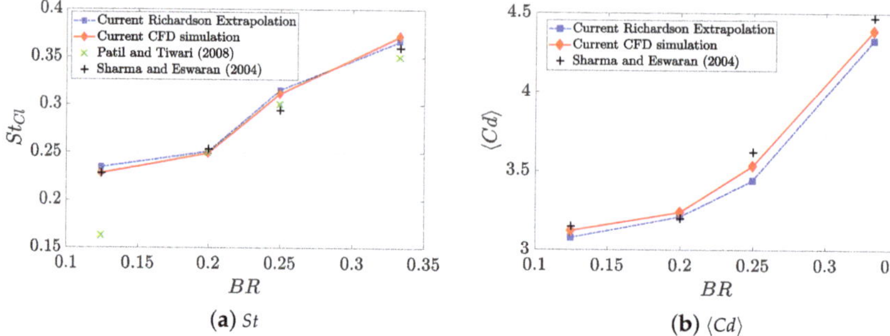

Figure 2. Validation of the CFD simulation with data from the literature: Patil and Tiwari (2008) [38] and Sharma and Eswaram (2004) [26]. The geometric configurations are confined flows around a square pillar ($AR = 1$) and a fully developed flow at $Re = 100$.

The parameter $GCI_{j+1,j}$ yields the discretisation uncertainty of each magnitude for the fine and medium grids with a safety factor of $Fs = 3$, i.e., for $j = 1,2$, since for the coarse mesh $j = 3$ there is no previous computation to compare the convergence with. In Table 1 can be seen that the fine grid performed very well, with uncertainty ranging between 2.1% and 0.5% for $\langle Cd \rangle$; and between 1.9% and 0.5% for St_{Cl}. Thus, the fine mesh, with $j = 1$, would be a very accurate option. However, the mesh must also be fine enough to solve the smallest thermal mixing length scales of θ. This is shown in the estimation of the Batchelor's length scale Γ [39,40], which is the ratio between the length scale of the smallest velocity structures Γ_{vel} and the square root of the Prandtl number: $\Gamma = \Gamma_{vel}/\sqrt{Pr}$. Since the size of the smallest structures in this laminar flow is of order H, then $\Gamma \sim 0.01$ H. This means that a finer mesh with a cell size smaller than ds_1 must be used. For that reason, the final mesh used to conduct the simulations was around 1/3 of the size of the fine mesh, i.e., $ds \simeq 0.004$ H (see Figure 3).

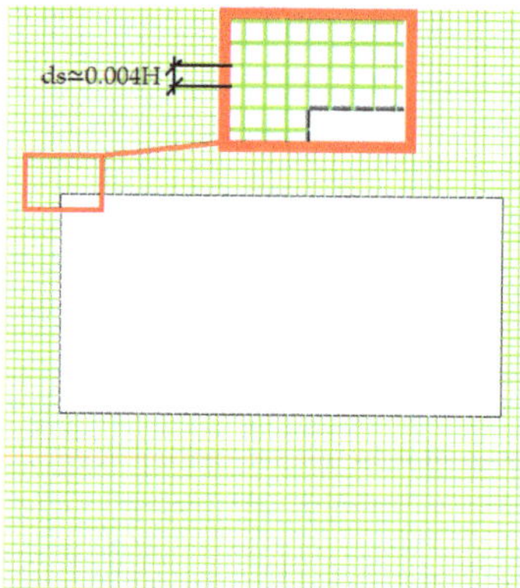

Figure 3. Detail of the uniform $ds \simeq 0.004$ H mesh around a pillar with $AR = 2$.

3. Characterisation of the Micro Heat Exchanger

In many works in the literature (for instance Reference [27]), variations in the blockage ratio BR are studied and the cylinder is kept as square geometry. Other researchers just varied the flow regime by means of the Reynolds number Re to see above which values an unsteady flow appears. However, no correlation models as rule of thumb are given to know in advance whether a configuration may or may not lead to vortex shedding for considered ranges, nor the full geometric dependence on the performance of quantities of interest. An exception is the study using arrays of fin pins in Reference [41], where a threshold criteria to predict which cases would lead to vortex shedding was provided based on the geometric variables (transversal and longitudinal pitch ratio and height aspect ratio), although for a constant Reynolds number.

In several investigations on the same problem geometry studied in the present work (for instance Reference [22]) it is frequent to see correlation models, but only related to the heat transfer characteristics on the surface heat transfer exchange (frequently Nusselt number as function of Reynolds number). No correlation models dependent on all AR, BR and Re conditions are found for the quantities of interest in the literature. Moreover, Nusselt number is out of our scope, since the present investigation is only focused on thermal mixing. If other heat sources are introduced, may not be feasible to differentiate whether convective heat transfer intensification is being improved by thermal mixing or by heating/cooling the flow wake as a consequence of the interaction with a solid at different temperature or imposed heat flux. In the present investigation, the full geometric and regime dependence on the performance of the vortex shedding-based micro heat exchanger will be given in the following.

3.1. Geometry Leading to Vortex Shedding

The most interesting feature of this device is the beneficial use of the vortex shedding flow wake to enhance the thermal mixing between the two fluids confined in the microchannel. For this purpose, the geometry must be adequate, as well as the flow rate. Given a Re, AR and BR, vortex shedding from the pillar may or may not take place. If there is no oscillatory pattern, the flow is actually steady. Thence, for a given pair of values (AR, BR), there is a Reynolds number above which the flow is unsteady and oscillatory flow detachment is present. This is denoted as critical Reynolds number Re_{cr}: $Re_{cr} = Re_{cr}(AR, BR)$. Below this value, the flow is steady.

This shows that it is crucial to determine the critical Reynolds according to the geometry in order characterise the micro heat exchanger. In Figure 4 it is illustrated the effect of Reynolds number and geometry on the generation of vortex shedding. It can be observed a limiting region of Re_{cr} to classify those geometries leading to oscillatory pattern. A correlation model can be thus found for Re_{cr} for characterisation. For each fixed value of AR, a different correlation Re_{cr}^{AR} is found as illustrated by the magenta solid lines in Figure 4, whose equations correspond to:

$$Re_{cr}^{AR} = \begin{cases} -400 \cdot BR + 330 \text{ if } AR = 1, \\ -200 \cdot BR + 230 \text{ if } AR = 0.5, \\ -200 \cdot BR + 190 \text{ if } AR = 0.25, \\ -200 \cdot BR + 170 \text{ if } AR = 0.125, \end{cases} \tag{10}$$

which are valid only for $Re \in [120, 200]$ and $BR \in [0.2, 0.5]$. The correlation for Re_{cr}^{AR} has a $Re_{cr}^{AR} = a^{AR}BR + b^{AR}$ linear form, being a^{AR} and b^{AR} coefficients which vary depending on AR values. With this approach, simple functions are used as correlation model; and only 1 case out of 80, the $Re = 120$, $AR = 0.5$ and $BR = 0.5$ case, would be clustered incorrectly. These coefficients can be now approximated to obtain a Re_{cr} non-linear correlation fully dependent on all the geometrical parameters within one single equation by finding the relationship of the coefficients above with respect to AR. Regression fits are obtained as shown in Figure 5 with a very accurate fit. Thus, finally a correlation for the Re_{cr} can be found as

$$Re_{cr} = aBR + b \begin{cases} a = -609.52AR^3 + 533.33AR^2 - 133.33AR - 190.48, \\ b = 183.7AR + 143.9, \end{cases} \tag{11}$$

which is valid only for $Re \in [120, 200]$, $AR \in [0.125, 1]$ and $BR \in [0.2, 0.5]$. The performance of this Re_{cr} correlation is shown in Figure 4 by means of dashed black lines.

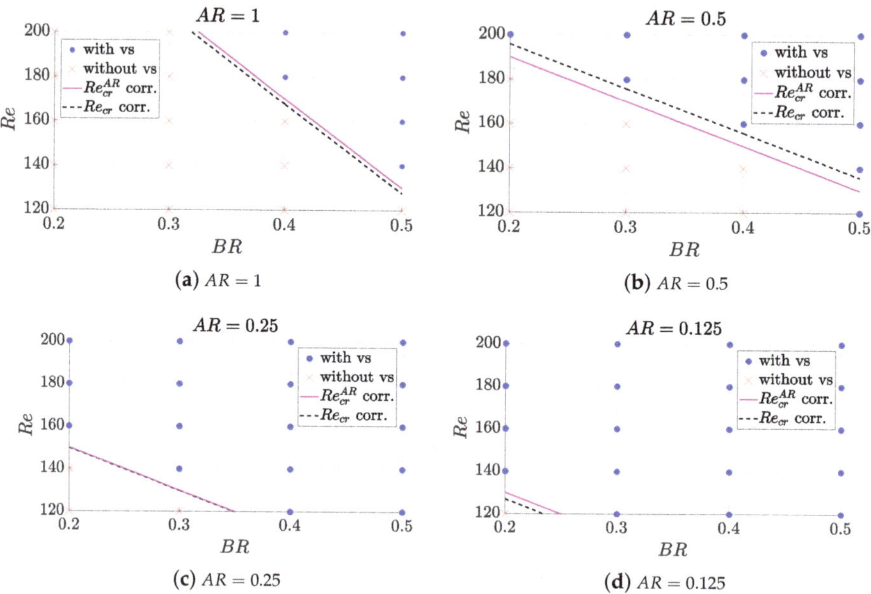

(a) $AR = 1$ **(b)** $AR = 0.5$ **(c)** $AR = 0.25$ **(d)** $AR = 0.125$

Figure 4. Impact of Reynolds number and geometry on the existence of vortex shedding (vs).

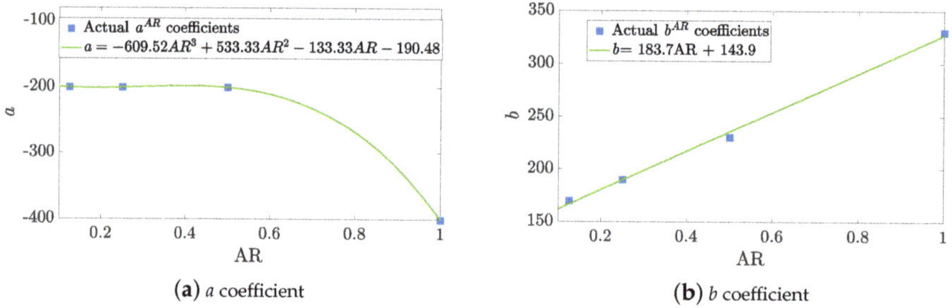

(a) *a* coefficient **(b)** *b* coefficient

Figure 5. Re_{cr} fitting coefficients.

3.2. Mechanics of the Flow Around the Rectangular Structure

When the flow impinges on the pillar, interesting structures are created. As a consequence, for Re values above the Re_{cr}, there is a signature oscillatory pattern for each configuration, which is shown on the Strouhal number. In Figure 6 is shown the behaviour of the Strouhal number for the lift coefficient St_{Cl} with increasing Re, BR and AR. When the configuration does not lead to vortex shedding, the St_{Cl} has zero value. An underlying linear behaviour is observed with respect to increasing BR, as already observed in Refenrence [27]. However, the effect of AR on the frequency is almost negligible. This is unexpected, because the aspect ratio obviously influences the existence of vortex shedding, but seems that this does not affect to the frequency value of the oscillation. This scenario also takes place with Re. If BR and AR are fixed and the Re is varied, it is observed that Re does not have any notable effect on the frequency either (although it affects the existence or not of the vortex detachment phenomenon, as seen in the Re_{cr} characterisation study). In short, within the values considered in this work for the three parameters that define each possible configuration, if a low non-zero frequency of oscillation for the wake is desired, the best practice would be to keep a BR as low as possible. If a high frequency of oscillation is desired, a higher BR should be used.

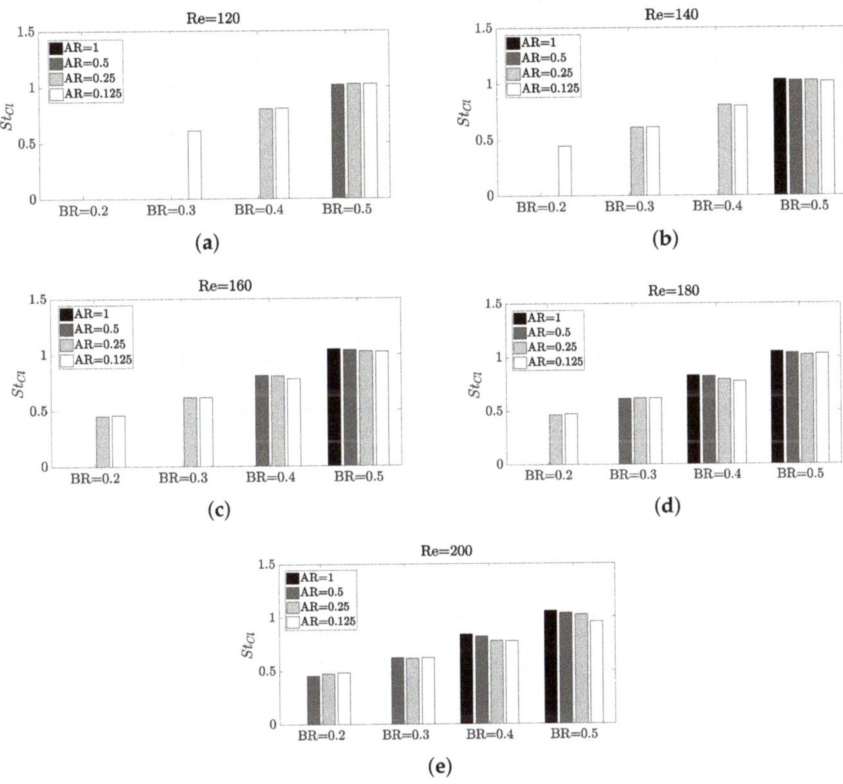

Figure 6. Impact of Reynolds number and geometry on Strouhal number in Cl.

Since the thermal mixing efficiency η is enhanced by the vortex shedding, a similar behaviour to St_{Cl} would be expected. However, by parameter exploration can be observed that the behaviour is very different. The parameter dependence pattern with η presents some sharp changes, as shown in Figure 7. The reason behind this nonexistent direct correlation between η and St_{Cl} is that it is not only important the frequency of oscillation, but also the amplitude, since a high frequency may be achieved under low amplitude of oscillation. In this sense, a certain configuration $[Re, AR, BR]$ has different impact on frequency and amplitude, as shown in Figure 8, where the amplitude of the oscillation is observed on the peak-to-peak value of Cl (Cl_{pp}). As confirmed by a comparison of Figure 6 with Figures 7 and 8, it is statistically correlated a high amplitude of oscillation with a high thermal mixing efficiency. Please note that the configurations with vortex shedding are indicated with '1' on top of the bars, and with '0' those which are steady. Thus, to find a configuration that enhances the amplitude of the frequency would lead to better thermal mixing results. The Pearson correlation coefficients R have been calculated to certify this, yielding a correlation between η and Cl_{pp} of a 0.9421, whereas the correlation between η and St_{Cl} was only a $R = 0.5728$ (weakly correlated). We recall that the correlation coefficient ranges between -1 and 1, so a value close to $R = 1$ means a high direct correlation. The correlation with St_{Cl} is in practice misleading, because the very low oscillatory cases have zero or close-to-zero frequency values, but mixing is still taking place. The correlation without these zero-value-frequency cases in the data set (now only 53 cases) has a correlation coefficient of

$R = -0.0366$ for St_{Cl} (the Cl_{pp} correlation coefficient is almost unperturbed when removing such data, with a $R = 0.9181$ value).

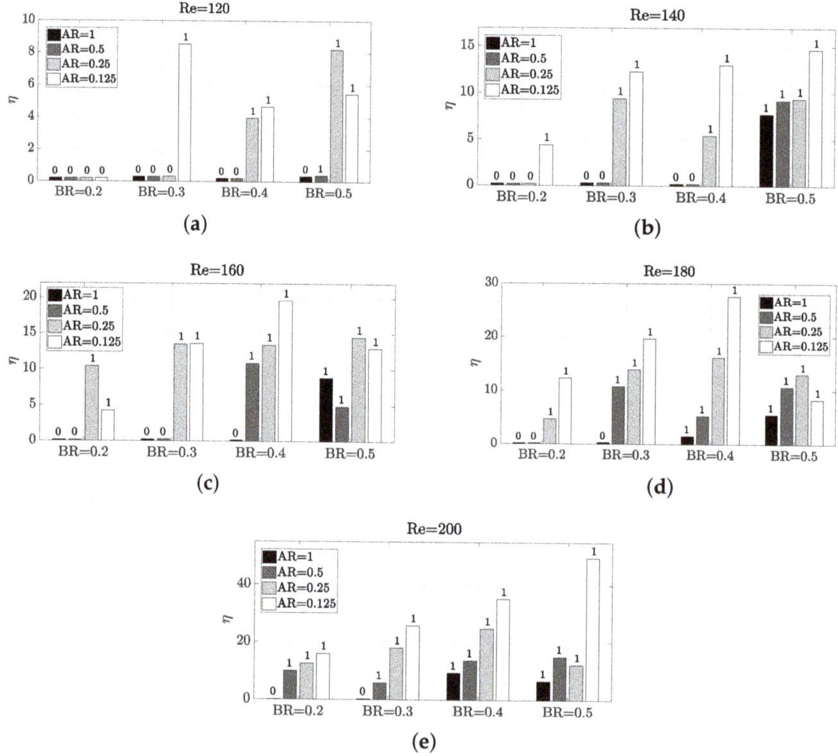

Figure 7. Impact of Reynolds number and geometry on thermal mixing efficiency η. The number on top of each bar indicates whether the configuration presents vortex shedding (1) or not (0).

The oscillatory motion of the fluid creates structures that enhance thermal mixing. As seen, the higher the amplitude of the vortex structures, the more enhanced the mixing is. In Figures 9–12 is shown the behaviour of three configurations, denoted as $maxEff$, $maxF$ and $minMEC$. $maxEff$ is the configuration which led to the maximum thermal mixing efficiency ($\eta = 49.68\%$) among the 80 cases simulated, $maxF$ is the configuration which led to the maximum frequency of oscillation ($St_{Cl} = 1.0522$), and $minMEC$ is the configuration that yielded the minimum mixing efficiency cost ($\phi = 0.0540$). Their characteristics are shown in Table 2.

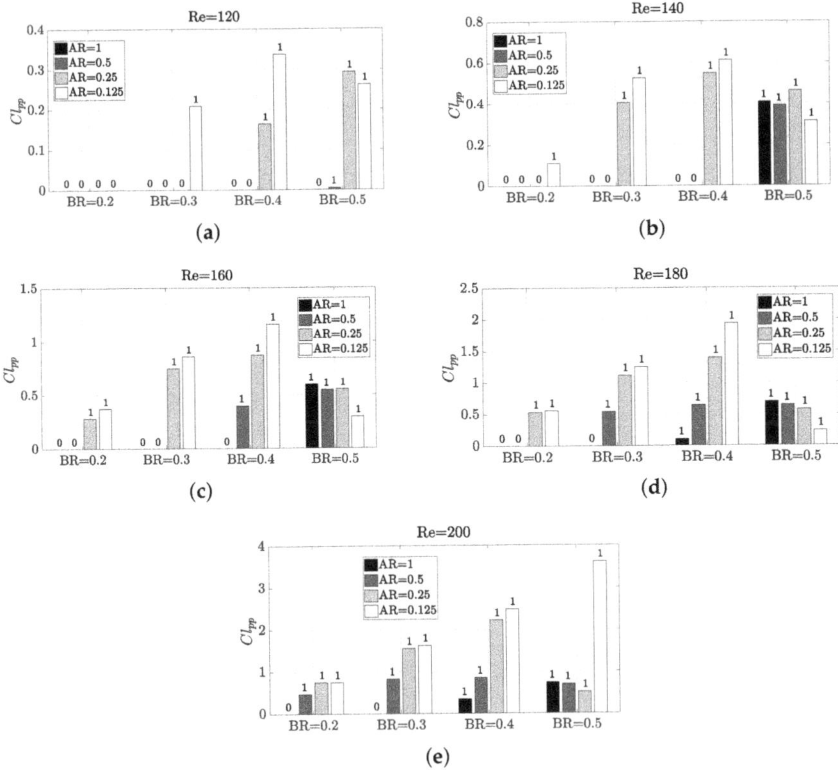

Figure 8. Impact of Reynolds number and geometry on the amplitude of the oscillation via Cl_{pp}. The number on top of each bar indicates whether the configuration presents vortex shedding (1) or not (0).

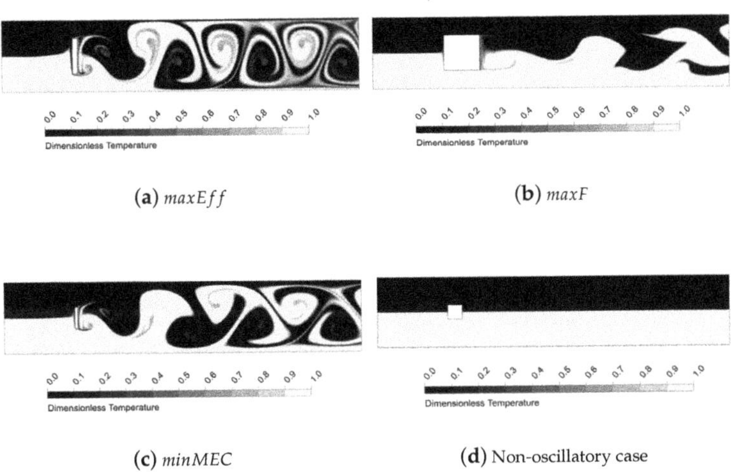

(a) $maxEff$ **(b)** $maxF$ **(c)** $minMEC$ **(d)** Non-oscillatory case

Figure 9. Thermal mixing performance for the micro heat exchangers $maxEff$, $maxF$, $minMEC$ and a non-oscillatory case ($AR = 1$, $BR = 0.2$ and $Re = 200$).

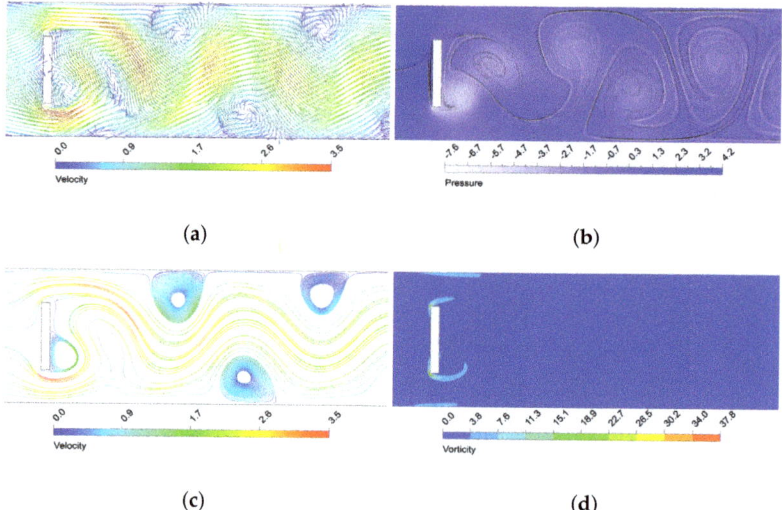

(**a**) (**b**)

(**c**) (**d**)

Figure 10. Performance of the $maxEff$ micro heat exchanger near the rectangular structure. (**a**) Dimensionless velocity vector field; (**b**) Dimensionless static pressure (with mixing structures on top); (**c**) Streamlines (dimensionless velocity coloured scale); (**d**) Dimensionless vorticity.

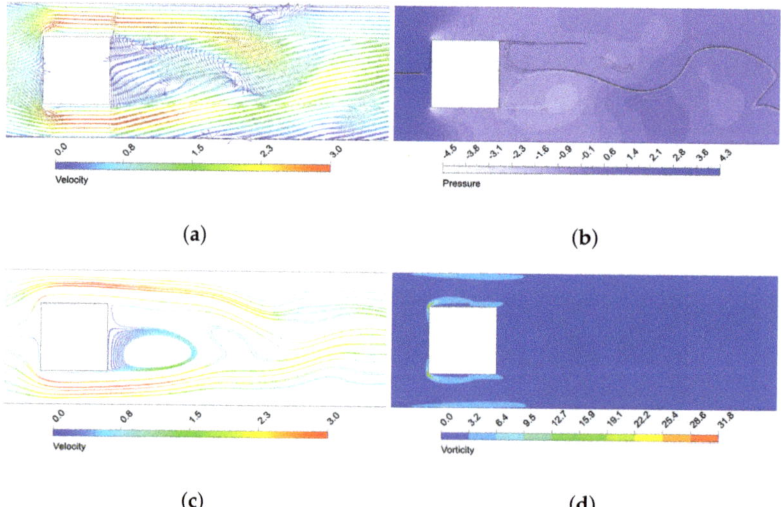

(**a**) (**b**)

(**c**) (**d**)

Figure 11. Performance of the $maxF$ micro heat exchanger near the rectangular structure. (**a**) Dimensionless velocity vector field; (**b**) Dimensionless static pressure (with mixing structures on top) (**c**) Streamlines (dimensionless velocity coloured scale); (**d**) Dimensionless vorticity.

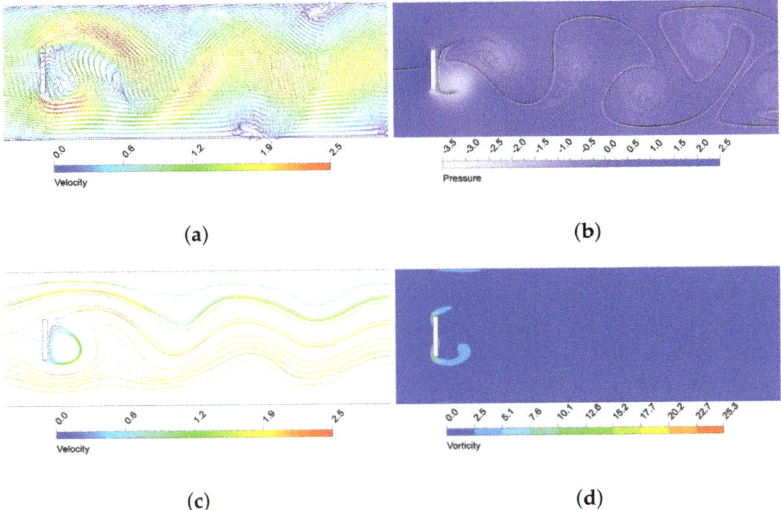

Figure 12. Performance of the *minMEC* micro heat exchanger near the rectangular structure. (**a**) Dimensionless velocity vector field; (**b**) Dimensionless static pressure (with mixing structures on top) (**c**) Streamlines (dimensionless velocity coloured scale); (**d**) Dimensionless vorticity.

Table 2. Efficient configurations for thermal mixing enhancement.

Configuration Label	AR	BR	Re	η	$\langle\Delta p\rangle$	St_{Cl}	ϕ
$maxEff$	0.125	0.5	200	49.6870%	3.30872	0.9549	0.0666
$maxF$	1	0.5	200	6.6167%	3.2020	1.0522	0.4839
$minMEC$	0.125	0.3	200	25.7284%	1.3886	0.6242	0.0540

By analysing the configurations in Figures 9–12, it can be observed that the highest velocities are achieved by the $maxEff$ configuration, which has a $BR = 0.5$. $maxF$ only differs with $maxEff$ on the aspect ratio, but this shows a dramatically effect on the maximum velocity achieved (reduced by a 15%). As can be observed in Figure 10b, in $maxEff$ the regions with higher velocity at the pillar object sides correspond to areas of very low pressure. This obvious fluid mechanics relation generates a suction effect on the high pressure upstream flow from the stagnation region, forming the beginning of the vortical structures that will potentially grow and travel downstream. This effect is strong for the vortex shedding thermal mixing: when the inlet flow impinges on the pillar structure, a portion of the hotter (colder) flow is able to pass through the upper (lower) side of the object (see the upstream curvature in the separation layer in Figure 9a). This occurs in an oscillatory manner, what enhances remarkably the thermal mixing and creates large vortical structures with low pressure cores. This situation does not take place for $maxF$, which has $AR = 1$ and thence is not a sudden compression-expansion-like channel. In this case, the oscillatory motion has higher frequency, but very low amplitude of oscillation, since no large pressure gradients near the object take place to produce the aforementioned suction effect of hot (cold) fluids. Despite the vorticity generated by the corners is still remarkable, this is dissipated once the flow leaves the sides of the object, so the impact on the vortex shedding is low and nearly parallel to the pillar sides, as opposite to smaller AR cases.

By comparison of $minMEC$ configuration with $maxEff$, where the only difference is in BR, one can see that the mixing is still good because large vortical structures are generated and propagated downstream. However, it can be observed a less influential role from the upper and lower walls of the microchannel: there is more separation between the thermal mixing flow and the channel walls (possibly because the upper and lower vortex shedding is taking place closer to the microchannel centreline, as BR is smaller). This generates less prominent recirculation zones near the microchannel wall in contrast to those that $maxEff$ produced (see Figure 10c), which improve vorticity (and hence mixing), but require additional pumping power to drag the flow downstream. Also, last but not least, the drag force is dramatically reduced, since the face surface size of the pillar where the flow impinges is reduced by a 40%. In the next section will be shown the relation between the drag forces and pumping power for each configuration.

3.3. Analysis of Forces on the Rectangular Structure

The forces along the micro heat exchanger define the pumping power P_p required to achieve the desired working conditions. The pumping power is plotted in Figure 13. As can be seen, the required pumping power is not very high. In fact, since this is strongly related to the pressure drop, a quick comparison with other pressure drop data from other microchannels reported in the literature is a valuable reference in Table 3. In such table, our dimensionless pressure loss ranges per unit length (made dimensionless with $\rho U^2/H$) are compared with those existing in a Y-shape microchannel [42] and a T-shape microchannel [43]. It is identified that among all the 80 cases simulated in the present work at various Reynolds and geometric configurations, all options had lower pressure loss than the Y and T-shape microchannels. For $maxEff$, $maxF$ and $minMEC$, their dimensionless pressure loss per unit length is 0.661744, 0.6404 and 0.27772, which are actually very low values.

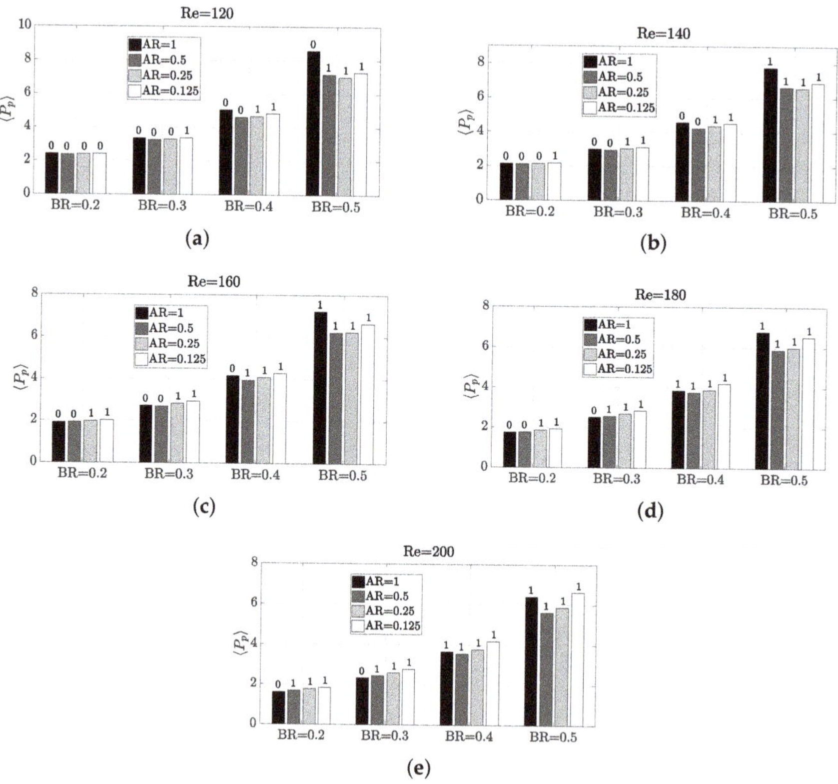

Figure 13. Impact of Reynolds number and geometry on the pumping power $\langle P_p \rangle$. The number on top of each bar indicates whether the configuration presents vortex shedding (1) or not (0).

Table 3. Comparison of pressure losses with different mixing microchannels.

	[42], $Re = 50$	[43], $Re = 200$	Current Research, $Re = [120, 200]$
$\Delta p/L$ [-]	0.93–3.59	1.23–23.52	0.16052–0.85624

To predict and characterise the performance of a system is of high value in engineering practice. For this reason, the relationship between the pumping power, the Reynolds number and the geometry is analysed. The following expression allows to quantify such underlying relationship:

$$\langle P_p \rangle = (\alpha_1 BR^{\alpha_2} AR^{\alpha_3} + \alpha_4) \frac{1}{(Re/100)^{\alpha_5}}, \tag{12}$$

where a vector for the fitting coefficients is defined as $\boldsymbol{\alpha} = [\alpha_1\ \alpha_2\ \alpha_3\ \alpha_4\ \alpha_5]$. This expression is an empirical correlation model, which intends to predict the value of the pumping power based on relations between the geometric and flow regime parameters. Since the exact model is not known beforehand, some assumptions must be made. In the literature is very frequent to find terms using relations among power functions to account for non-linear interactions [25,41,44,45]. There are many works in the literature that explored empirical correlation models for the drag coefficient Cd of objects. In such investigations (see for instance Reference [45] for a review), the Reynolds number was always modelled as inversely proportional to the drag coefficient, as demonstrated more than a century ago by Stokes [46] and Oseen [47]. As P_p and Cd are strongly related, the same modelling assumptions can be applied to P_p, leading to the correlation model suggested in Equation (12). Other researchers applied a similar logic to model the pressure drop and/or friction factor in the presence of arrays of pillars, as seen in References [23–25,41]. In Equation (12) the Reynolds number Re has been normalised with 100 to improve the numerical stability in the search of coefficients. It has been found by means of the Matlab *nlinfit* Least-Squares algorithm that the best coefficients are

$$\boldsymbol{\alpha} = [50.7966\ 3.0216\ 0.0312\ 2.1216\ 0.4520]\,. \tag{13}$$

This fit provided very accurate results, as shown in Figure 14. The correlation model yields a fitting error of 4.7202%, calculated for N samples as

$$error\ [\%] = \frac{1}{N}\sum_{i=1}^{N}\left(\frac{|Predicted_i - Numerical_i|}{Numerical_i}\right)\cdot 100. \tag{14}$$

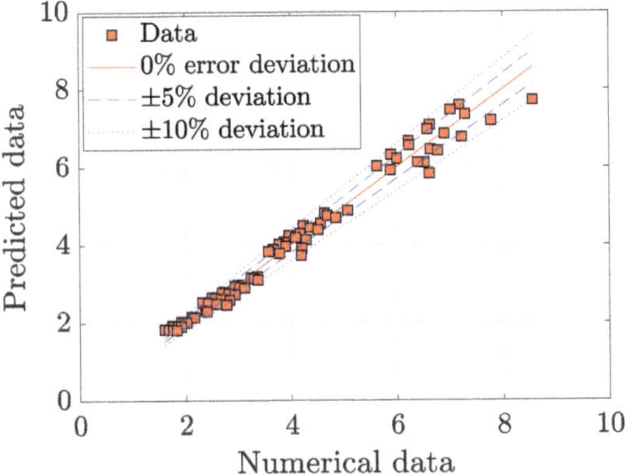

Figure 14. Fitting model for P_p.

Additionally, as mentioned above, $\langle P_p \rangle$ and $\langle Cd \rangle$ are related parameters. Their dependence is illustrated in Figure 15. It is shown that the correlation between these parameters is very high, which is evident since the higher the drag, the higher the pressure drop. However, it is less obvious the fact that these two parameters vary almost linearly, and when $AR < 0.5$, data is more deviated from a linear trend. Nevertheless, the correlation coefficients are nearly invariant ($R = 0.9912$ for any AR within the simulated range of values, $R = 0.9882$ for $AR < 0.5$, and $R = 0.9966$ for $AR \geq 0.5$). This shows that the pumping power needed is almost entirely due to the drag force on the pillar structure, and the contribution to the pressure drop by the microchannel walls is very weak in comparison with such drag. Those cases of $AR < 0.5$ that are more deviated from the overall linear trend are actually oscillatory cases of high amplitude. Thus, in these configurations the effect of the walls is contributing more to the pumping power requirement (the oscillatory flow "hits" the walls periodically), but still negligible.

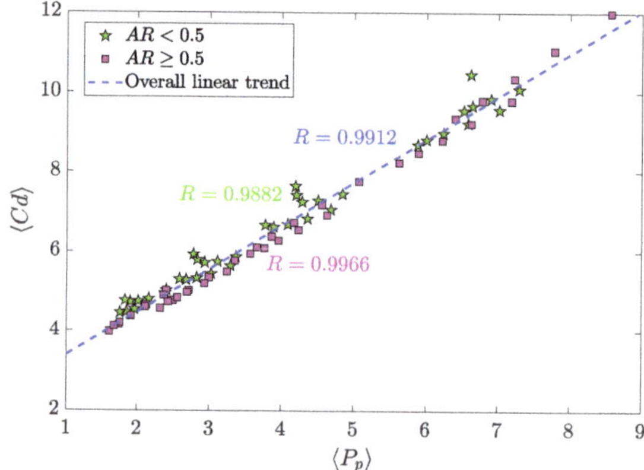

Figure 15. Pearson Correlation for $\langle P_p \rangle$ and $\langle Cd \rangle$. Please note that the correlation for each AR value has been obtained by including $AR = 0.5$ in both greater and lower case scenarios, to have a greater number of data in each set.

Due to their linear underlying relationship, a linear correlation model can be developed to characterise their dependence. The model has the form:

$$\langle Cd \rangle = \beta_1 \langle P_p \rangle + \beta_2, \tag{15}$$

and a vector with the fitting coefficients is defined as $\boldsymbol{\beta} = [\beta_1 \; \beta_2]$. The best fitting coefficients found are

$$\boldsymbol{\beta} = [1.0843 \; 2.3159], \tag{16}$$

with a fitting error of a 3.14%, as illustrated in Figure 16. In such figure can be observed that the data for $AR < 0.5$ had worse fitting performance than $AR \geq 0.5$, because of the reasons mentioned above. Finally, since a model dependent on the regime and geometrical conditions was found in Equation (12) for $\langle P_p \rangle$, this can be substituted in Equation (16) to make $\langle Cd \rangle$ only dependent on Re, AR and BR.

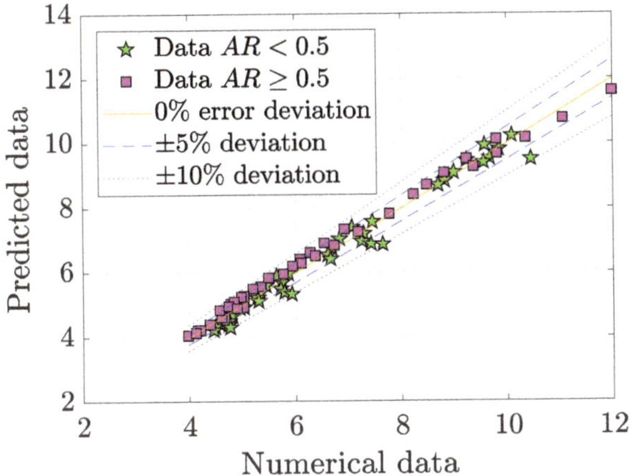

Figure 16. Model fit for $\langle Cd \rangle$ as a function of $\langle P_p \rangle$).

3.4. Mixing Behaviour of the Micro Heat Exchanger

As aforesaid, thermal mixing in the micro heat exchanger is influenced by the type of flow (Pr), flow regime (Re), and geometry (AR, BR, L). The opposite effect to achieve a good mixing is the increase in pumping power requirement, which is influenced by the same parameters. Therefore, the mixing efficiency cost is a good ratio to evaluate the trade-off problem of achieving a good mixing at low pumping power cost. This ratio will be calculated per unit length L, since the length of the microchannel also plays a role in the thermal mixing:

$$\phi_L = \frac{\phi}{L}. \tag{17}$$

Figure 17 details the operation of the micro heat exchanger in terms of ϕ_L. It is obvious that the lower the ϕ_L, the more cost efficient the heat exchanger is. Despite the underlying relations of P_p with Re, AR and BR were clear and a model was built in Equation (12), the relationship of η with these parameters is not that clear. Thence, in ϕ_L it is not either possible to find any pattern to characterise the performance. From Figure 17 can be observed that all but one configurations that present vortex shedding have ϕ_L below 0.5. The only single case which experiences vortex shedding and ϕ_L is high, is $Re = 120$, $AR = 0.5$ and $BR = 0.5$. For this configuration can be seen, despite that there is an oscillatory motion at high frequency (see Figure 6), and the amplitude is very low (see Figure 8). Also, the pumping power is large (see Figure 13). The observation that at $120 \leq Re \leq 160$, $BR = 0.4$ is the worst configuration possible is also interesting. This behaviour with $BR = 0.4$ is apparently attributed to the fact that along the "sub-channels" created at the sides of the pillar structure, if BR is further increased, the flow is accelerated due to the section reduction, and the oscillatory motion starts. This lack of oscillation is worsened as AR is increased, since the flow expansion is less abrupt and the flow still have some time to adapt to the walls. Values of $BR > 0.5$ are not considered in this study, since the increase in pumping power is significant. Therefore, a weak generalised aspect is that large values of AR need more cost for an efficient thermal mixing. This is especially notable for the higher Reynolds numbers considered, where the combination of low BR and high AR is very undesired.

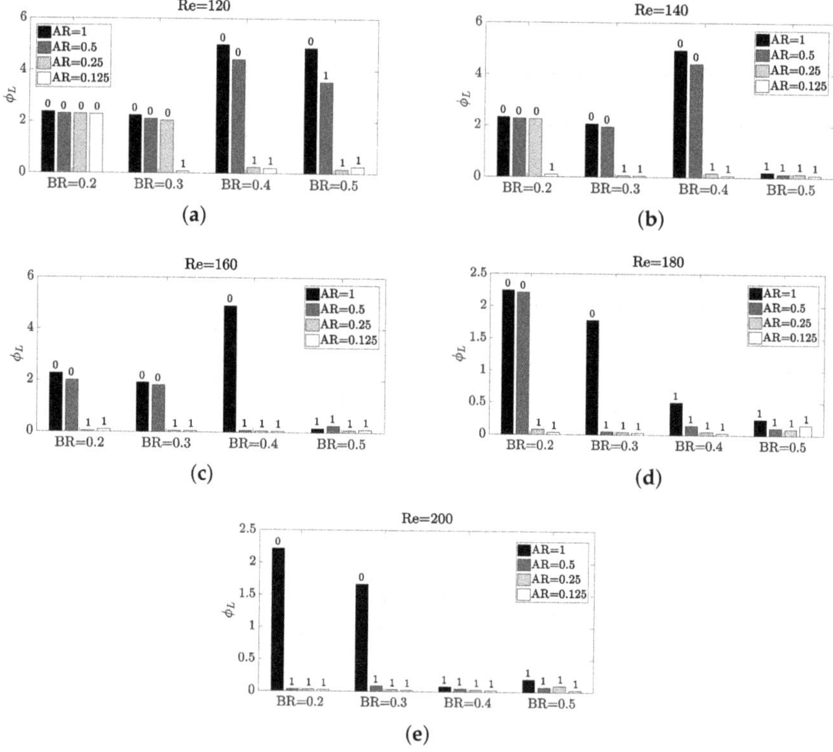

Figure 17. Impact of Reynolds number and geometry on ϕ_L. The number on top of each bar indicates whether the configuration presents vortex shedding (1) or not (0).

4. Conclusions

This paper investigated the performance of a microscale heat exchanger, which consists of a rectangular pillar structure in a microchannel, where two fluids at different temperature are mixing. This represents a low-cost passive thermal mixing microdevice very appropriate for microscale applications, since no moving parts are required for heat transfer enhancement. Besides the heat transfer problem, the mechanics of the vortex shedding have been characterised. Opposite to the existing works in the literature of single-object confined vortex shedding, which do not consider more than two design parameters, a large number of different configurations varying simultaneously the longitudinal aspect ratio, blockage ratio and Reynolds number have been simulated. By means of empirical models and analysis of relevant contours and plots, the underlying relations between these parameters (including critical Reynolds number values) have been analysed. One of the most interesting features observed in the mechanism is that, for configurations with at least moderate oscillation, a pressure suction-like effect takes place periodically, allowing a portion of the hotter (colder) flow to pass through the upper (lower) side around the object and enhancing the thermal mixing.

For an efficient mixing configuration must be taken into account that large values of AR need more pumping power cost to achieve an efficient thermal mixing, as shown by the mixing efficiency cost per unit length. The impact of BR is less clear: to increase its value is usually beneficial for the thermal mixing efficiency, but the pumping power is increased notably. Thus, it is not possible to generalise the conclusions to a given specific rule of thumb parameter configuration. In this investigation, the best combinations for the considered parameter ranges seem to take place mostly for large Re, large BR and small AR. However, this generalisation is not always true, as observed for instance in the thermal mixing efficiency. For this parameter, at $Re = 200$ the statement is true, but if $Re = 180$, a $BR = 0.5$ is less beneficial than a $BR = 0.4$. This shows the complexity of the micro heat exchanger mechanics and the need for a careful design and testing by engineers.

Author Contributions: Conceptualization, F.-J.G.-O. and J.O.-C.; methodology, F.-J.G.-O. and J.O.-C.; software, J.O.-C.; validation, F.-J.G.-O. and J.O.-C.; formal analysis, F.-J.G.-O.; investigation, F.-J.G.-O.; resources, F.-J.G.-O. and J.O.-C.; data curation, F.-J.G.-O.; writing—original draft preparation, F.-J.G.-O.; writing—review and editing, F.-J.G.-O. and J.O.-C.; visualization, F.-J.G.-O.; supervision, J.O.-C.; project administration, J.O.-C.; funding acquisition, J.O.-C. All authors have read and agreed to the published version of the manuscript.

Funding: This research was funded by the UMA18-FEDERJA-184 grant.

Conflicts of Interest: The authors declare no conflict of interest.

References

1. Cho, H.; Lee, H.; Park, C. Performance characteristics of an automobile air conditioning system with internal heat exchanger using refrigerant R1234yf. *Appl. Therm. Eng.* **2013**, *61*, 563–569. [CrossRef]
2. El-Baky, M.A.A.; Mohamed, M.M. Heat pipe heat exchanger for heat recovery in air conditioning. *Appl. Therm. Eng.* **2007**, *27*, 795–801. [CrossRef]
3. Jung, E.G.; Boo, J.H. Thermal analytical model of latent thermal storage with heat pipe heat exchanger for concentrated solar power. *Sol. Energy* **2014**, *102*, 318–332. [CrossRef]
4. Wei, X.; Joshi, Y. Stacked microchannel heat sinks for liquid cooling of microelectronic components. *J. Electron. Packag.* **2004**, *126*, 60–66. [CrossRef]
5. Ahuja, A.; Hendee, W. Thermal design of a heat exchanger for heating or cooling blood. *Phys. Med. Biol.* **1978**, *23*, 937. [CrossRef]
6. Pelosi, P.; Solca, M.; Ravagnan, I.; Tubiolo, D.; Ferrario, L.; Gattinoni, L. Effects of heat and moisture exchangers on minute ventilation, ventilatory drive, and work of breathing during pressure-support ventilation in acute respiratory failure. *Crit. Care Med.* **1996**, *24*, 1184–1188. [CrossRef]
7. Kratochvil, J.M. Ventilator with Air-to-Air Heat Exchanger and Pressure Responsive Damper. US Patent 5,435,377, 25 July 1995.
8. Cuce, P.M.; Riffat, S. A comprehensive review of heat recovery systems for building applications. *Renew. Sustain. Energy Rev.* **2015**, *47*, 665–682. [CrossRef]
9. Natsuhara, D.; Takishita, K.; Tanaka, K.; Kage, A.; Suzuki, R.; Mizukami, Y.; Saka, N.; Nagai, M.; Shibata, T. A Microfluidic Diagnostic Device Capable of Autonomous Sample Mixing and Dispensing for the Simultaneous Genetic Detection of Multiple Plant Viruses. *Micromachines* **2020**, *11*, 540. [CrossRef]
10. Huang, B.; Li, H.; Xu, T. Experimental Investigation of the Flow and Heat Transfer Characteristics in Microchannel Heat Exchangers with Reentrant Cavities. *Micromachines* **2020**, *11*, 403. [CrossRef]
11. Duan, Z.; Ma, H.; He, B.; Su, L.; Zhang, X. Pressure drop of microchannel plate fin heat sinks. *Micromachines* **2019**, *10*, 80. [CrossRef]
12. Xu, B.; Wong, T.N.; Nguyen, N.T.; Che, Z.; Chai, J.C.K. Thermal mixing of two miscible fluids in a T-shaped microchannel. *Biomicrofluidics* **2010**, *4*, 044102. [CrossRef] [PubMed]
13. Lobasov, A.; Shebeleva, A. Initial temperatures effect on the mixing efficiency and flow modes in T-shaped micromixer. *J. Phys. Conf. Ser.* **2017**, *899*, 022010. [CrossRef]
14. Mouheb, N.A.; Malsch, D.; Montillet, A.; Solliec, C.; Henkel, T. Numerical and experimental investigations of mixing in T-shaped and cross-shaped micromixers. *Chem. Eng. Sci.* **2012**, *68*, 278–289. [CrossRef]
15. Suh, Y.K.; Kang, S. A review on mixing in microfluidics. *Micromachines* **2010**, *1*, 82–111. [CrossRef]
16. Bier, W.; Keller, W.; Linder, G.; Seidel, D.; Schubert, K.; Martin, H. Gas to gas heat transfer in micro heat exchangers. *Chem. Eng. Process. Process Intensif.* **1993**, *32*, 33–43. [CrossRef]
17. Mallock, H.R.A. On the resistance of air. *Proc. R. Soc. Lond. Ser. A Contain. Pap. Math. Phys. Character* **1907**, *79*, 262–273.
18. Bénard, H. Formation de centres de giration a l'arrière d'un obstacle en movement. *Comptes Rendus Acad. Sci.* **1908**, *147*, 839–842.
19. Benard, H. Sur la zone de formation des tourbillons alternés derriere un obstacle. *CR Acad. Sci. Paris* **1913**, *156*, 1003–1005.

20. Von Karman, T. Über den Mechanismus des Widerstandes, den ein bewegter Körper in einer Flüssigkeit erfährt. *Nachrichten von der Gesellschaft der Wissenschaften zu Göttingen Mathematisch-Physikalische Klasse* **1911**, *1911*, 509–517.
21. Ortega-Casanova, J. On the onset of vortex shedding from 2D confined rectangular cylinders having different aspect ratios: Application to promote mixing fluids. *Chem. Eng. Process.-Process Intensif.* **2017**, *120*, 81–92. [CrossRef]
22. Rahnama, M.; Hadi-Moghaddam, H. Numerical investigation of convective heat transfer in unsteady laminar flow over a square cylinder in a channel. *Heat Transf. Eng.* **2005**, *26*, 21–29. [CrossRef]
23. Brunschwiler, T.; Michel, B.; Rothuizen, H.; Kloter, U.; Wunderle, B.; Oppermann, H.; Reichl, H. Interlayer cooling potential in vertically integrated packages. *Microsyst. Technol.* **2009**, *15*, 57–74. [CrossRef]
24. Prasher, R.S.; Dirner, J.; Chang, J.Y.; Myers, A.; Chau, D.; He, D.; Prstic, S. Nusselt number and friction factor of staggered arrays of low aspect ratio micropin-fins under cross flow for water as fluid. *J. Heat Transf.* **2007**, *129*, 141–153. [CrossRef]
25. Koşar, A.; Mishra, C.; Peles, Y. Laminar flow across a bank of low aspect ratio micro pin fins. *ASME J. Fluids Eng.* **2005**, *127*, 419–430. [CrossRef]
26. Sharma, A.; Eswaran, V. Effect of channel confinement on the two-dimensional laminar flow and heat transfer across a square cylinder. *Numer. Heat Transf. Part A Appl.* **2004**, *47*, 79–107. [CrossRef]
27. Turki, S.; Abbassi, H.; Nasrallah, S.B. Effect of the blockage ratio on the flow in a channel with a built-in square cylinder. *Comput. Mech.* **2003**, *33*, 22–29. [CrossRef]
28. Islam, S.U.; Zhou, C.; Shah, A.; Xie, P. Numerical simulation of flow past rectangular cylinders with different aspect ratios using the incompressible lattice Boltzmann method. *J. Mech. Sci. Technol.* **2012**, *26*, 1027–1041. [CrossRef]
29. Rahnama, M.; Hashemian, S.M.; Farhadi, M. Forced convection heat transfer from a rectangular cylinder: Effect of aspect ratio. In Proceedings of the 16th International Symposium on Transport Phenomena, ISTP-16, Prague, Czech Republic, 29 August–1 September 2005; pp. 1–5.
30. Kelkar, K.M.; Patankar, S.V. Numerical prediction of vortex shedding behind a square cylinder. *Int. J. Numer. Methods Fluids* **1992**, *14*, 327–341. [CrossRef]
31. Abbassi, H.; Turki, S.; Ben Nasrallah, S. Numerical Investigation of Forced Convection in a Horizontal Channel With a Built-In Triangular Prism. *J. Heat Transf.* **2002**, *124*, 571–573. [CrossRef]
32. Gallegos, R.K.B.; Sharma, R.N. Flags as vortex generators for heat transfer enhancement: Gaps and challenges. *Renew. Sustain. Energy Rev.* **2017**, *76*, 950–962. [CrossRef]
33. Shi, J.; Hu, J.; Schafer, S.R.; Chen, C.L.C. Numerical study of heat transfer enhancement of channel via vortex-induced vibration. *Appl. Therm. Eng.* **2014**, *70*, 838–845. [CrossRef]
34. Lee, J.B.; Park, S.G.; Sung, H.J. Heat transfer enhancement by asymmetrically clamped flexible flags in a channel flow. *Int. J. Heat Mass Transf.* **2018**, *116*, 1003–1015. [CrossRef]
35. Wang, Q.; Jaluria, Y. Unsteady mixed convection in a horizontal channel with protruding heated blocks and a rectangular vortex promoter. *Phys. Fluids* **2002**, *14*, 2113–2127. [CrossRef]
36. ANSYS. *ANSYS Fluent 12.0 User's Guide*; Ansys, Inc.: Canonsburg, PA, USA, 2009.
37. Roache, P.J. Perspective: a method for uniform reporting of grid refinement studies. *J. Fluids Eng.* **1994**, *116*, 405–413. [CrossRef]
38. Patil, P.P.; Tiwari, S. Effect of blockage ratio on wake transition for flow past square cylinder. *Fluid Dyn. Res.* **2008**, *40*, 753. [CrossRef]
39. Ottino, J.M. Mixing and chemical reactions a tutorial. *Chem. Eng. Sci.* **1994**, *49*, 4005–4027. [CrossRef]
40. Bothe, D.; Stemich, C.; Warnecke, H.J. Computation of scales and quality of mixing in a T-shaped microreactor. *Comput. Chem. Eng.* **2008**, *32*, 108–114. [CrossRef]
41. Rasouli, E.; Naderi, C.; Narayanan, V. Pitch and aspect ratio effects on single-phase heat transfer through microscale pin fin heat sinks. *Int. J. Heat Mass Transf.* **2018**, *118*, 416–428. [CrossRef]
42. Sarkar, S.; Singh, K.; Shankar, V.; Shenoy, K. Numerical simulation of mixing at 1–1 and 1–2 microfluidic junctions. *Chem. Eng. Process. Process Intensif.* **2014**, *85*, 227–240. [CrossRef]
43. Solehati, N.; Bae, J.; Sasmito, A.P. Numerical investigation of mixing performance in microchannel T-junction with wavy structure. *Comput. Fluids* **2014**, *96*, 10–19. [CrossRef]

44. Olsen, L.E.; Abraham, J.P.; Cheng, L.; Gorman, J.M.; Sparrow, E.M. Summary of forced-convection fluid flow and heat transfer for square cylinders of different aspect ratios ranging from the cube to a two-dimensional cylinder. In *Advances in Heat Transfer*; Elsevier: Amsterdam, The Netherlands, 2019; Volume 51, pp. 351–457.
45. Goossens, W.R. Review of the empirical correlations for the drag coefficient of rigid spheres. *Powder Technol.* **2019**, *352*, 350–359. [CrossRef]
46. Stokes, G.G. *On the Effect of the Internal Friction of Fluids on the Motion of Pendulums*; Pitt Press: Cambridge, UK, 1851; Volume 9.
47. Oseen, C. On Stokes' Formula and a Related Problem in Hydrodynamics. *Arkiv filr Mathematik Astronomi och Fysik* **1910**, *6*.

MDPI
St. Alban-Anlage 66
4052 Basel
Switzerland
Tel. +41 61 683 77 34
Fax +41 61 302 89 18
www.mdpi.com

Micromachines Editorial Office
E-mail: micromachines@mdpi.com
www.mdpi.com/journal/micromachines

www.ingramcontent.com/pod-product-compliance
Lightning Source LLC
LaVergne TN
LVHW071248170726
843515LV00006BA/1357

* 9 7 8 3 0 3 6 5 1 3 6 8 3 *